PROFESSIONAL
ELECTRICAL/ELECTRONIC
ENGINEER'S
LICENSE STUDY GUIDE

No. 742
$10.95

PROFESSIONAL ELECTRICAL/ELECTRONIC ENGINEER'S LICENSE STUDY GUIDE

BY EDWARD J. ROSS

TAB BOOKS
Blue Ridge Summit, Pa. 17214

FIRST EDITION

FIRST PRINTING—JUNE 1977

Copyright © 1977 by TAB BOOKS

Printed in the United States
of America

Library of Congress Cataloging in Publication Data

Ross, Edward J
    Professional electrical.

    Bibliography: p.
    Includes index.
    1. Electric engineering--Examinations, questions,
etc. 2. Electronics--Examinations, questions, etc.
I. Tab Books. II. Title.
TK169.R67   621.3'076   77-7495
ISBN 0-8306-7742-9
ISBN 0-8306-6742-3 pbk.

Contents

Preface

This book is prepared to enable the Electrical or Electronics Engineer to pass either the *Power Option* or the *Electronics Option* of the Professional Engineer (P.E.) exam. It contains problems and answers to every aspect of the electrical and electronics disciplines. Most of these problems were taken from past P.E. exams and are preceded by a brief review.

The first four chapters are general in scope. P.E. related information is contained in the first two chapters, which describe the contents of this book and how to use it to pass the P.E. exam. Ethics and Economics problems are contained in the next two; these are common to both the Electrical and Electronics Options.

There are two distinct and outstanding features of this book. First, it not only contains the old basic questions (several of which are given in almost every exam), but it also contains questions on the most advanced states of the art, such as digital techniques, transistors, and solid-state devices. Secondly, it utilizes mathematical methods such as inverse matrixes, determinants, and Laplace transforms to solve some of the complex problems. In many cases more than one solution, utilizing different methods, is used to solve a given problem. Nomographs are also provided as an alternate solution wherever possible. In all, this book is designed to accommodate the Engineer who has been out of school a while as well as the mathematically oriented savant.

Much credit must be given to the Penn State University, McKeesport Campus, where I have spent more than 20 years teaching evening classes and lately, P.E. courses. Without the above opportunities, this book could not have been written. Thanks also to my friends, fellow Westinghouse engineers, coinstructors, and students for their encouragement and for being either directly or indirectly a part of the effort contained herein. Finally, credits would not be complete without the deep appreciation to my wife and family for their understanding and minimal disturbances during the writing phases of this book.

Edward J. Ross, P.E.

Chapter 1

Introduction

This book is written for the purpose of instructing, guiding, and assisting the Electrical and Electronics Engineer in passing the Professional Engineer exam, which entitles the successor to a Professional Engineer's license. The P.E. lincense is becoming a prerequisite to many federal and state engineering positions. It is also required by most competent power companies. Whether this requirement will be expanded to mandatory licensing of the entire engineering society remains to be seen; society upgrading is the issue.

Possession of a P.E. license indicates that the owner has the education, experience, ability, and ethics with which to function as a registered professional. The ingredients needed to achieve this stature are contained in this book, which consist primarily of problems and answers, most of which were taken from previous P.E. exams. Also included are reviews to assist those who have been out of school a while, and a chapter on how to prepare for and to pass the exam.

The book is designed to span the gamut of age and experience. It suits the young engineer who is up to date with the latest state of the arts—who is familiar with transistors, semiconductors, op-amps, digital logic, etc. It will also satisfy the engineer who has been out of school a while and has not fully exercised his engineering talents. It may, in addition, help the practical self-taught engineer if he has a solid experience background; he may have to work harder, but hard work is not a stranger to a self-taught individual. Education and experience requirements, along with preparation and precedures, are all covered in Chapter 2.

The general breakdown of the entire contents of this book can be divided into four categories: (1) Ethics and economics problems, which must be taken by all P.E. candidates regardless of their discipline, are covered in Chapters 3 and 4. (2) Electronics and electrical problems of a general type that could appear in either Electrical or Electronic Option exams, and these are presented in Chapters 5 through 9. (3) Electronics Option problems contained in Chapters 10 through 16. (4) Power Option problems are covered in Chapters 17 through 20.

Electronics Option candidates would be mainly interested in categories (1), (2), and (3), whereas the Power Option types would pursue categories (1), (2), and (4). There are, however, areas of overlapping interests; for example, Chapter 14 on transmission lines and Chapter 18 on transformers. Although most of the problems are categorized here, they are detailed separately under each chapter heading in the Table of Contents.

Each chapter contains from 6 to 39 problems with answers. The chapter containing the lesser number of problems involve a special subject that may be an extension of another chapter. For example, Chapter 16 on Laplace transforms is in reality a special treatment of Chapter 15 on transients. Similarly, Chapter 12 on amplifiers can be considered an extension of Chapters 10 and 11, which cover transistors, semiconductors, and vacuum tube amplifiers.

At the opposite extreme, Chapter 17 contains 26 problems that cover a mixture of topics dealing with power distribution systems, line faults, heating and cooling, industrial plant power-factor correction, and other topics that are not suited for Chapters 18, 19, and 20.

Preceding the problems and answers in each chapter, there is a brief review of fundamentals, equations, graphs, tables, nomographs, and other pertinent information. The extent of each review depends on the subject. References are also provided in case a deeper review is needed.

The problems and answers are patterned after past P.E. exams, and it is reasonable to assume that future problems will be patterned after today's problems. Over the years, problems have become oriented toward semiconductors, op-amps, digital logic, higher math, and so on, but you can still expect a number of problems on basic theory and fundamentals. In any case, a thorough grasp of all the problems in this book, for a given option, will assuredly help wrest another P.E. license from the House of Examiners in your state.

10

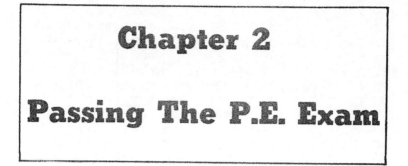

Chapter 2

Passing The P.E. Exam

The first step in passing the P.E. exam is to realize that the exam serves as a comprehensive technological measuring device that delves into every aspect of engineering. It is generated by sage and experienced engineers who approach the design of a P.E. problem in the same manner that a competent engineer would approach any design problem. It is an expertly planned, carefully executed, and well checked exam.

In knowing this, it should be evident to the electrical or electronic P.E. candidate that the exam may contain problems relating to any subject in any university's electrical curriculum, as well as to any industrial engineering project. As you can realize, that covers a lot of territory. How do you prepare for such an exam? By learning from others who have gone through it, and from the history of past exams. That is the crux of this chapter, and ultimately, of this book.

P.E. exams are given by every state in the U.S.A. The requirements vary somewhat between states and are constantly changing, usually toward more rigid requirements. For example, fewer states are now issuing P.E. licenses solely on the basis of long experience in a responsible or supervisory position. Those that do, require from 12 years to "long" years of experience, and the minimum age requirement varies from 35 to 50. This particular requirement is known as *eminence*, and also as the *grandfather clause*.

Tables 2-1 and 2-2 show the requirements and registration details of all the states. A copy of these details, along with

Table 2-1. Minimum Requirements for Registration.

STATE	PROFESSIONS REGISTERED	E.I.T. QUALIFICATIONS	PROFESSIONAL ENGINEER QUALIFICATIONS	CREDIT FOR EXPERIENCE		
				TEACHING	MILITARY	CONTRACTING
ALA.	ENGINEERS & L. S.	B(1)+C(2) or D(3)+C(2)	B(1)+D(3)+C(6) or D(5)+C(6)	E(1)+ E(1.1)	E(9)	No
ALAS.	ARCHITECTS, ENGINEERS & L. S.	N/A	B(1)+D(3)+C(6) or D(5)+C(6) or C(2)+ I(15)+F(12)	E(1)	E(9)	No
ARIZ.	ENGINEERS, L. S., ETC.	B(1)+C(2) or D(3)+ C(2)	B(1)+D(3)+C(2) or D(5)+C(6) or D(5)+ C(8)	E(6)	E(9)	E(11)
ARK.	ENGINEERS, L. S.	B(1)+C(2) or D(3)+ C(2)	B(1)+D(3)+C(6) or D(5)+C(6)	E(8)	E(10)	E(11)
CAL.	ENGINEERS, L. S.	C(2)	B(1)+D(1)+A(2)+C(2) or B(4)+D(3)+A(2) +C(2) or D(4)+A(2)+C(2)	E(3)	E(10)	E(10)
C. Z.	ENGINEERS	B(3)+C(2)	B(2)+D(3)+C(7) or B(6)+D(7)+C(6)		E(10)	E(10)
COLO.	ENGINEERS, L. S.	B(3)+C(2) or D(3)+ C(2)	B(3)+C(6)+D(3) or D(5)+C(6) or D(9)+ C(2)	E(1)	E(9)	E(7)
CONN.	ENGINEERS, L. S.	B(1)+C(2) or C(2)+ D(4)	B(1)+D(3)+C(6) or D(5)+I(23) or D(6) +C(6) or I(20)+C(9)	E(7)	E(7)	E(7)
DELA.	ENGINEERS, L. S.	B(1)+C(2)	B(1)+D(3)+C(6) or B(3)+D(5)+C(6) or I(16)	E(7)	E(7)	E(10)
D. C.	ENGINEERS	B(2)+C(2) or D(5)+ C(2)	B(2)+D(3)+C(6) or D(7)+C(6) or D(7)+ C(8)	E(7)	E(7)	E(7)
FLA.	ENGINEERS, L. S.	B(1)+C(2) or B(3)+ D(1)+C(2)	B(1)+D(3)+C(6) or D(6)+C(6)	E(2)	E(9)	E(10)
GA.	ENGINEERS, L. S.	B(1)+C(2) or D(5)+ C(2)	B(1)+D(3)+C(6) or D(7)+C(6) or B(1)+ D(9)+C(2) or D(9)+C(2)	E(2)	E(9)	E(10)
GUAM	ENGINEERS, L. S.,ARCH.	B(3)+C(2) or D(3)+ C(2)	B(3)+D(3)+C(6) or D(5)+C(6)	E(7)	E(10)	E(10)
HAW.	ENGINEERS, L. S.,ARCH.	B(4)+C(2) or D(10) +C(2)	B(4)+D(2)+C(6) or D(7)+C(6)	E(3)	E(9)	E(11)
IDAHO	ENGINEERS, L. S.	B(4)+C(2) or D(3)+ C(2)	B(1)+D(3)+C(6) or D(5)+C(6)	Yes	E(9)	E(9)
ILL. P. E.	ENGINEERS	B(2)+C(2) or D(3)+ C(2)	B(2)+D(3)+C(6) or D(5)+C(6)	E(2)	E(10)	E(11)
ILL. STRUCT.	STRUCTURAL		B(2)+D(3)+C(6) or D(4)+C(6) or D(7)+ F(6)	E(7)	E(7)	E(7)
IND.	ENGINEERS, L. S.	B(2)+C(2) or B(5)+ C(2)+D(1)	B(1)+D(3)+C(6) or D(7)+C(6)	E(1)	E(9)	No
IOWA	ENGINEERS, L. S.	B(1)+C(2) or D(5)+ C(2)	B(1)+D(3)+C(6) or D(7)+C(6) or I(5)	E(7)	E(7)	E(7)
KANS.	ENGINEERS, L. S.	B(2)+C(2) or D(3)+ C(2)	B(2)+D(3)+C(6) or D(5)+C(6) or B(2)+ D(7)+C(2)	E(1)	E(10)	E(7)
KY.	ENGINEERS, L. S.	B(1)+C(2) or C(2)+ D(3)	B(1)+C(6)+D(3) or B(1)+C(2)+C(9)+ D(3) or C(6)+D(5)	E(9)	E(10)	E(10)
LA.	ENGINEERS, L. S.	B(1)+C(2) or D(4)+ C(2)	B(1)+D(3)+C(2) or D(5)+C(6) or I(10)	E(2)	E(10)	E(10)
MAINE	ENGINEERS	B(1)+C(2) or D(5)+ C(2)	B(1)+D(3)+C(2) or D(7)+C(6) or I(6)+ C(8)	E(2)	E(9)	E(9)
MD.	ENGINEERS, L. S.	B(1)+C(2) or B(4)+ D(3)+C(2)	B(1)+D(3)+C(6) or B(4)+D(5)+C(6) or D(7)+C(2)	E(1)	E(9)	E(11)
MASS.	ENGINEERS, L. S.	B(1) or B(2)+D(3) or B(6)+D(7)	B(1)+D(3)+C(6) or B(2)/(3)+D(5)+ C(6) or D(7)+C(6)	E(1)	E(10)	No
MICH.	ENGINEERS, L.S.& ARCH.	B(1)/(2)+C(2) or D(3)+C(2)	B(1)/(2)+D(3)+C(6) or D(5)+C(6)	Yes	E(8)	No
MINN.	ENGINEERS, L.S.& ARCH.	B(1)+B(5)+C(2)	B(1)+D(3)+C(6) or B(1)+I(16)+C(2)	E(1)	E(10)	E(11)
MISS.	ENGINEERS, L. S.	B(2)+C(2) or D(3)+ C(2)	B(1)+D(3)+C(6) or D(5)+C(6)	Yes	E(10)	E(10)

12

Table 2-1. (continued).

ATE	LENGTH & TYPE ENGRS. EXAM.	PASSING GRADE	NEC RECOGNITION	POLICY ON EMINENCE	RECIPROCITY PRACTICES	REGISTRATION FEES PROF. ENGR. APPL.	CERT.	E. I. T. APPL.	CERT.	RECIPROCITY	RENEWAL*
R.	F(3)	G(7)	Yes	I(1)	J(6)	15.00	10.00	10.00	-----	25.00	10.00
AS.	F(3)	G(7)	Yes	I(1)	J(3)+F(12)	40.00	10.00	40.00	10.00	25.00+10.00	15.00
IZ.	F(3)	G(7)	H(3)	I(11)+C(8)	J(3)	15.00	10.00	10.00	-----	50.00	15.00
K.	F(3)	G(7)	Yes	I(5)	J(6)	30.00	15.00	10.00	-----	45.00	P.E.10.00 EIT 4.00
..	F(2)	G(7)	Yes		J(1)+J(5)	60.00	-----	40.00	-----	60.00	20.00
Z.	F(2)	G(6)	Yes	I(19)+(5)	J(3)	25.00	-----	10.00	-----	25.00	5.00
.O.	F(3)	G(7)	Yes	I(1)	J(6)	30.00	10.00	10.00	-----	25.00	6.00
IN.	F(3)	G(6)	H(3)	I(20)	J(6)	50.00	-----	25.00	-----	50.00	35.00
A.	F(3)	G(7)	Yes	I(16)	J(1)	30.00	-----	10.00	-----	30.00	6.00
C.	F(3)	G(3)	Yes	I(7)	J(6)	40.00	10.00	15.00	5.00	50.00	7.00
.	F(3)	G(7)	H(3)	No	J(1)+J(6)&(9)	35.00	-----	10.00	-----	35.00	15.00
	F(3)	G(7)	H(2)	I(1)	J(3)	15.00	-----	-----	5.00	15.00	10.00
M	F(3)	G(7)	Yes	I(10)+(22)+C(8)	J(6)	25.00	-----	10.00	-----	25.00	5.00
I.	F(2)	G(7)	Yes	I(1)	J(3)+C(9)	30.00	15.00	5.00	-----	45.00	15.00
HO	F(3)	G(7)	H(1)	No	J(3)+J(6)	40.00	10.00	10.00	5.00	50.00	13.00
.PE	F(3)	75	Yes	I(5)	J(6)	30.00	-----	15.00	-----	30.00	10.00*
.ST	C(6)	G(5)	H(1)	No	J(3)	30.00	-----	-----	-----	30.00	20.00*
.	F(11)	G(7)	H(5)+H(6)	I(24)	J(1), J(3)	5.00+ K(1)	7.50	-----	10.00	27.50+K(2)	15.00*
A	F(3)	G(7)	Yes	No	J(6)	25.00	10.00	10.00	-----	25.00	10.00
S.	F(2)	G(7)	Yes	No	J(6)	25.00	10.00-student 15.00 non-student	-----	-----	25.00	10.00
	F(2)	G(7)	Yes	I(1)	J(3)	25.00	15.00	15.00	-----	35.00	10.00
	F(3)	G(3)	Yes	I(10)+I(17)	J(6)	25.00	-----	10.00	-----	25.00	7.50
NE	F(2)	G(7)	Yes	60 + Emin.	J(3)	10.00	10.00	10.00	-----	20.00	3.00
	F(3)	G(7)	H(2)	I(7)	J(6)	20.00	15.00	10.00	-----	35.00	10.00
5.	C(6)	G(3)	Yes	I(10)+C(8)	J(5)	40.00+ 20.00 each examination	-----	20.00	-----	40.00+	15.00*
H.	F(2)	G(7)	H(4)	No	J(3)	30.00	40.00	30.00	-----	70.00	15.00
N.	F(3)	G(7)	Yes	No	J(3)	100.00	15.00	30.00	-----	100.00	15.00
5.	F(2)	G(7)	Yes	No	J(1) if reg. 10 yrs	25.00	-----	-----	10.00	25.00	8.00

:es subject to change

Table 2-1. (continued).

STATE	PROFESSIONS REGISTERED	E.I.T. QUALIFICATIONS	PROFESSIONAL ENGINEER QUALIFICATIONS	CREDIT FOR EXPERIENCE TEACHING	MILITARY	CONTRACTING
MO.	ENGINEERS, L.S.& ARCH.	B(3)+C(2) or D(3)+C(2)	D(5)+C(6) or B(3)+C(6)+D(3) or B(3)+I(10)+C(8)	Yes	E(9)	E(9)
MONT.	ENGINEERS, L.S.	B(1)+C(2) or D(3)+C(2)	B(1)+D'3)+C(6) or D(5)+C(6) or I(12)	Yes	E(10)	E(10)
NEBR.	ENGINEERS, ARCH.	B(1)+C(2) or D(3)+C(2)	B(1)+D(3)+C(6) or D(5)+C(6) or I(11)+C(8)	Yes	E(9)+E(12)	E(9)
NEV.	ENGINEERS, L.S.	B(2)+C(2) or D(3)+C(2)	B(2)+D(3)+C(6) or D(5)+C(6) or I(5)	Yes	E(9)	E(11)
N.H.	ENGINEERS	B(3)+C(2) or D(3)+C(2)	B(2)+D(3)+C(2) or D(5)+C(2) or I(15)	Yes	E(9)	No
N.J.	ENGINEERS, L.S.	B(1)+C(2) or D(3)+C(2)	B(1)+D(3)+C(6) or D(5)+C(6) or I(13)+C(8)	Yes	E(9)	No
N.M.	ENGINEERS, L.S.	B(2)+C(2) or D(4)+C(2)	B(2)+D(3)+C(6) or D(5)+C(6)	Yes	E(7)	E(7)
N.Y.	ENGINEERS, L.S.	B(1)+C(2) or B(6)+D(5)+C(2)	B(1)+D(3)+C(6) or B(6)+D(7)+C(6) or B(1)+I(5)+I(6)	E(10)	E(10)	E(10)
N..C.	ENGINEERS, L.S.	B(1)+C(2) or B(6)+D(4)+C(2)	B(1)+D(3)+C(6) or B(6)+D(6)+C(6)	E(1)	E(9)	E(4)
N.D.	ENGINEERS, L.S.	B(1)+C(2) or B(4)+C(2)	B(1)+D(3)+C(6)	D(1)	E(9)	E(9)
OHIO	ENGINEERS, L.S.	B(1)+C(2) or D(3)+C(2)	B(1)+D(3)+C(6) or D(5)+C(6)	E(1)	E(10)	E(10)
OKLA.	ENGINEERS, L.S.	B(1)+C(2) or B(4)+C(2)+D(1)	B(3)+D(3)+C(6) or D(5)+C(6) or I(16)	E(1)	E(9)	E(9)
ORE.	ENGINEERS, L.S.	B(3)+C(2) or D(3)+C(2)	B(3)+D(3)+C(6) or D(5)+C(6)	E(7)	E(10)	No
PA.	ENGINEERS, L.S.	B(3)+C(2) or D(3)+C(2)	B(1)+D(3)+C(6) or D(7)+C(6) or I(21)	Yes	E(10)	No
P.R.	ENGINEERS, L.S.& ARCH.	B(1)+C(6)	B(1)+D(3)+C(6)	Yes	E(9)	No
R.I.	ENGINEERS, L.S.	B(3)+C(2) or B(6)+D(5)+C(2)	B(3)+D(3)+C(2) or B(3)+D(7)+C(2) or B(6)+D(7)+C(2) or I(10)+C(2)	E(1)	Some E(12)	No
S.C.	ENGINEERS, L.S.	B(3)+C(2)	B(3)+D(3)+C(6) or D(7)+C(6) or I(12)+C(2)	E(9)	E(9)	E(11)
S.D.	ENGINEERS, L.S.& ARCH.	B(1)+C(2) or D(3)+C(2)	B(1)+D(3)+C(6) or D(5)+C(6)	E(2)	Some	No
TENN.	ENGINEERS, ARCH.	B(3)+C(2) or D(4)+C(2)	B(3)+D(3)+C(6) or D(5)+C(6) or I(5)/(10)	Yes	E(9)	E(10)
TEXAS	ENGINEERS	B(1)+C(2) or B(3)+C(2) or D(5)+C(2)	B(1)+D(3) or B(3)+D(3) or D(5)+C(6)	E(1)	Some E(12)	E(10)
UTAH	ENGINEERS, L.S.	B(1)+C(2) or D(3)+C(2)	B(1)+D(3)+C(6) or D(5)+C(6)	E(3)	E(7)	E(11)
VT.	ENGINEERS	B(2)+C(2) or B(6)+D(5)+C(2)	B(2)+D(3)+C(6) or D(7)+C(6) or I(10)/(22)+C(7)	E(2)	E(9)	E(9)
VA.	ENGINEERS, L.S.& ARCH.	B(1)+C(2) or D(4)+C(2)	B(1)+D(3)+C(6) or D(6)+C(6) or I(20)+F(6) or D(9)+C(2)	E(7)	E(7)	E(7)
V.I. (U.S.)	ENGINEERS, L.S.& ARCH.	B(2) or D(3)+C(2)	B(1)+D(1)+C(6) & Resident of V.I.(US) or D(5)+C(6) & Resident of V.I.(US)	Yes	E(9)	E(9)
WASH.	ENGINEERS, L.S.	B(1)+C(2) or D(3)+C(2)	B(1)+D(3)+C(6) or D(5)+C(6) or B(1)+I(11)+C(2)	E(4)	E(9)	No
W.V.	ENGINEERS	B(2)+C(2) or D(5)+C(2)	B(2)+D(3)+C(6) or D(5)+C(6)	Yes	E(9)	No
WISC.	ENGINEERS, L.S.,ARCH.	B(1)+C(2) or D(3)+C(2)	B(1)+D(3)+C(6) or D(5)+C(6) or I(12)	E(1)	E(9)	E(9)
WYO.	ENGINEERS, L.S.	B(1)+C(2) or D(3)+C(2)	B(1)+D(3)+C(6) or D(5)+C(6) or I(5)/(10)	E(4)	E(9)	E(9)

Table 2-2. (continued).

STATE	LENGTH & TYPE ENGRS. EXAM.	PASSING GRADE	NEC RECOGNITION	POLICY ON EMINENCE	RECIPROCITY PRACTICES	REGISTRATION FEES PROF. ENGR. APPL.	CERT.	E.I.T. APPL.	CERT.	RECIPROCITY	RENEWAL*
MO.	F(3)	G(7)	H(2)	I(5)	J(1), J(3)	35.00	-----	10.00	-----	35.00	10.00
MONT.	F(3)	G(7)	H(3)	J(3)	J(3) , J(5)	20.00	-----	20.00	-----	20.00	10.00
NEBR.	F(3)	G(7)	H(3)	I(11)+C(8)	J(3)	50.00	25.00	25.00	-----	75.00	10.00
NEV.	F(3)	G(7)	H(1)	I(5)	J(6)+C(9)	35.00	-----	10.00	-----	35.00	10.00
N. H.	F(3)	G(1)	Yes	I(15)+C(7)	J(6)	20.00	20.00	10.00	7.50	40.00	5.00
N. J.	F(3)	G(7)	H(1)	I(9)+C(8)	J(3)	40.00	-----	10.00	-----	40.00	5.00
N. M.	F(3) .	G(7)	Yes	I(2)+C(9)	J(6)	25.00	-----	15.00	-----	25.00	14.00
N. Y.	F(3)	75	H(1)	B(1)+I(5)+I(6)	J(10	40.00	-----	20.00	-----	40.00 P.E. 20.00 EIT	15.00*
N. C.	F(9)+C(6)+C(9)	G(7)	H(6)	No	J(3)	45.00	-----	10.00	-----	45.00	10.00
N. D.	F(3)	G(7)	Yes	I(1)	J(6)	25.00	-----	10.00	-----	25.00	10.00
OHIO	F(3)	G(7)	Yes	I(19)+C(8)+B(2)	J(6)+J(1)	15.00	15.00	15.00	-----	30.00	5.00
OKLA.	F(3)	G(7)	Yes	I(16)	J(6)	25.00	-----	10.00	-----	25.00	5.00
ORE.	F(3)	G(3)	Yes	I(1)	J(9)	10.00	10.00	5.00	5.00	25.00	10.00
PA.	F(3)	G(7)	H(1)	I(21)	J(6)	25.00	-----	15.00	-----	25.00	10.00*
P. R.	F(2)	G(7)	Yes	I(5)	B(1)+D(3)+C(8)	15.00	20.00	25.00	-----	35.00	-----
R. I.	F(3)	G(7)	Yes	I(10)+C(2)	J(6)	30.00	-----	10.00	-----	30.00	5.00
S. C.	F(3)	G(7)	Yes	I(12)+C(2)	J(6)	30.00	5.00	-----	10.00	25.00	10.00#
S. D.	F(3)	G(7)	Yes	No	J(5)+J(6)	30.00	5.00	5.00	-----	30.00	10.00
TENN.	F(3)	G(7)	Yes	I(5)+I(10)	J(9)	35.00	-----	10.00	-----	35.00	10.00
TEXAS	F(3)	G(7)	Yes	I(1)	J(6)	15.00	10.00	10.00	-----	10.00	10.00
UTAH	F(3)	G(7)	J(5)	I(12)	J(6)	30.00	-----	15.00	-----	30.00	3.00
VT.	F(3)	G(7)	H(5)+H(6)	I(5),(20)&(22)	J(5),(6) & (8)	45.00	-----	15.00	-----	30.00	5.00
VA.	F(3)	G(7)	H(3)	I(20)+F(6)	J(6)	30.00	-----	15.00	-----	30.00	10.00*
V.I. (U.S.)	F(3)	G(7)	Yes	I(12)	J(6)+resident of V. I. (U.S.)	50.00	-----	25.00	-----	50.00	25.00
WASH.	F(3)	G(7)	H(3)	No	J(6)	15.00	10.00	10.00	-----	15.00	7.50
W. V.	F(3)	G(7)	Yes	I(1)	J(3)+J(6)	10.00	15.00	10.00	-----	25.00	7.50
WISC.	F(3)	G(7)	H(3)	I(12)	J(6)	12.50	12.50	12.50	-----	75.00	25.00*
WYO.	F(2)+F(5)	G(7)	Yes	I(5)+I(10)	J(6)	15.00	-----	5.00	-----	15.00	8.00*

Prices subject to change.

Table. 2-2. Code Symbols for Table 2-1, Minimum Requirements for Registration

This table, current as of January 1973, was developed from data furnished by the Examining Boards of the states and territories. Each Board retains the right to change these conditions at any time.

A.
(1) Prerequisites are for professional engineering. Requirements for land surveying, in general, are the same but do vary in some states.
(2) Valid EIT status, 10 year limit.
(3) Land Surveyors work experience time is three years less than professional engineers.

B.
(1) ECPD school (Accredited by Engineers' Council for Professional Development)
(2) Approved school (Acceptable to Board)
(3) Board approved school of science and engineering
(4) 4 year engineering curriculum
(5) Degree
(6) High school

C.
(1) 6 hour examination
(2) 8 hour examination
(3) 12 hour examination
(4) 13 hour examination
(5) 14 hour examination
(6) 16 hour examination
(7) Examination as required by Board (Board may waive portions of examination)
(8) Oral or written examination (based on application)
(9) Personal interview

D.
(1) 2 years experience
(2) 3 years experience
(3) 4 years experience
(4) 6 years experience
(5) 8 years experience
(6) 10 years experience
(7) 12 years experience
(8) 13 years experience
(9) 15 years experience
(10) 9 years experience

E.
(1) Yes, if in responsible charge of engineering teaching.
  (1.1) If rank of assistant professor or higher.
(2) Yes, year for year
(3) Yes, 1 year
(4) 2 years maximum
(5) Yes, up to 3 years
(6) Yes, up to 5 years
(7) Yes, evaluated by Board
(8) Yes, same as other experience
(9) Yes, if of an engineering nature
(10) Only engineering experience
(11) No, unless planning and design involved
(12) Experience must be documented

F. All written examinations unless otherwise noted
(1) 4 hour sessions (see qualifications)
(2) Two 4 hour Fundamentals; Two 4 hour PE (branch)
(3) 8 hour Fundamentals and 8 hour Principles & Practice
(4) Three parts, 3 days
(5) 8 hour land surveying
(6) Oral examination, no time limit
(7) 4 hour basic, 4 hour general, 4 hour PE
(8) 3 hour basic, 3 hour specific
(9) According to qualification requirement
(10) Fundamental 8 hour general (closed book) Principles & Practice 8 hour (open book)
(11) Two 4 hour sessions on engineering fundamentals; 4 hours application of engineering principles; 1 hour misc. and 3 hours design problems in specialization
(12) Treatise or course on arctic engineering

other publications, can be obtained from the National Council of Engineering Examiners (NCEE)—see reference 49 in the Bibliography for the address. Table 2-2 explains the letter and numerals contained in Table 2-1. These tables will provide useful guidance throughout this chapter.

Most states offer professional certification of Engineers, Land Surveyors (L.S.), and Architects. But this book is directed toward the electrical and electronic disciplines, specifically.

Qualifications are listed for both P.E. candidates and Engineer-in-Training (E.I.T.) candidates. The E.I.T. is available to engineers with minimal experience and creditable education, as defined in the tables. Qualifications for P.E.

Table 2-2. (continued).

G.	(1) 70 average all parts, 60 minimum any part	(5) 75 average all parts, 60 minimum any part
	(2) Based on examination load	(6) 75 on each part
	(3) Pass or fail	(7) 70 on each 8 hour part
	(4) 70 full field, 75 limited field	
H.	(1) Supporting but not in itself qualifying (independent determination made in all cases)	(4) If registration is by written examination
	(2) Limited (accepted as evidence of qualification)	(5) When registration conforms to State (territory) law
	(3) Only for verification of record	(6) Yes, provided it is up-to-date
I.	(1) Not permitted under State Law	(14) Age 35, 16 years experience, 4 of which may be ECPD school
	(2) Rarely, must have national recognition	(15) Age 40, 15 years experience
	(3) Not eminence but under exemption	(16) Age 40, 20 years experience
	(4) Experience must be of high quality	(17) Age 45, 12 years responsible charge
	(5) Only on basis of outstanding professional accomplishment	(18) Age 45, 15 years experience
	(6) 15 years lawful practice	(19) Age 45, 20 years experience, 10 years responsible charge of outstanding engineering work
	(7) Recognized standing, 12 years experience	(20) Age 50, 20 years experience
	(8) 12 years progressive experience	(21) Age 50, 25 years experience, 15 years responsible charge
	(9) 15 years progressive experience	
	(10) 20 years progressive experience	(22) At least 10 years in responsible charge of important engineering work
	(11) 24 years progressive experience	(23) Age 40, ECPD Graduate, 8 hour examination
	(12) Age 35, 12 years experience, 5 years responsible charge	(24) Registered over 30 years and 15 years responsible charge
	(13) Age 35, 15 years experience	
J.	(1) Comity, not reciprocity	(6) Must meet requirement of State (territory) Law
	(2) Recognition, not reciprocity	(7) 45 minute to 3 hour examination
	(3) Recognize all states having equivalent requirements	(8) Fundamental part of examination may be waived
	(4) On equal basis (States which have agreement)	(9) Evaluation by Board
	(5) If first registration based on qualification by written examination	(10) Eligibility for N. Y. exam and already passed full equivalent
K.	(1) Examination fee of $15.00 (for each examination)	
	(2) Includes $5.00 application fee and $5.00 Certificate fee	
	(3) Student (B. S. Candidate) $8.00; others $20.00	

Renewal fee: *Biennial
up to $10.00 allowed

NEC - Committee on National Engineering Certification

Compiled by: NATIONAL COUNCIL OF ENGINEERING EXAMINERS
Uniform Laws and Procedures Committee

January 1973

candidates differ for engineers having varying amounts of industrial, teaching, military, and contracting experience. The minimum age in nearly all the states is 21 for those taking the E.I.T. exam and 21 to 25 for the P.E. exam. However, the minimum number of years work experience is zero in most states for the degreed E.I.T. candidate and about 4 years (in the majority of states) for the degreed P.E. candidate. For the nonaccredited degree, or the nondegreed, 6 to 12 years of experience is required (see Tables 2-1 and 2-2).

The P.E. exam is actually divided into two parts (the E.I.T. and the P.E.), both of which must be taken by the P.E. candidate. In some states, however, the E.I.T. portion can be waived by the examiners under special conditions. The exams are generally given on two different days. The E.I.T. exam usually consists of 8 hours on fundamentals, while the P.E. exam usually consists of 8 hours on principles and practices. Most states require a minimum grade of at least 70 on each test.

Not all professional Engineers are recognized by the National Engineering Certification (NEC) committee, nor do all states recognize eminence or reciprocity. Refer to Tables 2-1 and 2-2 for variations in requirements.

Reciprocity is usually recognized between states that require equal or better qualifications. Some examiners will call for a review of one's background and experience before an agreement on reciprocity is reached. Reciprocity fees vary from $10 to $100.

Registration fees are also listed in the tables. There are fees for application, certification, reciprocity, and renewal. The initial application fee usually entitles you to a second exam if you fail the first. For a third and fourth try, you must reapply.

Some states do not require a certification fee after passing—those that do usually have a lower application fee. Renewal fees must be paid either annually or biannually (every 2 years).

With all the above information at hand, let us take a hypothetical case: that is, apply for examination, prepare for it, and examine what it takes to pass it. The first step is to write to the State Board of Engineering Examiners, addressed to the capital city in the state or district for which registration is desired. (Puerto Rico, Virgin Islands, Guam, and the Canal Zone are also included.) You will be sent an application form with instructions covering the fees required, dates of the exams, and all necessary requirements for that state or district.

Most applications require more than just a resume. A state with a very thorough Board of Examiners will require most or all of the following:

- Your educational background.
- Transcript of grades.
- Your photograph.
- Your employment experience (with dates and names of supervisors). This is no time to be modest. Be sure to write your achievements using *active* terms, such as "I supervised" instead of "I was supervisor of"—or "We developed" instead of "We worked on." List your patents, licenses, published articles, teaching

experience, part time TV employment, awards and bonuses, technical society memberships, fraternities, and anything else you can think of.

- Three to five signatures of P.E. acquaintances. These engineers will be given your resume and they will be required to authenticate that portion of the resume with which they are familiar.

If your application is accepted, you will be notified where and when the E.I.T. and P.E. exams will be given. (The E.I.T. is usually given on a Friday and the P.E. on a Saturday.) Your thoughts about preparation, dusting off old and new text books, the library, reliance on friends and a refresher course, should then begin to gel in your mind.

Of the above thoughts, the one concerning a refresher course is a most commendable one for many reasons:

- The course is usually based upon previous P.E. exams.
- You will be attending and communicating with persons of the same interest and with whom you can discuss the problems and solutions.
- You will be taught by a P.E. who can answer your questions and work with you on troublesome areas.
- The instructor can provide references, hand-outs, "quickie-solution" nomographs, etc., for use at exam time.
- Time is of the essence. The candidate usually does not possess a lot of free time for studying. If done at home, where there is much distraction, you are not altogether sure what to study or how to organize your review.
- A refresher course is very efficient. You can get a high ratio of well guided technical refreshment per time spent.
- You can approach the exams with one of the most important assets—confidence.

If you insist on doing it alone, be sure to schedule your time and lesson plan. In the beginning this might seem trivial, but the deeper you get into your review, the more you will realize how expansive the task becomes.

At exam time, it is important to bring with you the items that will save you time, such as a calculator and slide rule. (The slide rule is a backup—just in case the calculator decides not to participate in the exam.) Your notes are the most valuable of all references, especially if you enrolled in a refresher course. Text books must be selected with care—*do not bring a trunkful of text books!* Remember, the time spent searching through books cannot be spent on the exam; the more books you have, the more confused you are apt to become, and the more time you will waste.

Do bring comprehensive handbooks containing tables, charts, trigonometric functions, logarithms, Laplace transforms, calculus tables, wire tables, and useful nomographs. Electronics Option candidates should bring a semiconductor manual, such as the type provided by most semiconductor manufacturers. For the Power Option you will need wire characteristics, tables with positive and negative sequence data, wire spacing information, NEC tables and diagrams, power nomograms, and a lighting manual (generally available from manufacturers).

In winding your way through the problems in this book, try to assume variations of the problems. Ask yourself if they can be performed in another way, for no two people necessarily react the same way to a given problem. You will notice that some problems do show alternate solutions. Keep a well organized record of your solutions because it will eventually be your most important reference at exam time.

Get a good rest the night before the exam and do not let yourself become nervous. Up to this point, you've done all you can, so relax.

During the exam, be sure to select a well lighted desk, one that is far away from outside noises and distractions. Be sure to follow these guidelines:

- Read through the *entire* exam before working any problems and before selecting your quota of problems.
- Always work the easiest problems first—it builds confidence and makes the most efficient use of your limited time.
- Be sure to draw a diagram of each problem. Jot down the given information and all the solutions to be found in lengthy problems. This prevents your omitting any portion of a problem.
- Be methodical: always identify your solutions or underscore the answer. The examiner should not have to search for the answer, nor guess at which equation was used.
- Make references, especially when using uncommon equations.
- Be prepared to make assumptions if insufficient data is given or if an obvious error exists in the question.

Clearly indicate these assumptions *before* you proceed to solve the problem.
- Never linger over a problem. Be certain you know both the problem and where to find reference data before you begin. If in doubt, set it aside and come back to it later.
- After completion of the exam, reread the problems and solutions to be sure all questions are understood and clearly answered.

It will take 2 to 3 months before you will know whether you passed or not. Usually, no news is good news, because the failures are notified first. Several months after favorable notification, a handsome Certificate of Registration will arrive from the Department of State. You can muse over what your feelings might be. Whatever they are, you can be certain of two of them. (1) You will be proud to be recognized as a competent professional. (2) You will know that you have earned it.

Good luck!

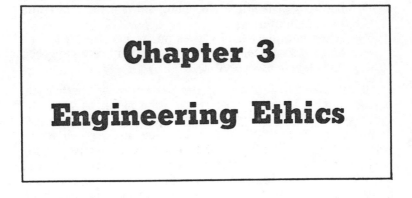

Chapter 3

Engineering Ethics

The following is the Code of Ethics for Engineers, reprinted by permission of the National Society of Professional Engineers, as revised, January 1974.

CODE OF ETHICS FOR ENGINEERS

Preamble—The Engineer, to uphold and advance the honor and dignity of the engineering profession and in keeping with high standards of ethical conduct:

- *Will be honest and impartial, and will serve with devotion his employer, his clients, and the public;*
- *Will strive to increase the competence and prestige of the engineering profession;*
- *Will use his knowledge and skill for the advancement of human welfare.*

Section 1—The Engineer will be guided in all his professional relations by the highest standards of integrity, and will act in professional matters for each client or employer as a faithful agent or trustee.

a. He will be realistic and honest in all estimates, reports, statements, and testimony.

b. He will admit and accept his own errors when proven wrong and refrain from distorting or altering the facts in an attempt to justify his decision.

c. He will advise his client or employer when he believes a project will not be successful.

d. He will not accept outside employment to the detriment of his regular work or interest, or without the consent of his employer.

e. He will not attempt to attract an engineer from another employer by false or misleading pretenses.

f. He will not actively participate in strikes, picket lines, or other collective coercive action.

g. He will avoid any act tending to promote his own interest at the expense of the dignity and integrity of the profession.

Section 2—The Engineer will have proper regard for the safety, health, and welfare of the public in the performance of his professional duties. If his engineering judgment is overruled by nontechnical authority, he will clearly point out the consequences. He will notify the proper authority of any observed conditions which endanger public safety and health.

a. He will regard his duty to the public welfare as paramount.

b. He shall seek opportunities to be of constructive service in civic affairs and work for the advancement of the safety, health and well-being of his community.

c. He will not complete, sign, or seal plans and/or specifications that are not of a design safe to the public health and welfare and in conformity with accepted engineering standards. If the client or employer insists on such unprofessional conduct, he shall notify the proper authorities and withdraw from further service on the project.

Section 3—The Engineer will avoid all conduct or practice likely to discredit or unfavorably reflect upon the dignity or honor of the profession.

a. The Engineer shall not advertise his professional services but may utilize the following means of identification:

(1) *Professional cards and listings in recognized and dignified publications, provided they are consistent in size and are in a section of the publication regularly devoted to such professional cards and listings. The information displayed must be restricted to firm name, address, telephone number, appropriate symbol, name of principal participants and the fields of practice in which the firm is qualified.*

(2) *Signs on equipment, offices and at the site of projects for which he renders services, limited to firm name, address, telephone number and type of services, as appropriate.*

(3) *Brochures, business cards, letterheads and other factual representations of experiencs, facilities, personnel and capacity to render service, providing the same are not misleading relative to the extent of participation in the projects cited, and provided the same are not indiscriminately distributed.*

(4) *Listings in the classified section of telephone directories, limited to name, address, telephone number and specialties in which the firm is qualified.*

b. *The Engineer may advertise for recruitment of personnel in appropriate publications or by special distribution. The information presented must be displayed in a dignified manner, restricted to firm name, address, telephone number, appropriate symbol, name of principal participants, the fields of practice in which the firm is qualified and factual descriptions of positions available, qualifications required and benefits available.*

c. *The Engineer may prepare articles for the lay or technical press which are factual, diginified and free from ostentations or laudatory implications. Such articles shall not imply other than his direct participation in the work described unless credit is given to others for their share of the work.*

d. *The Engineer may extend permission for his name to be used in commercial advertisements, such as may be published by manufacturers, contractors, material suppliers, etc., only by means of a modest dignified notation acknowledging his participation and the scope thereof in the project or product described. Such permission shall not include public endorsement of proprietary products.*

e. *The Engineer will not allow himself to be listed for employment using exaggerated statements of his qualifications.*

Section 4—The Engineer will endeavor to extend public knowledge and appreciation of engineering and its achievements and to protect the engineering profession from misrepresentation and misunderstanding.

a. *He shall not issue statements, criticisms, or arguments on matters connected with public policy which are inspired or paid for by private interests, unless he indicates on whose behalf he is making the statement.*

Section 5—The Engineer will express an opinion of an engineering subject only when founded on adequate knowledge and honest conviction.

a. *The Engineer will insist on the use of facts in reference to an engineering project in a group discussion, public forum or publication of articles.*

Section 6—The Engineer will undertake engineering assignments for which he will be responsible only when qualified by training or experience; and he will engage, or advise engaging, experts and specialists whenever the client's or employer's interests are best served by such service.

Section 7—The Engineer will not disclose confidential information concerning the business affairs or technical processes of any present or former client or employer without his consent.

a. While in the employ of others, he will not enter promotional efforts or negotiations for work or make arrangements for other employment as a principal or to practice in connection with a specific project for which he has gained particular and specialized knowledge without the consent of all interested parties.

Section 8—The Engineer will endeavor to avoid a conflict of interest with his employer or client, but when unavoidable, the Engineer shall fully disclose the circumstances to his employer or client.

a. The Engineer will inform his client or employer of any business connections, interests, or circumstances which may be deemed as influencing his judgment or the quality of his services to his client or employer.

b. When in public service as a member, advisor, or employee of a governmental body or department, an Engineer shall not participate in considerations or actions with respect to services provided by him or his organization in private engineering practice.

c. An Engineer shall not solicit or accept an engineering contract from a governmental body on which a principal or officer of his organization serves as a member.

Section 9—The Engineer will uphold the principle of appropriate and adequate compensation for those engaged in engineering work.

a. He will not undertake or agree to perform any engineering service on a free basis, except for civic, charitable, religious, or eleemosynary nonprofit orgainizations when the professional services are advisory in nature.

b. He will not undertake work at a fee or salary below the accepted standards of the profession in the area.

c. He will not accept remuneration from either an employee or employment agency for giving employment.

d. When hiring other engineers, he shall offer a salary according to the engineer's qualifications and the recognized standards in the particular geographical area.

e. If in sales employ, he will not offer, or give engineering consultation, or designs, or advice other than specifically applying to the equipment being sold.

Section 10—The Engineer will not accept compensation, financial or otherwise, from more than one interested party for the same service, or for services pertaining to the same work, unless there is full disclosure to and consent of all interested parties.

a. He will not accept financial or other considerations, including free engineering designs, from material or equipment suppliers for specifying their product.

b. He will not accept commissions or allowances, directly or indirectly, from contractors or other parties dealing with his clients or employer in connection with work for which he is responsible.

Section 11—The Engineer will not compete unfairly with another engineer by attempting to obtain employment or advancement or professional engagements by competitive bidding, by taking advantage of a salaried position, by criticizing other engineers, or by other improper or questionable methods.

a. The Engineer will not attempt to supplant another engineer in a particular employment after becoming aware that definite steps have been taken toward the other's employment.

b. He will not pay, or offer to pay, either directly or indirectly, any commission, political contribution, or a gift, or other consideration in order to secure work, exclusive of securing salaried positions through employment agencies.

c. He shall not solicit or submit engineering proposals on the basis of competitive bidding. Competitive bidding for professional engineering services is defined as the formal or informal submission, or receipt, of verbal or written estimates of cost or proposals in terms of dollars, man days of work required, percentage of construction cost, or any other measure of compensation whereby the prospective client may compare engineering services on a price basis prior to the time that one engineer, or one engineering organization, has been selected for negotiations. The disclosure of recommended fee schedules prepared by various engineering societies is not considered to constitute competitive bidding. An Engineer requested to submit a fee proposal or bid prior to the selection of an engineer or firm subject to the negotiation of a satisfactory contract, shall attempt to have the procedure changed to conform to ethical practices, but if not successful he shall withdraw from consideration for the proposed work. These principles shall be applied by the Engineer in obtaining the service of other professionals.

d. An Engineer shall not request, propose, or accept a professional commisson on a contingent basis under circumstances in which his professional judgment may be compromised, or when a contingency provision is used as a device for promoting or securing a professional commission.

e. While in a salaried position, he will accept part-time engineering work only at a salary or fee not less than that recognized as standard in the area.

f. An Engineer will not use equipment, supplies, laboratory, or office facilities of his employer to carry on outside private practice without consent.

Section 12—The Engineer will not attempt to injure, maliciously or falsely, directly or indirectly, the professional reputation, prospects, practice or employment of another engineer, nor will he indiscriminately criticize another engineer's work. If he believes that another engineer is guilty of unethical or illegal practice, he shall present such information to the proper authority for action.

a. An Engineer in private practice will not review the work of another engineer for the same client, except with the knowledge of such engineer, or unless the connection of such engineer with the work has been terminated.

b. An Engineer in governmental, industrial or educational employ is entitled to review and evaluate the work of other engineers when so required by his employment duties.

c. An Engineer in sales or industrial employ is entitled to make engineering comparisons of his products with products by other suppliers.

Section 13—The Engineer will not associate with or allow the use of his name by an enterprise of questionable character, nor will be become professionally associated with engineers who do not conform to ethical practices, or with persons not legally qualified to render the professional services for which the association is intended.

a. He will conform with registration laws in his practice of engineering.

b. He will not use association with a nonengineer, a corporation, or partnership, as a "cloak" for unethical acts, but must accept personal responsibility for his professional acts.

Section 14—The Engineer will give credit for engineering work to those to whom credit is due, and will recognize the proprietary interests of others.

a. Whenever possible, he will name the person or persons who may be individually responsible for designs, inventions, writings, or other accomplishments.

b. When an Engineer uses designs supplied to him by a client, the designs remain the property of the client and should not be duplicated by the Engineer for others without express permission.

c. Before undertaking work for others in connection with which he may make improvements, plans, designs, inventions, or other records which may justify copyrights or patents, the Engineer should enter into a positive agreement regarding the ownership.

d. Designs, data, records, and notes made by an engineer and referring exclusively to his employer's work are his employer's property.

Section 15—The Engineer will cooperate in extending the effectiveness of the profession by interchanging information and experience with other engineers and students, and will endeavor to provide opportunity for the professional development and advancement of engineers under his supervision.

a. He will encourage his engineering employees' efforts to improve their education.

b. He will encourage engineering employees to attend and present papers at professional and technical society meetings.

c. He will urge his engineering employees to become registered at the earliest possible date.

d. He will assign a professional engineer duties of a nature to utilize his full training and experience, insofar as possible, and delegate lesser functions to subprofessionals or to technicians.

e. He will provide a prospective engineering employee with complete information on working conditions and his proposed status of employment, and after employment will keep him informed of any changes in them.

Note: In regard to the question of application of the Code to corporations vis-a-vis real persons, business form or type should not negate nor influence conformance of individuals to the Code. The Code deals with professional services, which services must be performed by real persons. Real persons in turn establish and implement policies within business structures. The Code is clearly written to apply to the Engineer and it is incumbent on a member of NSPE to endeavor to live up to its provisions. This applies to all pertinent sections of the Code.

PROBLEM 3-1

An engineer, A, is asked to design, estimate the cost of, and construct a bridge, based on the surveys and data previously supplied by another engineer, B.

Is it proper or wise on the part of A to accept the responsibility involved without personally investigating or verifying the accuracy of the surveys and data supplied by B?

Answer

It would be unwise, in view of the responsibility to the client assumed by A, for him to proceed with the work without personally satisfying himself of the adequacy and accuracy of the data supplied by B, unless it is distinctly understood or stated in the terms of A's engagement that he should assume no responsibility for the correctness of the data supplied him. Such review and verification is desirable in order to protect both the client and the engineer from the consequences of erroneous data or misunderstood conditions.

PROBLEM 3-2

An engineer, A, received a circular letter from a patent attorney, B, asking to solicit or turn over to B legal work in patent cases, and offering to share B's fees with A in such cases. B gives in this letter the names of five engineers as business references.

(A) Would it be unethical for A to accept and act on B's proposition?

(B) Is it unethical for engineers to permit the use of their names as references in business enterprises or for the promotion of business schemes?

Answer

(A) The sharing of fees or profits as compensation for soliciting engagements is regarded among engineers as unethical. This does not apply to the division of fees in cases where two or more engineers are jointly employed or engaged upon the same work.

(B) It is regarded as at least imprudent for an engineer to permit the use of his name as a reference regarding the character and responsibility of persons or business concerns for advertising purposes. The practice might lead not only to personal embarrassment, but to discredit of the profession.

PROBLEM 3-3

An engineer is employed by a client to design, make plans for, and superintend the construction of an important work, his compensation to be based upon a percentage of the cost of the completed work. At a certain stage of the progress of the work, the client decides, or is compelled to abandon or suspend the work, through no fault of the engineer.

Is the engineer entitled to claim and receive the whole amount of the fee that would be due him if the work had gone on to completion?

Answer

Such contingencies should be provided for in the original agreement between engineer and client. In the absence of such a provision, and unless the circumstances should warrant a different action, and particularly where the abandonment occurs through no fault or bad faith on the part of the client, settlement upon an equitable basis for the work already done, and the expenses and obligations already incurred, would be fair and just to both parties.

PROBLEM 3-4

A city, by public advertisement, asked for proposals from engineers to perform certain professional services, the obvious purpose being to secure the services at the lowest possible cost.

Should engineers recognize or take part in such public competitions for services where it is to be presumed that the lowest bidder will be engaged for the services?

Answer

This method of procuring engineering services should be discouraged. It is not regarded as ethical for reputable engineers to enter such competitions. This does not, however, apply to competitions for designs for a specific structure where such competition is properly conducted and provision is made for reasonable compensation for rejected designs. The Institution of Civil Engineers of Great Britain, in a recent circular, urges that engineers should not enter into competition at the invitation of municipalities because of the lack of equity, dignity, and proper compensation offered.

PROBLEM 3-5

An engineer was employed by a client to make an investigation and report upon a specific matter relating to the client's business. In the course of the investigation the engineer made discoveries which, while foreign to the subject of his specific engagement, were so related to the business of the client as to be of great importance to him and, if disclosed, might seriously affect that business. These discoveries were mentioned in the engineers report. The engineer later, at his own expense, pursued the discoveries further and they appeared to be of great industrial value and scientific interest.

Does the present ethical relation of the engineer to his former client permit him to develop these discoveries for his personal use or profit, or to publish an account of them in the interest of science?

Answer

Insofar as the case presented involved questions of the proper relations between engineer and client, the Committee would reaffirm the broad, general principle that the engineer should not make use of information or discoveries, or the results therefrom, obtained while in the service of a client, in any manner adverse to the interests of the client.

PROBLEM 3-6

City and County authorities engaged an engineer to advise them in regard to providing increased navigation facilities and other improvements connected therewith. The engineer made necessary surveys and studies, recommended projects, made plans therefore, and secured the approval of them by the Department of Defense. The projects and plans were approved and adopted by the municipal authorities and the engineer was paid his fee for the service. Later he was recalled for further conference, represented the local authorities at public hearings, and when some changes were requested by the U. S. Assistant Engineer in charge, prepared revised plans and obtained the final approval of the Department of Defense. He was also asked to submit a proposition to carry out as engineer the construction work, which he did on the basis of a percentage of the cost. For these later services a merely nominal fee was charged in expectation that he would be retained to carry out the work. Shortly thereafter, the authorities, without notice to him, awarded the engineering work to a firm of local engineers upon exactly the same terms as named by the first engineer except for some minor changes as to time of payments.

Was the action of the firm of local engineers, in securing or accepting the engagement, in accordance with correct professional conduct?

Answer

Assuming the statement is substantially correct and full, the firm of local engineers must have known the facts of the situation, including the terms named by the prior engineer. The civil authorities must have acted with the consent, if not at the active solicitation, directly or indirectly, of the firm of local engineers. Under these circumstances it was grossly unprofessional for them to supplant the previous engineer. Section 8 of the Code of Ethics of the American Institute of Consulting Engineers, declares it to be "unprofessional and inconsistent with honorable and dignified bearing...to attempt to supplant a fellow engineer after definite steps have been taken toward his employment." While the public authorities concerned are not amenable to canons of professional ethics and while they should not be condemned without a hearing, the facts stated seem to warrant the conclusion that commonly accepted principles of fairness and just dealing were violated by them in the action taken, presumably without notice to the engineer whose previous work they had approved.

PROBLEM 3-7

Municipalities and other organizations sometimes, by public advertisement, invite competitive designs for engineering works.

(A) Is it considered proper professional conduct for engineers to enter into such competitions?

(B) If question (A) is answered in the affirmative, under what conditions may engineers enter such competitions?

Answer

(A) Reputable engineers generally discourage such competitions, believing that as a rule, they do not result in benefit to either the client or the engineer. But the practice is likely to continue in some lines of engineering work, and there appears to be no valid reason why engineers should not participate therein—provided that the conditions and the terms under which the competitions are to be conducted are fair and honorable.

(B) Some of the conditions and terms that make it permissible for engineers to compete in such cases are the following:

(1) That the competition shall be confined to engineers of good standing, preferably to a limited number designated by the client.

(2) That proper and adequate rules and regulations to control the competition shall be submitted by the client.

(3) That provision be made for payment to each competitor of a specified sum to cover the reasonable estimated cost of preparing a design.

(4) That the client shall specify his want or requirements in sufficient detail to enable the engineer to design intelligently.

(5) That the official body, board, or persons who are to pass upon the merits of the designs shall be designated in advance.

(6) That unsuccessful designs shall remain the property of the designer and be returned to him; and that the successful competitor shall be retained as chief or consulting engineer for the construction work.

PROBLEM 3-8

In what ways and to what extent is it considered proper for engineers to publicly advertise their professional business?

Answer

It is generally accepted in the profession that advertising by engineers should be restricted to modest, brief, and dignified forms such as announcements and professional cards in the public press. Professional cards may give the name and address of the person or firm, the branch of engineering and the specialities in which he practices and the names of the professional societies to which he belongs. Laudatory notices, articles, or accounts of professional work in the secular press should not be inspired or encouraged by the engineer.

PROBLEM 3-9

Would it be an infraction of professional ethics for a member of an engineering commission to divulge or publish the findings of the commission before its report has been agreed upon and submitted?

Answer

The results of the studies or actions of an engineering commission and its findings are to be regarded as the property of the commisson and not of its members individually. It would be improper as well as discourteous to their associates, and therefore unethical, for the individual members to divulge or publish any part of the results of findings of the commission in advance of the submission of its report, except with the official authorization or permission of the commission, and the consent of the principals involved.

PROBLEM 3-10

The Engineer-in-Chief of a corporation designed for it a structure and affixed his official signature to the plans therefore.

Is it ethical for his official successor to erase such signature and substitute his own in place thereof?

Answer

Assuming that the original plans are unchanged by the successor, it would be dishonorable, therefore a violation of professional ethics, to erase the designer's signature and substitute his own. He might properly endorse the plans "approved by" over his own signature. If changes in details, not requiring new drawings, are made in the original plans, the endorsement should be "approved as modified by."

PROBLEM 3-11

An engineer was retained by a client to advise him in reference to an engineering project and structure. The engineer submitted with his report complete drawings of the structure recommended by him. The client, with the permission of the engineer, submitted the plans to an engineering-contracting company, inviting a proposal to do the work. With such proposal the engineering-contracting company submitted a drawing identical with that of the engineer except that a few minor details and additions were included. Upon this drawing were stamped the words, "Copyrighted by XY" (the title of the company).

Was it a violation of professional ethics for XY to copyright the drawing under the conditions named?

Answer

The right of XY to copyright plans or documents originated and prepared by them cannot be questioned. This is also true of changes made by them in the designs and drawings prepared by others, provided it is made clear that the original plans are not their own, and that the changes or additions made by them upon which copyright is claimed are specifically described. In the case before us, it appears that this was not done. XY had no title by authorship, ownership, or assignment to the original drawings, and therefore no right to appropriate and copyright them. By copyrighting them the engineer himself is debarred from using them without their consent or compulsion. The copyright leads to the natural inference that the plans were designed and prepared wholly by them. To copyright the drawings under these conditions, whether intentionally or inadvertently, was morally wrong, unjust, and dicourteous to the engineer, and therefore a flagrant violation of professional ethics.

PROBLEM 3-12.

An engineer is appointed to a State Board which, by legislative enactment, is required to pass on plans, specifications, and designs of certain types of municipal improvements. The Board also has jurisdiction over, and is required to inspect and report upon, the operation of certain types of municipal plants. This engineer, while sitting on the Board and passing upon plans and operations of the plants mentioned, practices in the state in which the Board upon which he sits has jurisdiction, and prepares plans and specifications for plants which must be passed upon by the Board, and also installs and operates plants, the inspection and operation of which is a function of this Board to regulate.

Is it proper, or in accordance with professional ethics, for an engineer, while serving as a member of a public commission, to accept professional engagements that involve, or may involve, questions or matters that will come before the commission for official decision or action?

Answer

A fairly parallel question would be: Is it proper and in accordance with professional ethics for a Judge to consult with, advise, and prepare for a client, a suit at law that is to be tried before him as presiding judge? The answer is as obvious in one case as in the other. An engineer who accepts a public or quasi-public appointment under the conditions named in the statement is therby debarred, not only by professional ethics, but by moral obligations and sound public policy from accepting any engagement involving matters that may eventually come before him in his public capacity for consideration and decision. Evidence of intentional violation of this principle should be sufficient cause for discipline or removal of the offender.

PROBLEM 3-13

A consulting engineer has a client, A, for whom he is making certain important investigations, pending which an attorney, B, asks the engineer to accept a retainer as expert witness in a suit at law against A. The matter on which the engineer's testimony is wanted by B is totally unrelated to the matter being investigated for A.

(A) Is it proper for the engineer to accept the retainer from B?

(B) If the engineer may properly accept the retainer from B, should A be advised of the acceptance?

Answer

(A) While there is room for a difference of opinion from the strictly ethical point of view, the acceptance of B's retainer, under the conditions named, should be regarded as so questionable as to be unwise and improper.

(B) In any event, assuming that, as stated, the matters in regard to which the engineer is employed by A and sought to be retained by B have no connection with, relation to, or bearing upon each other, the engineer should not accept B's retainer without the full knowledge and consent of both A and B.

PROBLEM 3-14

A civil engineer, specializing in a certain branch of professional work, is requested through an explanatory circular, enclosing a blank form for reply, issued by the official board of an important public enterprise, to express his views as to the design and construction of a proposed public work in line of his special practice. No retainer or compensation is suggested and obviously none is intended. The request comes from a distant city and relates to an enterprise with which the engineer has no actual present or prospective professional or personal connection. He declined to reply in the manner requested but called attention to the fact that there are a number of experts in this country capable of giving them sound professional advice.

(A) Was such a reply warranted and proper from an ethical point of view, under the conditions named?

(B) Recognizing that engineers, like the members of other professions, owe to the public, as specially qualified good citizens, a certain amount of gratuitous advice and assistance in the planning and administration of works and affairs affecting the public welfare, what general principles or rules should guide their conduct in such matters?

Answer

(A) Such a reply was warranted under the circumstances recited.

(B) It is difficult, if not impossible, to sharply define the character or extent of the professional services that the engineer properly may or should contribute to the public without pecuniary compensation therefore. It may vary with conditions and circumstances in each individual case. The patriotic and public-spirited citizen will construe his duty in this respect liberally in favor of Nation, State, civic organizations, and community. Like members of other professions, the engineer is frequently called to serve on public commissions, committees, and civic organizations where his specialized knowledge and experience enable him to render services of great public value, for which no monetary compensation is provided or expected. Service of this character within proper bonds is to be commended and encouraged. As a general rule, gratuitous professional service may properly be limited to information, advice, and assistance of a general nature relating to the inception, principle involved, or the development of contemplated public enterprises. Specific recommendations, plans, details, specifications, and estimates for definite projects involving professional investigation and study should be regarded as the province of the formally retained engineer. These functions are essentially professional in character and involve special labor and responsibility for which the prudent engineer, acting in his capacity as a citizen only, may not care and should not be expected or required to assume without the usual professional recognition and compensation. The interests of the profession, as well as his own, suggest that his generous impulses be not allowed too great latitude, particularly in cases like the one cited, in which he has no public or community interest.

PROBLEM 3-15

An engineering firm, XY, inserted in the advertising pages of a prominent daily newspaper a display advertisement in large type, stating that the firm had been selected by a State Valuation Commission to ascertain the value of a public utility of the State and was then engaged upon the work.

(A) Should the inserting of such an advertisement in newspapers be considered as professional advertising?

(B) Was such an advertisement a violation of professional. ethics?

Answer

(A) Since the obvious purpose was to exploit the competency and high character of the firm as engineers, it was clearly a professional advertisement.

(B) Advertising of this character is held to be a violation of professional ethics by three leading professions in this country: Engineering, Law, and Medicine. The Code of Ethics of the American Society of Civil Engineers reads as follows: "It shall be considered unprofessional and inconsistent with honorable and dignified bearing for any member of the Society of Civil Engineers...to advertise in self-laudatory language, or in any other manner derogatory to the dignity of the profession." The Code of Ethics of the American Society of Mechanical Engineers contains this clause: "He should not advertise in an undignified, sensational, or misleading manner."

PROBLEM 3-16

A firm of engineers, XY, was retained some years since by a city to make a preliminary report upon a certain municipal improvement. Recently, the City Council decided to proceed with these improvements and authorized the Director of Public Service to employ XY as engineers to design the works, which he proceeded to do upon terms mutually agreed upon. Later, his attention was called to an opinion, rendered some years previously by the Attorney General of the State, to the effect that under the State laws and the City Charter the employment of technical assistance came under the provision of a law requiring that competitive bids must be asked for in all cases where the expenditure for services exceeded $500. Notwithstanding doubt as to the validity of this opinion, the city authorities thought it prudent to comply, and bids for the service were asked for and received. XY submitted a bid at substantially the same rate of compensation as that agreed upon prior to the public advertisement, and while lower bids were submitted the contract was awarded to XY. Prospective bidders were given a blank form of contract to be entered into between the Director of Public Service and the successful bidder, which XY was required to execute. This contract, as is usual in such cases, is so inequitable and unsatisfactory that, under ordinary conditions, high class, responsible engineers would probably refuse to submit a bid under its provisions.

Considering the exceptional circumstances in this case, should the action of XY in submitting a bid for the work be regarded as professionally unethical?

Answer

Regarding the practice of competitive bidding for engineering services the Committee reaffirms the principles stated in its answer to Problem 3-4: "This method of procuring engineering services should be discouraged . . . it is not regarded as ethical for reputable engineers to enter such competitions." This does not, however, apply to competitions for designs for a specific structure where such competition is properly conducted and provision is made for reasonable compensation for rejected designs. The conditions in this case were so unusual as to make it exceptional. Bids for service were asked for only to comply with alleged legal requirements, after XY had been retained in a proper manner; and XY submitted their bid at the same rate of compensation for the service as that in the earlier retainer. Under the circumstances the conduct of XY should not be held as unprofessional.

PROBLEM 3-17

In a case involving the respective rights of a water supply company owning the surface water rights over certain lands, and the owners of the lands, a (supposedly) friendly suit at law is instituted to determine the legal status of the parties. An engineer employed in the inception of the water supply project is asked by the land owners to accept a retainer as expert witness in their behalf. The engineer consulted with officials of the water company, who stated that they (the water company) had intended to retain him as expert witness in their case, but that there would be no objection whatever to his testifying on the other side, since his testimony would no doubt be the same in either case. After further conference with both parties and having their assurance that his action would be satisfactory to both, he accepted the retainer of the land owners.

Was the action of the engineer in accordance with correct professional ethics?

Answer

The circumstances in this case are somewhat unusual and differ in some respects from those stated in Problem 3-13, inasmuch as the suit at law is represented as being a "friendly" one to determine the legal status of the parties thereto, and that the retaining of the engineer by one of the parties was with full knowledge and consent of the other. The acceptance of the retainer under these conditions was permissible and should not be regarded as unethical.

PROBLEM 3-18

There was printed in the advertising pages of a leading engineering periodical a prominent advertisement reciting facts, relating to an important construction project, characterized as "a record job—constructed in record time, and within the estimated cost under a contract awarded to the ABC construction company." The advertisement concluded with the name of the construction company, "designers and constructors of hydroelectric and steam power plants, factories, warehouses, dry docks and water terminals," followed immediately and ending with the name of an engineering firm, XYZ, designated as "Engineers and Managers" with the names of the individual members of the firm. The form and wording warranted the inference that the firm and the corporation were identical. Investigation disclosed that this is substantially the case; the two concerns, while being different and distinct organizations, are practically under the same management and control, though one member of the engineering firm has no connection with the corporation. Such being the case, it is justifiable to conclude that the advertisement was authorized and approved by XYZ and that they were responsible therefore.

Under the conditions stated, should the advertisement be regarded as a violation of professional ethics by XYZ?

Answer

Yes. (Action approved November 23, 1951.)

PROBLEM 3-19

An engineer was retained by a client to make an examination and report for which it was agreed that a certain definite fee was to be paid. The work was completed and the report in duplicate rendered, and a partial payment on account of the fee made to the engineer; but the balance due as per agreement was not paid and remains unpaid after six years. The report having been lost or mislaid by the client, he now asks for a duplicate of data contained therein, which the engineer declines to supply until the original account shall be settled.

Is the engineer justified in refusing to supply the data asked for by the client until the original fee shall be paid in full?

Answer

Assuming the conditions and circumstances to be fully stated, the action of the engineer is justifiable. (Action approved January 10, 1952.)

PROBLEM 3-20

When may an engineer sign and seal work that is not his own?

Answer

Only when he has the written consent of the originator. He should also be aware that his signature is a written approval of the contents therein. If the work is already signed and turned over to the second engineer, the second engineer can co-sign or sign over a signature, assuming he is in agreement with it. He should never erase or remove another's signature. If changes are made by the second engineer, he should clearly mark the changes, date them, and sign the document "approved as modified by." (See also Problem 3-10.)

PROBLEM 3-21

An engineer, a graduate of a technical institution of the highest standing, had been in the employ of a firm of engineers for several years, representing them as resident engineer at the manufacturing plants of their different clients during engagements at each location of six months or more. On one engagement the type of service rendered to the client, a manufacturing firm, was management-engineering, advising and assisting the client in reducing cost and developing best methods of production, accounting, and sales. The service was contracted and rendered on the basis of service of the engineering firm and not of an individual member of the firm or its staff. While the resident engineer was necessarily in constant touch with the principals of the client organization, the job was thoroughly supervised by executives of the engineering firm by frequent visits to the plant, formulation of plans and policies, and control of the work progress. The work proceeded for a period of a little over six months with frequent expressions by the client of satisfaction with the results, and with an understanding—implied but not then expressed—that the engagement would be continued for an appreciable period. This was in fact recognized by all concerned. At this time the resident engineer was approached by the vice-president and large stockholder of the client company with an offer of direct employment, at a salary over 50% higher than the engineer was receiving from his firm, to begin at the expiration of one month's notice. He then accepted an invitation to confer with the president of the client company and at that conference accepted the offer. Not until after this did he mention it in any way to his own firm. In the conference which ensued between officers of the client company and the engineering company, the president of the client company (who closed the original agreement for service) stated that he had questioned whether the matter should not first be brought to the attention of the engineering company, but the engineer had reiterated that he was under no obligation to them. At this conference the statement was further made by the president of the client company that the work of the engineering firm had been satisfactory and that there was more that must be done. He virtually agreed in fact that there was six months more work and that the returns would be increasingly evident as this progressed. There appears to be no legal claim by the engineering company upon the client since one month's notice is sufficient for closure.

Was the engineer morally and ethically justified in:

(A) Dealing with the client without reference to his employers?

(B) Accepting a position without consulting his employers?

(C) Forcing the termination of his employer's service at appreciable loss in money and prestige to them and at a detriment to the client through less satisfactory service?

Answer

To these questions, negative answers were given on all three counts, because the action of the engineer was inconsistent with the best tenets of the profession.

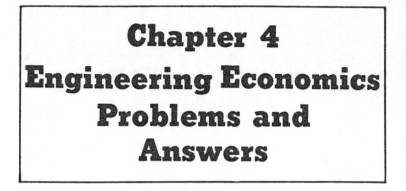

Chapter 4
Engineering Economics Problems and Answers

All the problems in this chapter deal with engineering economics and require the use of one or more of the six interest tables, labeled B-1 through B-6. Preceding these interest tables are two cross-reference tables, table A-1 and B-1, both of which categorize the various types of economics problems. Depending on your skills and interests, you may find one reference table easier to use than the other.

The two reference tables differ primarily concerning single payments versus multiple payments. Table A-1 separates the various interest formulas and categorizes them into one of three economics symbols commonly used in economics books. Table A-2 also categorizes the problems, but shows the correct interest table to use in this chapter as well as the pertinent interest formula.

In all cases, the P.E. student needs only to carefully read the economics problem and to decide on the category and formula (from one of the cross-reference tables) and then the correct interest table to be used. There are a great number of problems of all types included in this chapter, but with the proper equation and interest table, you can quickly solve most of them.

PROBLEM 4-1
(A) Show how the yearly cost of owning a machine is approximately equal to its straight-line yearly depreciation plus the average yearly interest on its cost.

Table B-1. Values of $(1 + i)^n$ Compound Amount Factor (S^n); $A_0 P(1 + i)^n$

n	2 %	2½ %	3 %	3½ %	4 %	4½ %	5 %	6 %	7 %	8 %
1	1.020	1.025	1.030	1.035	1.040	1.045	1.050	1.060	1.070	1.080
2	1.040	1.051	1.061	1.071	1.082	1.092	1.103	1.124	1.145	1.166
3	1.061	1.077	1.093	1.109	1.125	1.141	1.158	1.191	1.225	1.260
4	1.082	1.104	1.126	1.148	1.170	1.193	1.216	1.262	1.311	1.360
5	1.104	1.131	1.159	1.188	1.217	1.246	1.276	1.338	1.403	1.469
6	1.126	1.160	1.194	1.229	1.265	1.302	1.340	1.419	1.510	1.587
7	1.149	1.189	1.230	1.272	1.316	1.361	1.407	1.504	1.606	1.714
8	1.172	1.218	1.267	1.317	1.369	1.422	1.477	1.594	1.718	1.851
9	1.195	1.249	1.305	1.363	1.423	1.486	1.551	1.689	1.838	1.999
10	1.219	1.280	1.344	1.411	1.480	1.553	1.629	1.791	1.967	2.159
11	1.243	1.312	1.384	1.460	1.539	1.623	1.710	1.898	2.105	2.332
12	1.268	1.345	1.426	1.511	1.601	1.696	1.796	2.012	2.252	2.518
13	1.294	1.379	1.469	1.564	1.665	1.772	1.886	2.133	2.410	2.720
14	1.319	1.413	1.513	1.619	1.732	1.852	1.980	2.261	2.579	2.937
15	1.346	1.448	1.558	1.675	1.801	1.935	2.079	2.397	2.759	3.172
16	1.373	1.485	1.605	1.734	1.873	2.022	2.183	2.540	2.952	3.426
17	1.400	1.522	1.653	1.795	1.948	2.113	2.292	2.693	3.159	3.700
18	1.428	1.560	1.702	1.857	2.026	2.208	2.407	2.854	3.380	3.996
19	1.457	1.599	1.754	1.923	2.107	2.308	2.527	3.026	3.617	4.316
20	1.486	1.639	1.806	1.990	2.191	2.412	2.653	3.207	3.870	4.661
21	1.516	1.680	1.860	2.059	2.279	2.520	2.786	3.400	4.141	5.034
22	1.546	1.722	1.916	2.132	2.370	2.634	2.925	3.604	4.430	5.437
23	1.577	1.765	1.974	2.206	2.465	2.752	3.072	3.820	4.741	5.871
24	1.608	1.809	2.033	2.283	2.563	2.876	3.225	4.049	5.072	6.341
25	1.641	1.854	2.094	2.363	2.666	3.005	3.386	4.292	5.427	6.848
26	1.673	1.900	2.157	2.446	2.772	3.141	3.556	4.549	5.807	7.396
27	1.707	1.948	2.221	2.532	2.883	3.282	3.733	4.822	6.214	7.988
28	1.741	1.996	2.288	2.620	2.999	3.430	3.920	5.112	6.649	8.627
29	1.776	2.046	2.357	2.712	3.119	3.584	4.116	5.418	7.114	9.317
30	1.811	2.098	2.427	2.807	3.243	3.745	4.322	5.743	7.612	10.06
31	1.848	2.150	2.500	2.905	3.373	3.914	4.538	6.088	8.145	10.87
32	1.885	2.204	2.575	3.007	3.508	4.090	4.765	6.453	8.715	11.74
33	1.922	2.259	2.652	3.112	3.648	4.274	5.003	6.841	9.325	12.68
34	1.961	2.315	2.732	3.221	3.794	4.466	5.253	7.251	9.978	13.69
35	2.000	2.373	2.814	3.334	3.946	4.667	5.516	7.686	10.68	14.79
40	2.208	2.685	3.262	3.959	4.801	5.816	7.040	10.29	14.97	21.72
45	2.438	3.038	3.782	4.702	5.841	7.248	8.985	13.76	21.00	31.92
50	2.692	3.437	4.384	5.585	7.107	9.033	11.47	18.42	29.46	46.90
55	2.972	3.889	5.082	6.633	8.646	11.26	14.64	24.65	41.32	68.91
60	3.281	4.400	5.892	7.878	10.52	14.03	18.68	32.99	57.95	101.3
65	3.623	4.978	6.830	9.357	12.80	17.48	23.84	44.14	81.27	148.8
70	4.000	5.632	7.918	11.11	15.57	21.78	30.43	59.08	114.0	218.6
75	4.416	6.372	9.179	13.20	18.95	27.15	38.83	79.06	159.9	321.2
80	4.875	7.210	10.64	15.68	23.05	33.83	49.56	105.8	224.2	472.0
85	5.383	8.157	12.34	18.62	28.04	42.16	63.25	141.6	314.5	693.5
90	5.943	9.229	14.30	22.11	34.12	52.54	80.73	189.5	441.1	1,019.0
95	6.562	10.44	16.58	26.26	41.51	65.47	103.0	253.5	618.7	1,497.0
100	7.245	11.81	19.22	31.19	50.50	81.59	131.5	339.3	867.7	2,200.0

(B) Explain why depreciation cannot be deducted for income tax purposes if a reserve for replacement is set up out of income.

Answer

(A) This is only one of the methods used. The argument is that: (1) the straight-line depreciation is a convenient method of determining the amount of capital consumed in annual

Table B-2. Values of $\dfrac{1}{(1+i)^n}$; —Present Worth Factor (p^n) = $\dfrac{A}{(1+i)^n}$

n	2%	2½%	3%	3½%	4%	4½%	5%	6%	7%	8%
1	0.9804	0.9756	0.9709	0.9662	0.9615	0.9569	0.9524	0.9434	0.9346	0.9259
2	0.9612	0.9518	0.9426	0.9335	0.9246	0.9157	0.9070	0.8900	0.8734	0.8573
3	0.9423	0.9286	0.9151	0.9019	0.8890	0.8763	0.8638	0.8396	0.8163	0.7938
4	0.9238	0.9060	0.8884	0.8714	0.8548	0.8386	0.8227	0.7921	0.7629	0.7350
5	0.9057	0.8839	0.8626	0.8420	0.8219	0.8025	0.7835	0.7473	0.7130	0.6806
6	0.8880	0.8623	0.8375	0.8135	0.7903	0.7679	0.7462	0.7050	0.6663	0.6302
7	0.8706	0.8413	0.8131	0.7860	0.7599	0.7348	0.7107	0.6651	0.6227	0.5835
8	0.8535	0.8207	0.7894	0.7594	0.7307	0.7032	0.6768	0.6274	0.5820	0.5403
9	0.8638	0.8007	0.7664	0.7337	0.7026	0.6729	0.6446	0.5919	0.5439	0.5002
10	0.8203	0.7812	0.7441	0.7089	0.6756	0.6439	0.6139	0.5584	0.5083	0.4632
11	0.8043	0.7621	0.7224	0.6849	0.6496	0.6162	0.5847	0.5268	0.4751	0.4289
12	0.7885	0.7436	0.7014	0.6618	0.6246	0.5897	0.5568	0.4970	0.4440	0.3971
13	0.7730	0.7254	0.6810	0.6394	0.6006	0.5543	0.5303	0.4688	0.4150	0.3677
14	0.7579	0.7077	0.6611	0.6178	0.5775	0.5400	0.5051	0.4423	0.3878	0.3405
15	0.7430	0.6905	0.6419	0.5969	0.5553	0.5167	0.4810	0.4173	0.3624	0.3152
16	0.7284	0.6736	0.6232	0.5767	0.5339	0.4945	0.4581	0.3936	0.3387	0.2919
17	0.7142	0.6572	0.6050	0.5572	0.5134	0.4732	0.4363	0.3714	0.3166	0.2703
18	0.7002	0.6412	0.5874	0.5384	0.4936	0.4528	0.4155	0.3503	0.2959	0.2502
19	0.6864	0.6255	0.5703	0.5202	0.4746	0.4333	0.3957	0.3305	0.2765	0.2317
20	0.6730	0.6103	0.5537	0.5026	0.4564	0.4146	0.3769	0.3118	0.2584	0.2145
21	0.6598	0.5954	0.5375	0.4856	0.4388	0.3968	0.3589	0.2942	0.2415	0.1987
22	0.6468	0.5809	0.5219	0.4692	0.4220	0.3797	0.3418	0.2775	0.2257	0.1839
23	0.6342	0.5667	0.5067	0.4533	0.4057	0.3634	0.3256	0.2618	0.2109	0.1703
24	0.6217	0.5529	0.4919	0.4380	0.3901	0.3477	0.3101	0.2470	0.1971	0.1577
25	0.6095	0.5394	0.4776	0.4231	0.3751	0.3327	0.2953	0.2330	0.1842	0.1460
26	0.5976	0.5262	0.4637	0.4088	0.3607	0.3184	0.2812	0.2198	0.1722	0.1352
27	0.5859	0.5134	0.4502	0.3950	0.3468	0.3047	0.2678	0.2074	0.1609	0.1252
28	0.5744	0.5009	0.4371	0.3817	0.3335	0.2916	0.2551	0.1956	0.1504	0.1159
29	0.5631	0.4887	0.4243	0.3687	0.3207	0.2790	0.2429	0.1846	0.1406	0.1073
30	0.5521	0.4767	0.4120	0.3563	0.3083	0.2670	0.2314	0.1741	0.1314	0.0994
31	0.5412	0.4615	0.4000	0.3442	0.2965	0.2555	0.2204	0.1643	0.1228	0.0920
32	0.5306	0.4538	0.3883	0.3326	0.2851	0.2445	0.2099	0.1550	0.1147	0.0852
33	0.5202	0.4427	0.3770	0.3213	0.2741	0.2340	0.1999	0.1462	0.1072	0.0789
34	0.5100	0.4319	0.3660	0.3105	0.2636	0.2239	0.1904	0.1379	0.1002	0.0730
35	0.5000	0.4214	0.3554	0.3000	0.2534	0.2143	0.1813	0.1301	0.0937	0.0676
40	0.4529	0.3724	0.3066	0.2526	0.2083	0.1719	0.1420	0.0972	0.0668	0.0460
45	0.4102	0.3292	0.2644	0.2127	0.1712	0.1380	0.1113	0.0727	0.0476	0.0313
50	0.3715	0.2909	0.2281	0.1791	0.1407	0.1107	0.0872	0.0543	0.0339	0.0213
55	0.3365	0.2572	0.1968	0.1508	0.1157	0.0888	0.0683	0.0406	0.0242	0.0145
60	0.3048	0.2273	0.1697	0.1269	0.0951	0.0713	0.0535	0.0303	0.0173	0.0099
65	0.2761	0.2009	0.1464	0.1069	0.0781	0.0572	0.0419	0.0227	0.0123	0.0067
70	0.2500	0.1776	0.1263	0.0900	0.0642	0.0459	0.0329	0.0169	0.0088	0.0046
75	0.2265	0.1569	0.1089	0.0758	0.0528	0.0368	0.0258	0.0127	0.0063	0.0031
80	0.2051	0.1337	0.0940	0.0638	0.0434	0.0296	0.0202	0.0095	0.0045	0.0021
85	0.1858	0.1226	0.0811	0.0537	0.0357	0.0237	0.0158	0.0071	0.0032	0.0014
90	0.1683	0.1084	0.0699	0.0452	0.0293	0.0190	0.0124	0.0053	0.0023	0.0010
95	0.1524	0.0958	0.0603	0.0381	0.0241	0.0153	0.0097	0.0039	0.0016	0.0007
100	0.1380	0.0846	0.0520	0.0321	0.0198	0.0123	0.0076	0.0029	0.0012	0.0005

operations, and (2) the interest or expected return is most reasonably based on the *average* book values during the life of the asset.

(B) If a *reserve* is set up out of *income* it is assumed that the *net* income and consequent profit would be reduced by the

Table B-3. Value of $i = \dfrac{(1 + i)^n - 1}{i}$ $S = D\left[\dfrac{(1 + i)^n - 1}{i}\right]$

n	2 %	2½ %	3 %	3½ %	4 %	4½ %	5 %	6 %	7 %	8 %
1	1.000	1.000	1.000	1.000	1.000	1.000	1.000	1.000	1.000	1.000
2	2.020	2.025	2.030	2.035	2.040	2.045	2.050	2.060	2.070	2.080
3	3.060	3.076	3.091	3.106	3.122	3.137	3.153	3.184	3.215	3.246
4	4.122	4.153	4.184	4.215	4.246	4.278	4.310	4.375	4.440	4.506
5	5.204	5.256	5.309	5.362	5.416	5.471	5.526	5.637	5.751	5.867
6	6.308	6.388	6.468	6.550	6.633	6.717	6.802	6.975	7.153	7.336
7	7.434	7.547	7.662	7.779	7.898	8.019	8.142	8.394	8.654	8.923
8	8.583	8.736	8.892	9.051	9.214	9.380	9.549	9.897	10.26	10.64
9	9.755	9.955	10.16	10.37	10.58	10.80	11.03	11.49	11.98	12.49
10	10.95	11.20	11.46	11.73	12.01	12.29	12.58	13.18	13.82	14.49
11	12.17	12.48	12.81	13.14	13.49	13.84	14.21	14.97	15.78	16.65
12	13.41	13.80	14.19	14.60	15.03	15.46	15.92	16.87	17.89	18.98
13	14.68	15.14	15.62	16.11	16.63	17.16	17.71	18.88	20.14	21.50
14	15.97	16.52	17.09	17.68	18.29	18.93	19.60	21.02	22.55	24.21
15	17.29	17.93	18.60	19.30	20.02	20.78	21.58	23.28	25.13	27.15
16	18.64	19.38	20.16	20.97	21.82	22.72	23.66	25.67	27.89	30.32
17	20.01	20.86	21.76	22.71	23.70	24.74	25.84	28.21	30.84	33.75
18	21.41	22.39	23.41	24.50	25.65	26.86	28.13	30.91	34.00	37.45
19	22.84	23.95	25.12	26.36	27.67	29.06	30.54	33.76	37.38	41.45
20	24.30	25.54	26.87	28.28	29.78	31.37	33.07	36.79	41.00	45.76
21	25.78	27.18	28.68	30.27	31.97	33.78	35.72	39.99	44.87	50.42
22	27.30	28.86	30.54	32.33	34.25	36.30	38.51	43.39	49.01	55.46
23	28.84	30.58	32.45	34.46	36.62	38.94	41.43	47.00	53.44	60.89
24	30.42	32.35	34.43	36.67	39.08	41.69	44.50	50.82	58.18	66.76
25	32.03	34.16	36.46	38.95	41.65	44.57	47.73	54.86	63.25	73.11
26	33.67	36.01	38.55	41.31	44.31	47.57	51.11	59.16	68.68	79.95
27	35.34	37.91	40.71	43.76	47.08	50.71	54.67	63.71	74.48	87.35
28	37.05	39.86	42.93	46.29	49.97	53.99	58.40	68.53	80.70	95.34
29	38.79	41.86	45.22	48.91	52.97	57.42	62.32	73.64	87.35	104.0
30	40.57	43.90	47.58	51.62	56.08	61.01	66.44	79.06	94.46	113.3
31	42.38	46.00	50.00	54.43	59.33	64.75	70.76	84.80	102.1	123.3
32	44.23	48.15	52.50	57.33	62.70	68.67	75.30	90.89	110.2	134.2
33	46.11	50.35	55.08	60.34	66.21	72.76	80.06	97.34	118.9	146.0
34	48.03	52.61	57.73	63.45	69.86	77.03	85.07	104.2	128.3	158.6
35	49.99	54.93	60.46	66.67	73.65	81.50	90.32	111.4	138.2	172.3
40	60.40	67.40	75.40	84.55	95.03	107.0	120.8	154.8	199.6	259.1
45	71.89	81.52	92.72	105.8	121.0	138.8	159.7	212.7	285.7	386.5
50	84.58	97.48	112.8	131.0	152.7	178.5	209.3	290.3	406.5	573.8
55	98.59	115.6	136.1	160.9	191.2	227.9	272.7	394.2	575.9	848.9
60	114.1	136.0	163.1	196.5	238.0	289.5	353.6	533.1	813.5	1,253.
65	131.1	159.1	194.3	238.8	295.0	366.2	456.8	719.1	1,147.	1,847.
70	150.0	185.3	230.6	288.9	364.3	461.9	588.5	967.9	1,614.	2,720.
75	170.8	214.9	272.6	348.5	448.6	581.0	756.7	1,301.	2,270.	4,003.
80	193.8	248.4	321.4	419.3	551.2	729.6	971.2	1,747.	3,189.	5,887.
85	219.1	286.3	377.9	503.4	676.1	914.6	1,245.	2,343.	4,479.	8,656.
90	247.2	329.2	443.3	603.2	828.0	1,145.	1,595.	3,141.	6,287.	12,724.
95	278.1	377.7	519.3	721.8	1,013.	1,433.	2,041.	4,209.	8,824.	18,702.
100	312.2	432.5	607.3	862.6	1,238.	1,791.	2,610.	5,638.	12,382.	27,485.

amount of deposit into the reserve. Now if depreciation were *also* treated as an operating cost, profit would again be reduced by the amount of depreciation, which amounts to a *double* deduction for depreciation.

Table B-4. Values of $= \dfrac{i}{(1 + i)^n - 1}$ $\quad D = S\left[\dfrac{i}{(1 + i)^n - 1}\right]$

n	2%	2½%	3%	3½%	4%	4½%	5%	6%	7%	8%
1	1.0000	1.0000	1.0000	1.0000	1.0000	1.0000	1.0000	1.0000	1.0000	1.0000
2	0.4951	0.4938	0.4926	0.4914	0.4902	0.4890	0.4878	0.4854	0.4831	0.4808
3	0.3268	0.3251	0.3235	0.3219	0.3204	0.3188	0.3172	0.3141	0.3111	0.3080
4	0.2426	0.2408	0.2390	0.2373	0.2355	0.2337	0.2320	0.2286	0.2252	0.2219
5	0.1922	0.1903	0.1884	0.1865	0.1846	0.1828	0.1810	0.1774	0.1739	0.1705
6	0.1585	0.1566	0.1546	0.1527	0.1508	0.1489	0.1470	0.1434	0.1398	0.1363
7	0.1345	0.1325	0.1305	0.1285	0.1266	0.1247	0.1228	0.1191	0.1156	0.1121
8	0.1165	0.1145	0.1125	0.1105	0.1085	0.1066	0.1047	0.1010	0.0975	0.0940
9	0.1025	0.1005	0.0984	0.0965	0.0945	0.0926	0.0907	0.0870	0.0835	0.0801
10	0.0913	0.0893	0.0872	0.0852	0.0833	0.0814	0.0795	0.0759	0.0724	0.0690
11	0.0822	0.0801	0.0781	0.0761	0.0742	0.0723	0.0704	0.0668	0.0634	0.0601
12	0.0746	0.0725	0.0705	0.0685	0.0666	0.0647	0.0628	0.0593	0.0559	0.0527
13	0.0681	0.0660	0.0640	0.0621	0.0601	0.0583	0.0565	0.0530	0.0497	0.0465
14	0.0626	0.0605	0.0585	0.0566	0.0547	0.0528	0.0510	0.0476	0.0443	0.0413
15	0.0578	0.0558	0.0538	0.0518	0.0499	0.0481	0.0463	0.0430	0.0398	0.0368
16	0.0537	0.0516	0.0496	0.0477	0.0458	0.0440	0.0423	0.0390	0.0359	0.0330
17	0.0500	0.0479	0.0460	0.0440	0.0422	0.0404	0.0387	0.0354	0.0324	0.0296
18	0.0467	0.0447	0.0427	0.0408	0.0390	0.0372	0.0356	0.0324	0.0294	0.0267
19	0.0438	0.0418	0.0398	0.0379	0.0361	0.0344	0.0328	0.0296	0.0268	0.0241
20	0.0412	0.0391	0.0372	0.0354	0.0336	0.0319	0.0302	0.0272	0.0244	0.0219
21	0.0388	0.0368	0.0349	0.0330	0.0313	0.0296	0.0280	0.0250	0.0223	0.0198
22	0.0366	0.0347	0.0328	0.0309	0.0292	0.0276	0.0260	0.0231	0.0204	0.0180
23	0.0347	0.0327	0.0308	0.0290	0.0273	0.0257	0.0241	0.0213	0.0187	0.0164
24	0.0329	0.0309	0.0291	0.0273	0.0256	0.0240	0.0225	0.0197	0.0172	0.0150
25	0.0312	0.0293	0.0274	0.0257	0.0240	0.0224	0.0210	0.0182	0.0158	0.0137
26	0.0297	0.0278	0.0259	0.0242	0.0226	0.0210	0.0196	0.0169	0.0146	0.0125
27	0.0283	0.0264	0.0246	0.0229	0.0212	0.0197	0.0183	0.0157	0.0134	0.0115
28	0.0270	0.0251	0.0233	0.0216	0.0200	0.0185	0.0171	0.0146	0.0124	0.0105
29	0.0258	0.0239	0.0221	0.0205	0.0189	0.0174	0.0160	0.0136	0.0115	0.0096
30	0.0247	0.0228	0.0210	0.0194	0.0178	0.0164	0.0151	0.0127	0.0106	0.0088
31	0.0236	0.0217	0.0200	0.0184	0.0169	0.0154	0.0141	0.0118	0.0098	0.0081
32	0.0226	0.0208	0.0190	0.0174	0.0160	0.0146	0.0133	0.0110	0.0091	0.0075
33	0.0217	0.0199	0.0182	0.0166	0.0151	0.0137	0.0125	0.0103	0.0084	0.0069
34	0.0208	0.0190	0.0173	0.0158	0.0143	0.0130	0.0118	0.0096	0.0078	0.0063
35	0.0200	0.0182	0.0165	0.0150	0.0136	0.0123	0.0111	0.0090	0.0072	0.0058
40	0.0166	0.0148	0.0133	0.0118	0.0105	0.0093	0.0083	0.0065	0.0050	0.0039
45	0.0139	0.0123	0.0108	0.0095	0.0083	0.0072	0.0063	0.0047	0.0035	0.0026
50	0.0118	0.0103	0.0089	0.0076	0.0066	0.0056	0.0048	0.0034	0.0025	0.0017
55	0.0101	0.0087	0.0073	0.0062	0.0052	0.0044	0.0037	0.0025	0.0017	0.0012
60	0.0088	0.0074	0.0061	0.0051	0.0042	0.0035	0.0028	0.0019	0.0012	0.0008
65	0.0076	0.0063	0.0051	0.0042	0.0034	0.0027	0.0022	0.0014	0.0009	0.0005
70	0.0067	0.0054	0.0043	0.0035	0.0028	0.0022	0.0017	0.0010	0.0006	0.0004
75	0.0059	0.0047	0.0037	0.0029	0.0022	0.0017	0.0013	0.0008	0.0004	0.0003
80	0.0052	0.0040	0.0031	0.0024	0.0018	0.0014	0.0010	0.0006	0.0003	0.0002
85	0.0046	0.0035	0.0026	0.0020	0.0015	0.0011	0.0008	0.0004	0.0002	0.0001
90	0.0041	0.0030	0.0023	0.0017	0.0012	0.0009	0.0006	0.0003	0.0002	0.0001
95	0.0036	0.0027	0.0019	0.0014	0.0010	0.0007	0.0005	0.0002	0.0001	0.0000
100	0.0032	0.0023	0.0017	0.0012	0.0008	0.0006	0.0004	0.0002	0.0001	0.0000

PROBLEM 4-2

A steam boiler is purchased on the basis of guaranteed performance. A test indicates that the actual operating cost will be $300 more per year than guaranteed. If the expected life

Table B-5. Values of $\dfrac{(1 + i)^n - 1}{i(1 + i)^n}$ $R = D\left[\dfrac{(1 + i)^n - 1}{i(1 + i)^n}\right]$

n	2%	2½%	3%	3½%	4%	4½%	5%	6%	7%	8%
1	0.980	0.975	0.970	0.966	0.961	0.957	0.952	0.943	0.935	0.926
2	1.942	1.927	1.913	1.900	1.886	1.873	1.859	1.833	1.808	1.783
3	2.884	2.856	2.829	2.802	2.775	2.749	2.723	2.673	2.624	2.577
4	3.808	3.762	3.717	3.673	3.630	3.588	3.546	3.465	3.387	3.312
5	4.713	4.646	4.580	4.515	4.452	4.390	4.329	4.212	4.100	3.993
6	5.601	5.508	5.417	5.329	5.242	5.158	5.076	4.917	4.767	4.623
7	6.472	6.349	6.230	6.115	6.002	5.893	5.786	5.582	5.389	5.206
8	7.325	7.170	7.020	6.874	6.733	6.596	6.463	6.210	5.971	5.747
9	8.162	7.971	7.786	7.608	7.435	7.269	7.108	6.802	6.515	6.247
10	8.983	8.752	8.530	8.317	8.111	7.913	7.722	7.360	7.024	6.710
11	9.787	9.514	9.253	9.002	8.760	8.529	8.306	7.887	7.499	7.139
12	10.58	10.26	9.954	9.663	9.385	9.119	8.863	8.384	7.943	7.536
13	11.35	10.98	10.63	10.30	9.986	9.683	9.394	8.853	8.358	7.904
14	12.11	11.69	11.30	10.92	10.56	10.22	9.899	9.295	8.745	8.244
15	12.85	12.38	11.94	11.52	11.12	10.74	10.38	9.712	9.108	8.559
16	13.58	13.06	12.56	12.09	11.65	11.23	10.84	10.11	9.447	8.851
17	14.29	13.71	13.17	12.65	12.17	11.71	11.27	10.48	9.763	9.122
18	14.99	14.35	13.75	13.19	12.66	12.16	11.69	10.83	10.06	9.372
19	15.68	14.98	14.32	13.71	13.13	12.59	12.09	11.16	10.34	9.604
20	16.35	15.59	14.88	14.21	13.59	13.01	12.46	11.47	10.59	9.818
21	17.01	16.18	15.42	14.70	14.03	13.40	12.82	11.76	10.84	10.02
22	17.66	16.77	15.94	15.17	14.45	13.78	13.16	12.04	11.06	10.20
23	18.29	17.33	16.44	15.62	14.86	14.15	13.49	12.30	11.27	10.37
24	18.91	17.88	16.94	16.06	15.25	14.50	13.80	12.55	11.47	10.53
25	19.52	18.42	17.41	16.48	15.62	14.83	14.09	12.78	11.65	10.67
26	20.12	18.95	17.88	16.89	15.98	15.15	14.38	13.00	11.83	10.81
27	20.71	19.46	18.33	17.29	16.33	15.45	14.64	13.21	11.99	10.94
28	21.28	19.96	18.76	17.67	16.66	15.74	14.90	13.41	12.14	11.05
29	21.84	20.45	19.19	18.04	16.98	16.02	15.14	13.59	12.28	11.16
30	22.40	20.93	19.60	18.39	17.29	16.29	15.37	13.76	12.41	11.26
31	22.94	21.40	20.00	18.74	17.59	16.54	15.59	13.93	12.53	11.35
32	23.47	21.85	20.39	19.07	17.87	16.79	15.80	14.08	12.65	11.43
33	23.99	22.29	20.77	19.39	18.15	17.02	16.00	14.23	12.75	11.51
34	24.50	22.72	21.13	19.70	18.41	17.25	16.19	14.37	12.85	11.59
35	25.00	23.15	21.49	20.00	18.66	17.46	16.37	14.50	12.95	11.65
40	27.36	25.10	23.11	21.36	19.79	18.40	17.16	15.05	13.33	11.92
45	29.49	26.83	24.52	22.50	20.72	19.16	17.77	15.46	13.61	12.11
50	31.42	28.36	25.73	23.46	21.48	19.76	18.26	15.76	13.80	12.23
55	33.17	29.71	26.77	24.36	22.11	20.25	18.63	15.99	13.94	12.32
60	34.76	30.91	27.68	24.94	22.62	20.64	18.93	16.16	14.04	12.38
65	36.20	31.96	28.45	25.52	23.05	20.95	19.16	16.29	14.11	12.42
70	37.50	32.90	29.12	26.00	22.39	21.20	19.34	16.38	14.16	12.44
75	38.68	33.72	29.70	26.41	23.68	21.40	19.48	16.46	14.20	12.46
80	39.74	34.45	30.20	26.75	23.92	21.57	19.60	16.51	14.22	12.47
85	40.71	35.10	30.63	27.04	24.11	21.70	19.68	16.55	14.24	12.48
90	41.59	35.67	31.00	27.28	24.27	21.80	19.75	16.58	14.25	12.49
95	42.38	36.17	31.32	27.48	24.40	21.88	19.81	16.60	14.26	12.49
100	43.10	36.61	31.60	27.66	24.50	21.95	19.85	16.62	14.27	12.49

of the boiler is 20 years and money is worth 6%, what deduction from the purchase price would compensate the purchaser for the extra operating cost?

Table B-6. Values of $\dfrac{i(1+i)^n}{(1+i)^n - 1}$ $\qquad D = R\left[\dfrac{i(1+i)^n}{(1+i)^n - 1}\right]$

n	2 %	2½ %	3 %	3½ %	4 %	4½ %	5 %	6 %	7 %	8 %
1	1.0200	1.0250	1.0300	1.0350	1.0400	1.0450	1.0500	1.0600	1.0700	1.0800
2	0.5150	0.5188	0.5226	0.5264	0.5302	0.5340	0.5378	0.5454	0.5531	0.5608
3	0.3468	0.3501	0.3535	0.3569	0.3603	0.3638	0.3672	0.3741	0.3811	0.3880
4	0.2626	0.2658	0.2690	0.2723	0.2755	0.2787	0.2820	0.2886	0.2952	0.3019
5	0.2122	0.2152	0.2184	0.2215	0.2246	0.2278	0.2310	0.2374	0.2439	0.2505
6	0.1785	0.1815	0.1846	0.1877	0.1908	0.1939	0.1970	0.2034	0.2098	0.2163
7	0.1545	0.1575	0.1605	0.1635	0.1666	0.1697	0.1728	0.1791	0.1856	0.1921
8	0.1365	0.1395	0.1425	0.1455	0.1485	0.1516	0.1547	0.1610	0.1675	0.1740
9	0.1225	0.1255	0.1284	0.1314	0.1345	0.1376	0.1407	0.1470	0.1535	0.1601
10	0.1113	0.1143	0.1172	0.1202	0.1233	0.1264	0.1295	0.1359	0.1424	0.1490
11	0.1022	0.1051	0.1081	0.1111	0.1141	0.1172	0.1204	0.1268	0.1334	0.1401
12	0.0946	0.0975	0.1005	0.1035	0.1066	0.1097	0.1128	0.1193	0.1259	0.1327
13	0.0881	0.0910	0.0940	0.0971	0.1001	0.1033	0.1065	0.1130	0.1197	0.1265
14	0.0826	0.0855	0.0885	0.0916	0.0947	0.0978	0.1010	0.1076	0.1143	0.1213
15	0.0778	0.0808	0.0838	0.0868	0.0899	0.0931	0.0963	0.1030	0.1098	0.1168
16	0.0737	0.0766	0.0796	0.0827	0.0858	0.0890	0.0923	0.0990	0.1059	0.1130
17	0.0700	0.0729	0.0760	0.0790	0.0822	0.0854	0.0887	0.0954	0.1024	0.1096
18	0.0667	0.0697	0.0727	0.0758	0.0790	0.0822	0.0855	0.0924	0.0994	0.1067
19	0.0638	0.0668	0.0698	0.0729	0.0761	0.0794	0.0827	0.0896	0.0968	0.1041
20	0.0612	0.0641	0.0672	0.0704	0.0736	0.0769	0.0802	0.0872	0.0944	0.1019
21	0.0588	0.0618	0.0649	0.0680	0.0713	0.0746	0.0780	0.0850	0.0923	0.0998
22	0.0566	0.0596	0.0627	0.0659	0.0692	0.0725	0.0760	0.0830	0.0904	0.0980
23	0.0547	0.0577	0.0608	0.0640	0.0673	0.0707	0.0741	0.0813	0.0887	0.0964
24	0.0529	0.0559	0.0590	0.0623	0.0656	0.0690	0.0725	0.0797	0.0872	0.0950
25	0.0512	0.0543	0.0574	0.0607	0.0640	0.0674	0.0710	0.0782	0.0858	0.0937
26	0.0497	0.0528	0.0559	0.0592	0.0626	0.0660	0.0696	0.0769	0.0846	0.0925
27	0.0483	0.0514	0.0546	0.0579	0.0612	0.0647	0.0683	0.0757	0.0834	0.0914
28	0.0470	0.0501	0.0533	0.0566	0.0600	0.0635	0.0671	0.0746	0.0824	0.0905
29	0.0458	0.0489	0.0521	0.0554	0.0589	0.0624	0.0660	0.0736	0.0814	0.0896
30	0.0446	0.0478	0.0510	0.0544	0.0578	0.0614	0.0651	0.0726	0.0806	0.0888
31	0.0436	0.0467	0.0500	0.0534	0.0569	0.0604	0.0641	0.0718	0.0798	0.0881
32	0.0426	0.0458	0.0490	0.0524	0.0559	0.0596	0.0633	0.0710	0.0791	0.0875
33	0.0417	0.0449	0.0482	0.0516	0.0551	0.0587	0.0625	0.0703	0.0784	0.0869
34	0.0408	0.0440	0.0473	0.0508	0.0543	0.0580	0.0618	0.0696	0.0777	0.0863
35	0.0400	0.0432	0.0465	0.0500	0.0536	0.0573	0.0611	0.0690	0.0772	0.0858
40	0.0366	0.0398	0.0433	0.0468	0.0505	0.0543	0.0583	0.0665	0.0750	0.0839
45	0.0339	0.0373	0.0408	0.0445	0.0483	0.0522	0.0563	0.0647	0.0735	0.0826
50	0.0318	0.0353	0.0389	0.0426	0.0466	0.0506	0.0548	0.0634	0.0725	0.0817
55	0.0301	0.0337	0.0373	0.0412	0.0452	0.0494	0.0537	0.0625	0.0717	0.0812
60	0.0288	0.0324	0.0361	0.0401	0.0442	0.0485	0.0528	0.0619	0.0712	0.0808
65	0.0276	0.0313	0.0351	0.0392	0.0434	0.0477	0.0522	0.0614	0.0709	0.0805
70	0.0276	0.0304	0.0343	0.0385	0.0427	0.0472	0.0517	0.0610	0.0706	0.0804
75	0.0259	0.0297	0.0337	0.0379	0.0422	0.0467	0.0513	0.0608	0.0704	0.0802
80	0.0252	0.0290	0.0331	0.0574	0.0418	0.0464	0.0510	0.0606	0.0703	0.0802
85	0.0246	0.0285	0.0326	0.0370	0.0415	0.0461	0.0508	0.0604	0.0702	0.0801
90	0.0240	0.0280	0.0323	0.0367	0.0412	0.0459	0.0506	0.0603	0.0702	0.0801
95	0.0236	0.0276	0.0319	0.0364	0.0410	0.0457	0.0505	0.0602	0.0701	0.0800
100	0.0232	0.0273	0.0316	0.0362	0.0408	0.0456	0.0504	0.0602	0.0701	0.0800

Answer

Basis for adjustment (R) is present worth of excessive operating cost (D) for 20 years at 6%. Use interest Table B-5 for annuity factor:

$$R = (300)\left[\frac{(1.06)^{20} - 1}{(0.06)(1.06)^{20}}\right]$$

$$= (300)(11.47)$$

$$= \$3441$$

PROBLEM 4-3

A new community requires one duct for power. It is estimated that two ducts will be required in 5 years, three ducts in 15 years, and four ducts in 25 years. The cost of installing one duct at a time is $1700 per 1000 feet and the cost of installing four ducts at one time is $4400 per 1000 feet. With money at 6%, is it more economical to install one duct at a time as needed or to install four ducts now?

Table A-1. Compound Interest Factors Assuming Uniform Payments at End of Each Period.

	NAME	ABBREVIATION	FORMULA	GRANT'S SYMBOL	THUESEN'S SYMBOL	MATH SYMBOL
SINGLE PAYMENTS	Single Payment Compound Amount Factor	SPCAF	$(1 + i)^n$	caf'-i-n	(SP)i-n (xxxxx)	
	Single Payment Present Worth (or value) Factor	SPPWF	$\dfrac{1}{(1 + i)^n}$	pwf'-i-n	(PS)i-n (xxxxx)	V^n
UNIFORM MULTIPLE PAYMENTS	Capital Recovery Factor	CRF	$\dfrac{i(1i)^n}{(1 + i)^n - 1}$	crf-i-n	(RP)i-n (xxxxx)	$\dfrac{1}{a_{\overline{n}|}}$
	Uniform Series Present Worth Factor	USPWF	$\dfrac{(1 + i)^n - 1}{i(1 + i)^n}$	pwf-i-n	(PR)i-n (xxxxx)	$a_{\overline{n}|}$
	Sinking Fund Deposit Factor	SFDF	$\dfrac{i}{(1 + i)^n - 1}$	ssf-i-n	(RS)i-n	$\dfrac{1}{S_{\overline{n}|}}$
	Uniform Series Compound Amount Factor	USCAF	$\dfrac{(1 + i)^n - 1}{i}$	caf-i-n	(SR)i-n (xxxxx)	$S_{\overline{n}|}$
	Arithmetic Series Factor	ASF	$\dfrac{1 - n(SFDF)}{i}$			
	Arithmetic Series Present Worth (or value) Factor	ASPWF	ASF×USPWF			

Table A-2. Engineering Economics Interest Table Reference

TIME FACTOR	SINGLE PAYMENT — USING COMPOUND AMOUNT FACTOR	SINGLE PAYMENT — USING PRESENT WORTH FACTOR	UNIFORM MULTIPLE PAYMENTS — USING COMPOUND AMOUNT FACTOR	UNIFORM MULTIPLE PAYMENTS — USING THE SINKING FUND FACTOR	UNIFORM MULTIPLE PAYMENTS — USING THE PRESENT WORTH FACTOR	UNIFORM MULTIPLE PAYMENTS — USING CAPITAL RECOVERY FACTOR
Present / End of year 1 / End of year 2 / End of year 3 / End of year N − 1 / End of year N	(cash flow diagram)	(cash flow diagram)	(cash flow diagram)	(cash flow diagram)	(cash flow diagram)	(cash flow diagram)
Actual Factor to use	$(1 + i)^n$	$\dfrac{1}{(1 + i)^n}$	$\dfrac{(1 + i)^n - 1}{i}$	$\dfrac{i}{(1 + i)^n - 1}$	$\dfrac{(1 + i)^n - 1}{i(1 + i)^n}$	$\dfrac{i(1 + i)^n}{(1 + i)^n - 1}$
Symbolic factor	SP i-n (xxxxx)	PS i-n (xxxxx)	SR i-n (xxxxx)	RS i-n (xxxxx)	PR i-n (xxxxx)	RP i-n (xxxxx)
Use table	B − 1	B − 2	B − 3	B − 4	B − 5	B − 6
Numerical values	>1	<1	>1	<1	>1	<1
If this value is known:	P, i, n	S, i, n	R, i, n	S, i, n	R, i, n	P, i, n
Solve for	S	P	S	R	P	R
Use equation	S = P(xxxxx)	P = S(xxxxx)	S = R(xxxxx)	R = S(xxxxx)	P = R(xxxxx)	R = P(xxxxx)

Where P = present worth
S = compound amount
i = annual interest rate
n = number of years
R = equal payments

50

Answer

Use endowment method in which summations of the present worths (Table B-2) of all separate investments are compared:

FOR 1000 FT	PRESENT WORTH OF INVESTMENT	
Time of Investment	Four Ducts Now	One Duct at a Time
Now	$4400	$1700
In 5 years		(1700) (0.7473) = $1270
In 15 years		(1700) (0.4173) = $ 710
In 25 years		(1700) (0.2330) = $ 396
Total present worth	$4400	$4076

It is therefore better to install one duct at a time. This decision is also favored by intangibles—the risk of forecasting the future needs, and the possibilities of unforseen obsolescence and maintenance.

PROBLEM 4-4

During a slack period a manufacturer can sell 3000 articles per month which is 2/3 of the capacity of the factory. The investment in the factory is $2,000,000 which depreciates at the rate of 5%. Other fixed expenses are $20,000 per year. Maintenance costs vary from 1% at zero output to 4% at full output. If labor and material for the article cost $1 and the article must be sold for $1.20, should the factory stay in production or shut down?

Answer

ALTERNATIVES:	SHUT DOWN	TWO-THIRDS CAPACITY
Output/year	0	36,000
Depreciation at 5%	100,000	100,000
Fixed expense	20,000	20,000
Maintenance, fixed	20,000	20,000
Maintenance, variable	0	40,000
Labor and material	0	36,000
Total operating cost	$140,000	$216,000
Income at $1.20 each	0	43,200
Loss/year	$140,000	$172,800

To shut down means the lower loss. Actually they had better go out of business because they can't come near breaking even at 100% capacity. Possibly the investment was misstated and should have been $200,000, which would make the results more reasonable, but still not enough for profitable operations. There is only 20¢ margin between selling price and prime cost!

PROBLEM 4-5

A snow-loading machine costing $10,000 requires four operators at $12.00 per day. The machine can do the work of 50 hand shovelers at $7.00 per day. Fuel, oil and maintenance for the machine amount to $30 per day. If the life of the machine is 8 years and interest on money is 6%, how many days of snow removal per year are necessary to make purchase of the machine economical?

Answer

ALTERNATIVES:	HAND WORK	MACHINE
Investment	0	$10,000
Depreciation (straight line)	0	1,250
Labor for X days	$350X$	$48X$
Fuel, maintenance, etc.	0	$30X$
Expected return (on first cost)	0	600
Total cost + return	$350X$	$1850 + 78X$

The systems are equal when:

$$350X = 1850 + 78X$$
$$272X = 1850$$
$$X = 6.8 \text{ days/year}$$

PROBLEM 4-6

On the basis of the following data, is it more economical to repair an engine or to replace it with a new one? The old engine cost $1400 ten years ago. Repairs will cost $500 and will extend its usefulness for five years. A new engine will cost $1240 and last ten years. Annual cost for fuel, lubricants, and repairs will be $75 more for the repaired engine than for the new engine. Assume an interest rate of 6%. Neglect salvage values.

Answer

ALTERNATIVES:	KEEP OLD	BUY NEW
Investment (in-place value)	$500	$1240
Expected life (years)	5	10
Depreciation	$100	124
Additional maintenance	75	0
Expected return at 6%	30	74.40
Total cost + return	$205	$ 198.40
Advantage/year		$ 6.60

To most managers, $6.60 would be an inadequate margin to favor additional investment unless intangibles leaned that way. I assume that the old machine had been "written off."

PROBLEM 4-7

A municipality wishes to raise funds for improvements by issuing 5½% bonds. There is $20,000 available per year for interest payments and retirement of the bonds at 110. What may be the amount of the bond issue if all the bonds are to be retired in 20 years?

Answer

Assume $100 par value of bonds and that a sinking fund which also earns income at 5 1/2% will require appropriate annual deposits. Let X equal the amount of issue at par value. Then the sinking fund deposit:

$$D = 1.10X[0.055/(0.055)^{20} - 1]$$
$$D = (1.10X) (0.0287) = 0.03157X$$

Interest/year:

$$I = 0.055X$$

Available amount for sinking fund and interest is $20,000. Therefore:

$$
\begin{aligned}
D + I &= 20,000 \\
&= 0.03157X + 0.055X \\
&= 0.08657X \\
X &= 20,000/0.08657 \\
&= \$231,000 \text{ amount of issue}
\end{aligned}
$$

This same solution applies if bonds were retired annually in amounts equivalent to a sinking fund deposit because the rates of interest were assumed to be the same.

PROBLEM 4-8

A trestle will cost $10,000 if built of untreated timber and will have an estimated life of 25 years. If treated lumber is used, the cost will be $15,000, the estimated life will be increased to 50 years, and the annual maintenance expense reduced by $100. If money costs six percent, which material should be used?

Answer

ALTERNATIVES:	UNTREATED	TREATED
Investment	$10,000	$15,000
Life (years)	25	50
Difference in maintenance	100	
Depreciation (straight line)	400	300
Expected return at 6%	600	900
Annual cost + interest	$ 1,100	$ 1,200

The foregoing method favors untreated lumber. The *exact* method will do likewise. The *average interest* method, which reduces interest approximately 50%, will then favor the treated lumber.

PROBLEM 4-9

A shop can make an article on either an ordinary machine or an automatic machine. Which machine would you use to make 3000 articles if cost data are as follows:

ALTERNATIVES:	ORDINARY MACHINE	AUTOMATIC MACHINE
Cost of set-up	$5.00	$15.00
Production per hour	1 20	1 50
Wages per man per hour	$1.00	$1.00
Machines supervised per man	1	4
Overhead per hour*	$0.40	$0.70

Answer

Comparison to be based on cost per unit of output.

ALTERNATIVES:	ORDINARY	AUTOMATIC	AUTOMATIC** (alternate solution)
Machines used	1	4	1
Production/hour	120	600	150
Man hrs/job	25	5	5
Set-up cost	5.00	60.00	15.00
Labor (direct)	25.00	5.00	5.00
Overhead	10.00	14.00*	14.00
Total cost	$ 40.00	$ 79.00	$ 34.00
Unit cost	$0.0133	$ 0.02633	$ 0.01133

*It is assumed that the overhead is a machine hour rate which is $2.80/hour for 4 automatics. Also assumed that $15 is set-up cost for *one* machine. Using *one* machine would show lower cost, and if the operator could tend other work at the same time, it would pay.

**Assuming one automatic and one operator and the operator is running other machines on other work at the time this job is running on one machine.

PROBLEM 4-10

Dies for a set of stampings cost $2400; the stampings cost 8 cents a set, and the cost of assembly is 2 1/2 cents a set. A die for an aluminum casting to replace the stamping assembly costs $1600; the rough casting costs 11 cents, and the machining cost is 3 cents. At what production are the costs equal, and what is the total unit cost at this point?

Answer

If X equals the number of units produced:

ALTERNATIVES:	STAMPING	CASTING
Tool cost	$2400	$1600
Material and labor	$0.105X$	$0.14X$
Total cost	$0.105X + 2400$	$0.14X + 1600$

Note that no interest or charge for the use of capital is included since no time or interest rate is given.

(A) Costs are equal when:

$$0.105X + 2400 = 0.14X + 1600$$
$$0.035X = 800$$
$$X = 22,850 \text{ units}$$

(B) At this production total cost is:

$$(0.105)(22,850) + 2400 = \$4800$$
$$\text{Unit cost} = 4800/22,850$$
$$= \$0.21 \text{ each}$$

PROBLEM 4-11

A part has a finished value of $4.80. Of this cost, 40% is material, 30% labor, 20% overhead, and 10% profit. If material goes up 25%, the pay rate goes up 20%, labor production rate per hour goes down 15%, and overhead is the same. What is the new value to clear 10% profit?

Answer

	OLD RATES*	NEW RATES**
Material	$1.92	$2.50
Labor	1.44	2.03
Overhead	.96	.96
Profit	.48	.10X
Selling price	$4.80	X

*Old rates all calculated from given percentages of $4.80.
**New rates calculated from old rates. If labor output goes down 15% it means that labor hours go up by 1/0.85 or 1.177 (i.e., 17.7%), while the cost per hour goes up 20%. Hence new labor cost is:

$$\$1.44 \ (1.177) \ (1.20) = 2.03$$

From above new rates:

$$X = 2.50 + 2.03 + 0.96 + 0.10X$$
$$0.90X = 5.49$$
$$X = \$6.10 \text{ selling price}$$

PROBLEM 4-12

A man borrows $3000 and agrees to repay the debt in 12 equal yearly installments at $370. Of this amount, $250 is supposed to represent amortization and $120 to represent 4% interest. What actual interest will the man be paying? Give the answer to the nearest whole percent and state whether the exact answer is more or less.

Answer

There is no direct algebraic solution. It has to be done by solving for an interest factor and using tables. For uniform annual payments (interest and amortization combined), the annual amount is:

$$\$370 = 3000 \left[\frac{i}{(1 + i)^{12} - 1} \right] + 3000i$$

or $0.1233 = f_1 + i$, in which f_1 is the sinking fund factor.

Now f_1 and i can be combined thus:

$$\frac{i}{(1 + i)^n - 1} + i = \frac{i + (1 + i)^n - i}{(1 + i)^n - 1}$$

$$= \frac{i(1 + i)^n}{(1 + i)^n - 1}$$

$$= f_2$$

which is the familiar annuity factor (see tables). Thus, $f_2 = 0.1233$. For 12 yrs the corresponding value of i is 6.5%.

PROBLEM 4-13

Two methods are available for recovering ore. One method recovers 75 tons per 100 tons treated at a cost of $3.00 per ton recovered. The other method recovers 80 tons per 100 tons treated at a cost of $3.25 per ton recovered. If the value of the recovered ore is $6.00 per ton, which method of recovery should be used? At what value of the recovered ore would it be economical to change the method of recovery?

Answer

PROCESSING 100 TONS:	LOW COST	HIGH COST
Tons ore recovered	75	80
Operating cost	$225.00	$260.00
Selling price at $6.00	450.00	480.00
Profit	$225.00	$220.00

This favors the low cost method. The two methods are equal when:

$$75X - 225 = 80X - 260$$
$$5X = 35$$
$$X = \$7.00 \text{ per ton}$$

PROBLEM 4-14

A syndicate wishes to purchase an oil well that estimates indicate will produce a net income of $200,000 per year for 30 years. What should the syndicate pay for the well if, out of this net income, a return of 10% on the investment is desired and a sinking fund will be established at 3% interest to recover the investment?

Answer

By Hoskold's formula, $P =$ value of investment, $r =$ rate of return, and $R =$ net income. Thus:

$$P = R\left[\cfrac{1}{r + \cfrac{i}{(1 + i)^n - 1}}\right]$$

where $R = \$200,000$
$i = .03$
$r = .10$
$n = 30$

Then,

$$P = 200,000\left[\frac{1}{0.10 + 0.0210}\right] = 200,000/0.1210 = \$1,653,000$$

57

PROBLEM 4-15

It is necessary to replace the ties on a railroad line. Untreated ties, costing $1.50 in place, have a life of 6 years. What expenditure per tie is warranted for creosoting if the life of the tie is extended to 9 years? Assume an interest rate of 6%.

Answer

If X equals the cost of treatment:

ALTERNATIVES:	TREATED TIES	UNTREATED
Investment	1.50 + X	1.50
Life (years)	9	6
Depreciation (straight line)	(1.50 + X)/9	0.25
Expected return at 6%	(0.06) (1.50 + X)	0.09

The two methods are equal in economy when:

$$(1.50 + X)/9 + 0.09 + 0.06X = 0.25 + 0.09$$
$$0.1111X + 0.1667 + 0.09 + .06X = .34$$
$$0.1711X = .0833$$
$$X = \$0.487 \text{ each tie}$$

PROBLEM 4-16

Two methods, A and B, of conveying water are being studied. Method A requires a tunnel; first cost $150,000; life, perpetual; annual operation and upkeep, $2000. Method B requires a ditch plus a flume. First cost of ditch is $50,000; life, perpetual; annual operation and upkeep, $2000. First cost of flume is $30,000; life, 10 years; salvage value, $5000; annual operation and upkeep, $4000. Compare the two methods for perpetual service, assuming an interest rate of 5%.

Answer

ANNUAL COST & INTEREST	A	B
Straight-line depreciation	0	2500*
Upkeep	2000	6000
Interest on first cost at 5%	7500	4000
	$9500	$12,500

*Depreciation is on flume only.

Answer: Select method "A."

PROBLEM 4-17

A new timber floor is to be placed on a bridge. If untreated timber is used, the cost will be $10,000 and the estimated life 15 years. If treated timber is used, the cost will be $16,000, but the

58

estimated life will be 25 years and the yearly maintenance cost will be reduced by $100. Is the extra expenditure for treated timber justified? A 6% return is expected on the investment.

Answer

ALTERNATIVES:	UNTREATED	TREATED
Investment	$10,000	$16,000
Life (years)	15	25
Depreciation (straight line)	667	640
Added maintenance	100	
Return on first cost at 6%	600	960
Comparative cost	$ 1,367	$ 1,600

The answer favors the untreated timber.

PROBLEM 4-18

A 4%, $1000 bond, with interest paid annually, will mature in 10 years. What is its market value if 3% is considered a fair return?

Answer

Value now is based on combined present worths of annual interest and principal on data of retirement.

Present worth, bond interest (40) (8.530) = $ 341.20
Present worth, principal (1000) (0.7441) = 744.10
Market value today $1085.30

(For annuity and present worth factors, see interest tables.)

PROBLEM 4-19

A $1,000,000 issue of 3%, 15-year bond was sold at 95. If miscellaneous initial expenses of the financing were $20,000 and a yearly expense of $2000 is incurred, what is the true cost to the nearest 0.5% that the company is paying for the money it borrowed?

Answer

Gross receipts from sale of bonds
(1,000,000) (0.95) = $950,000
Expense of financing   20,000
Net return from bond issue $930,000

Annual cost of interest (1,000,000) (0.03) = 30,000
Annual expense of administration   2,000

Annual cost of handling bond issue $32,000
Sinking fund 1,000,000 [$(1.03)^{15}$ − 1/0.03] = 53,800

Total annual obligation   = $85,800

Now let i = *actual* interest rate; then:

$$930,000 = 85,800 \left[\frac{i(1 + i)^n}{(1 + i)^n - 1} \right] = (85,800) f$$

$$f = 930,000/85,800 = 10.83$$

From annuity tables for 15 yrs, this is equivalent to 4.38%.

PROBLEM 4-20

The terms of a $10,000 mortgage call for 10 equal yearly installments that will amortize the mortgage and pay 5% interest on the unpaid balance. What is the unpaid balance of the mortgage after the fifth payment is made?

Answer

The fixed annual payment is the same as the first year's obligation and is calculated as follows:

Amortization of mortgage 10,000 $[0.05/(1.05)^{10} - 1]$ = $795.00
Interest (10,000) (.05) = 500,00
Total $1295.00

By short method the total amount of mortgage retired in 5 years is

$$795 \left[\frac{(1.05)^5 - 1}{0.05} \right] = 795 \, (5.526) = \$4390$$

so that the unpaid balance is 10,000 − 4390 = $5610.

An alternate but longer method is:

YEAR	MORT-GAGE	INTER-EST*	PAY ON MORTGAGE*	BALANCE OF MORTGAGE
1	10,000	500	795	9,205
2	9,205	460	835	8,370
3	8,370	419	876	7,491
4	7,491	375	920	6,571
5	6,571	329	966	5,605

*Note that the sum of the interest and mortgage payment is fixed at $1295. The 5-year balance is $5,605.

PROBLEM 4-21

A certain floor surfacing in a factory has to be replaced every five years at a cost of $1500. How long should a floor surfacing costing twice as much last to justify the larger expenditure if a 6% return is required on the investment?

Answer

ALTERNATIVES:	ORIGINAL	IMPROVED
Investment	$1500	$3000
Life (years)	5	X
Depreciation	$ 300	3000/X
Expected return at 6%	90	180
Annual charge	$ 390	180 + 3000/X

The two methods are equal when

$$390 = 180 + 3000/X$$
$$(390 - 180)X = 3000$$
$$X = 14.3 \text{ years}$$

PROBLEM 4-22

Compare the following types of construction on the basis of annual cost, using interest at 5%:

	TIMBER TRESTLE	STEEL BRIDGE
First cost	$4000	$12,000
Life (years)	15	60
Salvage value	none	$1,000
Annual maintenance	$500	$300

Answer

	TIMBER	STEEL
Depreciation (straight line)	$267	$ 183
Maintenance	500	300
Expected return at 5%	200	600
Annual charge	$967	$1083

This favors the timber trestle. The *average interest* method will favor the steel; but this looks a long way ahead. The *exact* method (with sinking fund depreciation) will favor the steel by a narrow margin. I think the average business man would favor the demonstrated calculation. After years the steel bridge might be inadequate. The long-range calculation is risky.

PROBLEM 4-23

A firm has decided to purchase for $5000 a machine that will require replacement every five years. The salvage value will be $500. What will be the annual cost to the firm if the interest rate is 4%?

Answer

Investment	$5000
Life (years)	5
Salvage value	$ 500
Depreciation (straight line)	$ 900
Expected return at 5%	250
Annual charge	$1150

PROBLEM 4-24

Two 25-hp motors are being considered for purchase by a factory. The first costs $200 and has an efficiency of 85%. The second costs $150 and has an efficiency of 82%. If all charges, such as depreciation, insurance, maintenance, etc., amount annually to 15% of the original cost, and current costs 2.4 cents per kilowatt-hour, how many hours of full-load operation per year are necessary to justify purchase of the more expensive motor?

Answer

The critical factor is cost of current based on the kilowatt-hours required by the two motors.

$$kWh = hp (0.746)/efficiency$$

Then for A, annual cost of electricity is

$$(0.024) (X) (25) (0.746)/0.85 = 0.527X$$

For B,

$$(0.024) (X) (25) (0.746)/0.82 = 0.546X$$

Summarizing the costs:

	A	B
Fixed charges at 15%	$30.00	$22.50
Electricity	+ 0.527X	+0.546X
Total annual charges	$30.00 + 0.527X	$22.50 + 0.546X

The two motors are equal when:

$$30.00 + 0.527X = 22.50 + 0.546X$$
$$0.019X = 7.50$$
$$X = 395 \text{ hrs/year}$$

PROBLEM 4-25

A $1000 bond pays 4% and matures in 20 years. How much would you pay for the bond to realize 3% on your investment?

Answer

Bond valuation is the sum of two present worths:

$$\text{P.W. principal} = 1000 \left[\frac{1}{(1.03)^{20}} \right]$$

$$= 1000 \, (.5537)$$

$$= \$553.70$$

$$\text{P.W. annual interest} = 40 \left[\frac{(1.03)^{20} - 1}{0.03 \, (1.03)^{20}} \right]$$

$$= 40 \, (14.88)$$

$$= 595.20$$

The total of the two is $1148.90.

PROBLEM 4-26

A company requires 800,000 kWh of electric energy per year, which can be purchased at 1.3 cents per kWh. Will it pay the company to build a power plant of the above capacity under the following assumptions: First cost, $60,000; annual operation and maintenance $7500; life, 15 years; salvage value, $5000; insurance, 1%; cost of money, 4%?

Answer

Present cost of power: $(0.013) \ (800,000) = \$10,400/\text{year}$. The proposed set-up is then

Investment	$60,000
Life (years)	15
Salvage value	$ 5,000
Depreciation (straight line)	$ 3,667
Operating cost	7,500
Insurance and interest	3,000
Total annual charges	$14,167

which exceeds the present cost of power.

PROBLEM 4-27

A syndicate has purchased a mine for $1,000,000, which yields a net income of $100,000 per year. It is estimated that the mine will produce for 20 years. Part of each year's net income is set aside at 4% so that the original investment will be intact when the mine is exhausted. What annual rate of net income (after depreciation) is the syndicate realizing on its investment?

Answer
Use Hoskold's formula:

Sinking fund deposit at 4%	1,000,000 (0.0336) = $33,600
Expected return	1,000,000i
Net income	1,000,000i + 33,600

Then,

$$\$100,000 = 1,000,000i + 33,600$$
$$i = 66,400/1,000,000$$
$$= 0.0664, \text{ or } 6.64\%$$

PROBLEM 4-28
A lot was bought in January, 1930 for $1200. Taxes were paid in advance in January of each year as follows: $20 in 1930, $20 in 1931, $25 in 1932, $25 in 1933, $25 in 1934, and $20 in 1935. At the end of 1935 the lot was sold for $1500. What was the return on the investment, assuming simple interest? Outline a method of solution if compound interest is assumed.

Answer
The result of the ultimate sale is a capital gain of
1500−$1200 = $300.
(a) For a simple solution, assume that the capital gain accumulates annually on a straight line basis. Thus:

Average capital gain	300/6 = $50.00/yr
Average taxes	(20 + 20 + 25 + 25 + 20)/6 = 22.50/yr
Average net gain per year	$27.50
Annual yield on investment	27.50/1200 = 2.29%

(b) Yield by compound interest cannot be calculated algebraically or directly from interest tables. If i is assumed to be the yield realized, this rate can be used to evaluate the future worth of all expenditures (first cost and taxes) which should be matched by the selling price of the lot. Solve by cut and try. Use three or more values of i and draw a curve to determine the value at which the future worth (in 6 years) equals $1500.

For example:

FUTURE WORTH		AT 3% INTEREST	AT 2% INTEREST
Investment	6	$(1200)(1.194) = 1432.80$	$(1200)(1.126) = 1351.20$
Taxes:	1st yr 6	$(20)(1.194) = 23.88$	$(20)(1.126) = 22.52$
	2nd yr 5	$(20)(1.159) = 23.18$	$(20)(1.104) = 22.08$
	3rd yr 4	$(25)(1.126) = 28.15$	$(25)(1.082) = 27.05$
	4th yr 3	$(25)(1.093) = 27.33$	$(25)(1.061) = 26.53$
	5th yr 2	$(25)(1.061) = 26.52$	$(25)(1.040) = 26.00$
	6th yr 1	$(20)(1.030) = 20.60$	$(20)(1.020) = 20.40$
Total worth in 6 yrs.		$1582.46	$1495.78

Which shows that i is slightly more than 2%.

PROBLEM 4-29

A new boiler has just been installed. It is expected that there will be no maintenance charges until the end of the 11th year, when $100 will be spent on the boiler and $100 will be spent at the end of each successive year until the boiler is scrapped at the age of 25 years. What sum of money set aside at this time at 3% will take care of all maintenance expenses for the boiler?

Answer

The fund needed at the beginning of the 11th year is the present worth of the succeeding maintenance costs. This is:

$$100 \left[\frac{(1.03)^{15} - 1}{0.03(1.03)^{15}} \right] = \$1194$$

The present worth of this amount 10 years hence at 3% is:

$$1194 \left[\frac{1}{(1.03)^{10}} \right] = (1194)(.7441)$$

$$= \$888, \text{ set aside now}$$

PROBLEM 4-30

An automobile costs $3000. It is run the same distance each year. Annual expenses and maintenance costs increase $50 each year. The trade-in value is $1800 at the end of the first year and decreases $100 each year. When should the car be traded in if cost of transportation is to be a minimum?

Answer

The figures given for this problem are unreasonable. A $1200 depreciation the first year is out of line with $100 depreciation for each succeeding year. The replacement should be made when the annual cost (including depreciation) on the old car exceeds the first years cost of operating a new car. Let the first year's operating cost be C, and the age X. Then:

$$1200 + C = 100 + C + 50X$$
$$50X = 1100$$
$$X = 22 \text{ years}$$

If the first cost were $2000, then:

$$200 + C = 100 + C + 50X$$
$$X = 2 \text{ years}$$

PROBLEM 4-31

Two routes, A and B, for a highway between two towns are being considered. Route A is 12 miles long and will cost $1,000,000. Route B is 14 miles long and will cost $650,000. Rebuilding of each road will be required every 15 years at $25,000 per mile. Annual maintenance costs are $500 per mile. Is the shorter route justified under the following assumptions?

Estimated traffic is 200 trucks per day and 800 passenger cars per day. Cost of gas and oil is 10 cents per mile for trucks, 2 cents per mile for passenger cars. Speed of traffic is 40 miles per hour. The value of an hour saved is $1.00 for a truck, and negligible for a passenger car. Money costs 4%.

Answer

ENDOWMENT METHOD:	ROUTE A	ROUTE B
(1) Annual depreciation		
Sinking fund method	$14,980	$17,480
Maintenance	6,000	7,000
(2) Cost of mileage (cars)	70,100	81,700
(3) Cost of mileage (trucks)	87,600	102,200
(4) Cost of time (trucks)	21,900	25,550
Total annual cost	$200,580	$233,930
(5) Capitalized annual cost	5,014,500	5,848,250
Immediate investment	1,000,000	650,000
(6) Total capitalized cost	$6,014,500	$6,498,250

The comparison favors route A, with the lower capitalized cost.

PROBLEM 4-32

A $10,000 mortgage is being paid off in 20 equal yearly installments, each installment being part amortization and part interest at 6% on the unpaid balance. The mortgage provides for paying off the mortgage in a lump sum at any time with an amount equal to the unpaid balance plus a charge of 1% of the unpaid balance. What would have to be paid to discharge the mortgage after 10 payments have been made?

Answer

First year amortization:

$$10,000 \left[\frac{0.06}{(1.06)^{20} - 1} \right] = \$272$$

Short method: Amortization in 10 years

$$272 \left[\frac{(1.06)^{10} - 1}{0.06} \right]$$

	= $3,585
Mortgage	10,000
Unpaid balance	$6,415
Plus 1%	642
Amount due	$7,057

For longer method see Problem 4-20.

PROBLEM 4-33

A man paid $1100 for a $1000 bond that pays $40 per year. In 20 years the bond will be redeemed for $1050. What net rate of interest will the man obtain on his investment?

Answer

Present worth of redeemed bond:

$$\frac{1050}{(1 + i)^{20}} = 1050/f_1$$

Present worth of interest:

$$40 \left[\frac{(1 + i)^{20} - 1}{i(1 + i)^{20}} \right] = 40f_2$$

Total value of bond now:

$$\$1100 = 1050/f_1 + 40f_2$$

The required value of i cannot be solved algebraically. To solve by plotting curve, select several values of i, obtain f_1 and f_2 from interest tables and calculate present worth. At the point where total present worth equals $1100 is the correct interest rate. For example:

PRESENT WORTH OF:	2.5%	3%	3.5%	4%
Redeemed bond	640	582	527.50	479
Interest	623	595	568.50	543
Total present worth	$1263	$1177	$1096	$1022

Which indicates that i is slightly less than 3.5%.

PROBLEM 4-34

How much would the owner of a building be justified in paying for a sprinkler system that would save 1500 a year in insurance premiums? The system has to be renewed every 20 years and has a salvage value equal to 10% of its initial cost. Money is worth 5%

Answer

The net annual saving will be:

Saving in insurance	$500
Less depreciation (straight line)	$(0.90)(X)/20 = 0.045X$
Net saving	$500 - 0.045X$

The present worth of annual savings will be:

$$(500 - 0.045X) \left[\frac{(1.05)^{20} - 1}{0.05 \, (1.05)^{20}} \right] = (500 - 0.045X) \, (12.46)$$

$$= 6230 - 0.561X$$

The first cost of the sprinkler system is X:

$$X = 6230 - 0.561X$$
$$1.561X = 6230$$
$$X = \$3990$$

PROBLEM 4-35

Compare two types of bridges, A and B, on the basis of capitalized cost at 5% interest. Bridge A has an estimated life of 25 years; initial cost, $50,000; renewal cost, $35,000; annual maintenance, $500; repairs every 5 years, $2000; salvage value, $5000. Bridge B has an estimated life of 50 years; initial cost, $75,000; renewal cost $75,000; annual maintenance cost, $100; repairs every 5 years, $1000; salvage value, $10,000. The initial cost can be paid out of available funds. All other expenses will be defrayed by sinking funds.

Answer

In order to use capitalized cost, the repairs and renewals will be reduced to an annual cost as if they were treated as sinking fund depreciation—sinking fund since we are dealing with time value of money. The capitalized annual cost is C/i (see Problem 4-31).

ALTERNATIVES:	A	B
Depreciation	630	312
Repairs (annual)	362	181
Maintenance	500	100
Annual cost	$1,492	$593
Capitalized	29,840	11,860
First cost	50,000	75,000
Total capitalized cost	$79,840	$86,860

This favors the proposal for bridge A.

PROBLEM 4-36

Two companies, A and B, manufacture the same article. Company A, relying mostly on machines, has fixed expenses of $12,000 per month and a direct cost of $8 per unit. Company B, using more hand work, has fixed expenses of $4,000 and a direct cost of $20 per unit.

(A) At what monthly production rate will the total cost per unit be the same for the two companies:

(B) If a unit sells for $30, how many units must each company produce to clear expenses?

(C) How do profits compare for 1000 units per month?

Answer

COMPANY:	A	B
Fixed costs	$12,000	$4,000
Direct costs	8X	20X
Total cost/month	$12,000 + 8X	$4,000 + 20X
Income/month	30X	30X

(A) Total costs will be the same when:

$$12,000 + 8X = 4,000 + 20X$$
$$12X = 8,000$$
$$X = 667 \text{ units/month}$$

(B) Company A will break even when:

$$12,000 + 8X = 30X$$
$$X = 12,000/22$$
$$= 545/\text{month}$$

Company B will break even when:

$$4000 + 20X = 30X$$
$$X = 4000/10$$
$$= 400/\text{month}$$

(C) At 1000 units/month, profit will be:

For A:

$$(30)(1000) - (12,000) - (8)(1000) = \$10,000$$

For B:

$$(30)(1000) - (4000) - (20)(1000) = \$6,000$$

PROBLEM 4-37

Is it advisable for a railroad to replace a timber trestle by a concrete culvert? The estimated costs are the following:

	TRESTLE	CULVERT
First cost	$2600	$10,000
Annual maintenance	120	30
Life in years	9	100

Make the comparison on the basis of annual costs using 5% interest.

Answer

	TRESTLE	CULVERT
Depreciation (sinking fund)	$236	$ 4.00
Maintenance	120	30.00
Interest	130	500.00
Cost plus interest	$586	$534.00

The figures favor the culvert. If straight-line depreciation were used the trestle would appear the best.

PROBLEM 4-38

A father wishes to develop a fund for his newborn son's college education. The fund is to pay $2000 on the 18th, 19th, 20th, and 21st birthdays of the son. The fund will be built up by the deposit of a fixed sum on the son's first to 17th birthdays, inclusive. If the fund earns 2%, what should the yearly deposit into the fund be?

Answer

Fund needed by 17th birthday:

$$(2000)\left[\frac{(1.02)^4 - 1}{0.02(1.02)^4}\right] = (2000)(3.808) = \$7616$$

Annual deposit to create this fund:

$$(7616) \left[\frac{0.02}{(1.02)^{17} - 1} \right] = (7616)\ (.0500)$$

$$= \$380.80/\text{year}$$

PROBLEM 4-39

A motion-picture theater, with an average daily summer attendance of 1200 admissions at an average price of 50 cents, is considering the installation of air conditioning at a cost of $20,000. Air conditioning will be operated for a period of 20 weeks. By what per cent must the average daily summer attendance rise in order to justify the investment? Assume the following conditions: Daily energy requirement 300 kWh at 3 cents per kWh; daily water requirements, 2000 cu ft at 20 cents per 100 cu ft; taxes and insurance, 2 per cent; annual maintenance and operations, $200; expected life of installation, 10 years; investment charges, 6 percent.

Answer

All charges for the air conditioning will be treated as annual costs although they will apply to only 20 weeks of operations.

Income (20 weeks):

$$(1200)\ (7)\ (20)\ (0.50) = \$84,000$$

Let X represent the increased attendance. Then income after air conditioning is $84,000 + (84000)\ (X)$ and increment income is $84000X$.

Increment cost is:

Depreciation	$20,000/10 =$	$2,000
Energy cost	$(300)\ (7)\ (20)\ (0.03) =$	1,260
Water cost	$(2000)\ (7)\ (20)\ (0.0020) =$	560
Taxes	$(0.02)\ (20,000) =$	400
Maintenance		200
Expected return	$(0.06)\ (20,000) =$	1200
Annual cost plus expected return		$5,620

The system will pay for itself when:

$$84000\ X = 5620$$
$$X = 0.067, \text{ a } 6.7\% \text{ increase}$$

Chapter 5
Electrical Networks

Four-terminal networks, pads, fixed attenuators, etc., are used wherever frequency-independent impedance matching or fixed attenuation is needed. Such networks are usually resistive pads consisting of L, H, T, π, or lattice structures. In most cases these can be represented as balanced or un-balanced T or π networks. At times, however, it becomes necessary to insert L and C components to achieve a frequency-sensitive attenuation characteristic.

The problems contained in this chapter deal with an assortment of basic networks. Problems concerning resonance and RLC networks are covered in Chapter 6, while frequency-shaping networks and filters are detailed in Chapter 7.

The remainder of this review section consists of helpful equations relating to impedance matching and attenuation networks. A polar-to-rectangular coordinates conversion chart (Fig. 5-1) is included and is considered to be a valuable time-saving implement. This chart can be used in solving a variety of vector problems frequently encountered by the P.E. candidate, including many such problems in this book. If you are not already familiar with Feightner's chart, you are strongly urged to become so.

T NETWORKS

Basic Formulas

$$R_1 = \frac{Z_1 (K + 1) - 2\sqrt{KZ_1 Z_2}}{K - 1}$$

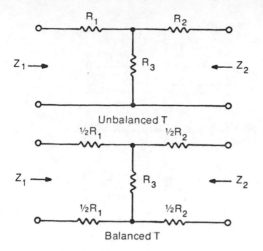

Unbalanced T

Balanced T

Fig. C-5-1.

$$R_2 = \frac{Z_2 (K + 1) - 2\sqrt{K Z_1 Z_2}}{K - 1}$$

$$R_3 = \frac{2\sqrt{K Z_1 Z_2}}{K - 1}$$

If $Z_1 = Z_2$:

$$R_1 = R_2 = \frac{Z_1 (\sqrt{K} - 1)}{\sqrt{K} + 1}$$

$$R_3 = \frac{2Z_1 \sqrt{K}}{K - 1}$$

where K is the ratio of the input power to the output power of the network.

The minimum possible value for K is:

$$K_{MIN} = \frac{2Z_1}{Z_2} - 1 + 2 \sqrt{\frac{Z_1}{Z_2}\left(\frac{Z_1}{Z_2} - 1\right)}$$

Other Formulas *(ref. 31)*

$$R_1 = \frac{Z_1 (A^2 + 1)}{A^2 - 1} - R_3$$

$$R_2 = \frac{Z_2 \, (A^2 + 1)}{A^2 - 1} - R_3$$

$$R_3 = \frac{2 \, A\sqrt{Z_1 \, Z_2}}{A^2 - 1}$$

where A is the required attenuation ratio in decibels.
If $R_1 = R_2$:

$$R_1 = R_2 = \frac{R(A - 1)}{A + 1}$$

$$R_3 = \frac{2 \, RA}{A^2 - 1}$$

For symmetrical pads, $R = Z_1 = Z_2$. Attenuation in decibels of a T pad:

$$\alpha_T = 20 \log \left[\frac{R_1 \, (R + R_1 + R_3) + R_3 \, (R + R_1)}{R \, R_3} \right]$$

PI NETWORKS

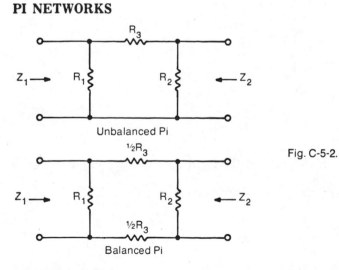

Unbalanced Pi

Balanced Pi

Fig. C-5-2.

Basic Formulas   *(ref. 1)*

$$R_1 = \frac{(K - 1) \, Z_1 \, \sqrt{Z_2}}{(K + 1) \, \sqrt{Z_2} - 2\sqrt{K \, Z_1}}$$

74

$$R_2 = \frac{(K - 1) \, Z_2 \, \sqrt{Z_1}}{(K + 1) \, \sqrt{Z_1} - 2\sqrt{K} \, Z_2}$$

$$R_3 = \frac{(K - 1) \, \sqrt{Z_1 \, Z_2}}{2\sqrt{K}}$$

If $Z_1 = Z_2$:

$$R_1 = R_2 = \frac{Z_1 \, (\sqrt{K} + 1)}{\sqrt{K} - 1}$$

$$R_3 = \frac{Z_1 \, (K - 1)}{2\sqrt{K}}$$

where K is the ratio of the input power to the output power of the network.

The minimum possible value for K is:

$$K_{\text{MIN}} = \frac{2Z_1}{Z_2} - 1 + 2\sqrt{\frac{Z_1}{Z_2}\left(\frac{Z_1}{Z_2} - 1\right)}$$

Other Formulas *(ref. 31)*

$$\frac{1}{R_1} = \frac{A^2 + 1}{Z_1 \, (A^2 - 1)} - \frac{1}{R_3}$$

$$\frac{1}{R_2} = \frac{A^2 + 1}{Z_2 \, (A^2 - 1)} - \frac{1}{R_3}$$

$$R_3 = \frac{(A^2 - 1) \, \sqrt{Z_1 \, Z_2}}{2A}$$

where A is the required attenuation ratio in decibels.

If $R_1 = R_2$:

$$R_1 = R_2 = \frac{R \, (A + 1)}{A - 1}$$

$$R_3 = \frac{R \, (A^2 - 1)}{2A}$$

For symmetrical pads, $R = Z_1 = Z_2$. Attenuation in decibels of a π pad:

$$\alpha_\pi = 20 \log \left[\frac{R_3 \ (R + R_1) R R_1}{R R_1} \right]$$

ITERATIVE LATTICE NETWORKS

(ref. 2)

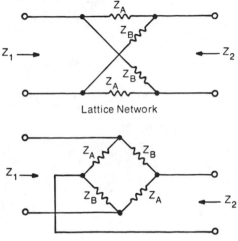

Lattice Network

Fig. C-5-3.

Lattice Network (planar representation)

$$Z_I = Z_1 = Z_2 = \sqrt{Z_A \ Z_B}$$

If the phase angle θ and Z_I are known:

$$Z_A = Z_I \ \tanh\left(\frac{\theta}{2}\right)$$

$$Z_B = \frac{Z_I}{\tanh(\theta/2)}$$

Note: The reference 2 also contains 3-element reactive networks for impedance matching.

PARALLEL RESISTANCE

The accompanying chart is handy for estimating parallel resistances, parallel inductances, and series capacitances. Two examples are worked out on the chart to illustrate its use.

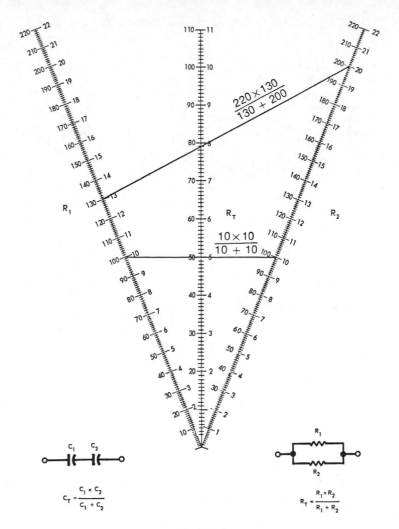

$$\frac{220 \times 130}{130 + 200}$$

$$\frac{10 \times 10}{10 + 10}$$

R_1 R_T R_2

$$C_T = \frac{C_1 \times C_2}{C_1 + C_2}$$

$$R_T = \frac{R_1 \times R_2}{R_1 + R_2}$$

Fig. C-5-4.

USE OF FEIGNTNER'S GRAPHIC VECTOR SOLUTION CHART

Conversion from polar to rectangular coordinates, and back, are easily achieved with Feightner's chart (Fig. 5-1). It is essentially a set of rectangular coordinates superimposed on a set of polar coordinates. As a bonus, the polar lines are extended to provide sine, cosine, and tangent functions. Simply read the function for the angle shown.

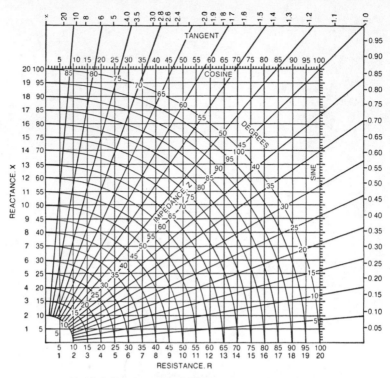

Fig. 5-1. Feightner's graphic vector solution chart.

Assume the rectangular coordinates, $R + jX = 30 + j45$. Move vertically along the 30Ω line until it intersects with the 45Ω line. Read an angle of approximately 56.5° on angle line and follow the polar impedance arc and read about 54Ω. Solution: $30 + j45 = 54 \angle 56.5°$. If the rectangular equation were $30 - j45$, $Z = 54 \angle -56.5°$.

Convert from $Z = 80 \angle 35°$ to rectangular. On the polar chart find 35°. At its intersection with 80Ω on the arced polar line, move down vertically to read $R = 65$ and to the left to read $X = 45$. Solution: $80 \angle 35 = 65 + j45$. If $Z = 80 \angle -35°$, then $R - jX = 65 - j45$. Use the outer scales the same way as inner scales with no modification in angles.

PROBLEM 5-1

For the four-terminal network shown:

(A) Find the four values of characteristic impedance and identify them.

(B) Sketch four or more different networks that are potentially equivalent.

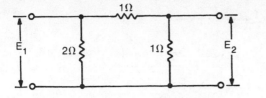

Problem 5-1.

Answer *(ref, 1, p. 20-8)*

(A) The four characteristic impedances are:

Input resistance (open circuit output):

$$Z_{11} = \frac{R_1 \ (R_2 \ + \ R_3 \ )}{R_1 \ + \ (R_2 \ + R_3 \ )} = \frac{2 \times 2}{2 + 2} = 1\Omega$$

Input resistance (short circuit output):

$$Z_{12} = \frac{R_1 \ R_2}{R_1 \ + \ R_2} = \frac{2 \times 1}{2 + 1} = 0.667\Omega$$

Output resistance (open circuit input):

$$Z_{22} = \frac{R_3 \ (R_1 \ + \ R_2 \ )}{R_3 \ + \ (R_1 \ + \ R_2 \ )} = \frac{1 \times 3}{1 + 3} = 0.75\Omega$$

Output resistance (short circuit input):

$$Z_{21} = \frac{R_3 \ R_2}{R_3 \ + \ R_2} = \frac{1 \times 1}{1 + 1} = 0.5\Omega$$

(B) The nature of these characteristic impedances are illustrated in Fig. 5-2.

PROBLEM 5-2

In the circuit shown, $Z_1 = 200 + j51$, $Z_2 = 30 - j90$, and $Z_3 = 80 + j90$. Find Z_{IN} and Z_{OUT} and give your answer in polar and rectangular form.

79

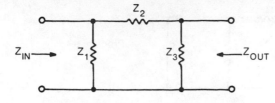

Problem 5-2.

Answer

Using the polar – rectangular conversion chart (Fig. 5-1):

$$Z_1 = 210 \angle 15$$
$$Z_2 = 95 \angle -72$$
$$Z_3 = 121 \angle 48$$

The input impedance (See Fig 5-2) is:

$$Z_{IN} = Z_{11} = \frac{Z_1 (Z_2 + Z_3)}{Z_1 + (Z_2 + Z_3)}$$

where $Z_2 + Z_3 = (30 + j90) + (80 + j90)$
$$= 110 + j0 = 110 \angle 0$$
$$Z_1 + Z_2 + Z_3 = (110 + j0) + (200 + j51)$$
$$= 310 + j51 = 320 \angle 10$$

Therefore

$$Z_{IN} = \frac{(210 \angle 15)(110 \angle 10)}{320 \angle 10}$$

$$= 72 \angle 5, \text{ or } 73 + j4 \text{ ANS}$$

The output impedance is:

$$Z_{OUT} = Z_{22} = \frac{(Z_1 + Z_2) Z_3}{(Z_1 + Z_2) + Z_3}$$

where $\quad Z_1 + Z_2 = (200 + j51) + (30 - j90)$
$$= 230 - j39, \text{ or } 235 \angle -10$$

Therefore

$$Z_{OUT} = \frac{(235 \angle -10)(121 \angle 48)}{320 \angle 10}$$

$$= 88.9 \angle 28, \text{ or } 94 + j17 \text{ ANS}$$

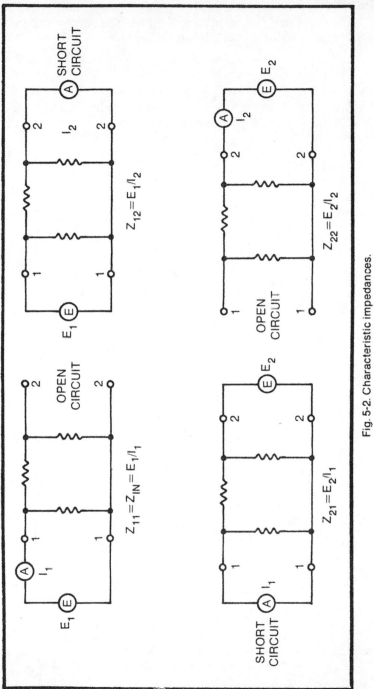

Fig. 5-2. Characteristic impedances.

PROBLEM 5-3

By use of the inverse matrix Z^{-1}, find the current through each resistor in the circuit shown.

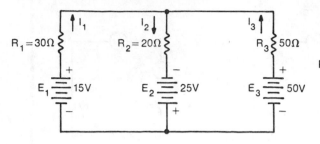

Problem 5-3.

Answer   (*ref. 5, p. 845*)

Loop equations:

$$-I_1 + I_2 - I_3 = 0V$$
$$30I_1 + 20I_2 \quad\;\; = +25V + 15V = 40V$$
$$0I_1 + 20I_2 + 50I_3 = 50V + 25V = 75V$$

Determinant:

$$Z = \begin{vmatrix} -1 & +1 & -1 \\ 30 & +20 & 0 \\ 0 & +20 & +50 \end{vmatrix}$$

$$= +(-10000) + 0 + (-600) - 0 - 1500 - 0$$
$$= -3100$$

Coefficient of Z:

$$\text{coef. } Z = \begin{bmatrix} + \begin{vmatrix} 20 & 0 \\ 20 & 50 \end{vmatrix} & - \begin{vmatrix} 30 & 0 \\ 0 & 50 \end{vmatrix} & + \begin{vmatrix} 30 & 20 \\ 0 & 20 \end{vmatrix} \\[2em] - \begin{vmatrix} 1 & -1 \\ 20 & 50 \end{vmatrix} & + \begin{vmatrix} -1 & -1 \\ 0 & 50 \end{vmatrix} & - \begin{vmatrix} -1 & +1 \\ 0 & 20 \end{vmatrix} \\[2em] + \begin{vmatrix} +1 & -1 \\ 20 & 0 \end{vmatrix} & - \begin{vmatrix} -1 & -1 \\ 30 & 0 \end{vmatrix} & + \begin{vmatrix} -1 & +1 \\ 30 & 20 \end{vmatrix} \end{bmatrix}$$

$$\text{coef. } Z = \begin{bmatrix} (1000 - 0) & - & (1500 - 0) & + & (600 - 0) \\ -(50 + 20) & + & (-50 - 0) & - & (-20 - 0) \\ +(30 + 20) & - & (\;\;0 + 30) & + & (\;\;30 + 20) \end{bmatrix}$$

82

$$\text{coef. } Z = \begin{bmatrix} 1000 - 1500 + 600 \\ -70 - 50 + 20 \\ +20 - 30 - 50 \end{bmatrix}$$

Inverse matrix:

$$Z^{-1} = \frac{1}{|Z|} \left(\text{coef. } Z^{T} \right)$$

where

$$(\text{coef. } Z)^{T} = \begin{bmatrix} 1000 & -70 & +20 \\ -1500 & -50 & -30 \\ +600 & +20 & -50 \end{bmatrix}$$

Thus:

$$Z^{-1} = \begin{vmatrix} \dfrac{1000}{-3100} & \dfrac{-70}{-3100} & \dfrac{+20}{-3100} \\[2mm] \dfrac{-1500}{-3100} & \dfrac{-50}{-3100} & \dfrac{-30}{-3100} \\[2mm] \dfrac{+600}{-3100} & \dfrac{+20}{-3100} & \dfrac{-50}{-3100} \end{vmatrix}$$

Register currents:

$$[I] = [Z^{-1}] \, [V]$$

$$= \begin{bmatrix} -\dfrac{1000}{3100} & +\dfrac{70}{3100} & -\dfrac{20}{3100} \\[2mm] +\dfrac{1500}{3100} & +\dfrac{50}{3100} & +\dfrac{30}{3100} \\[2mm] -\dfrac{600}{3100} & -\dfrac{20}{3100} & +\dfrac{50}{3100} \end{bmatrix} \begin{bmatrix} 0 \\[2mm] 40 \\[2mm] 75 \end{bmatrix}$$

$$= \begin{bmatrix} -\dfrac{1000}{3100}\,(0) + \dfrac{70}{3100}\,(40) - \dfrac{20}{3100}\,(75) \\[3mm] +\dfrac{1500}{3100}\,(0) + \dfrac{50}{3100}\,(40) + \dfrac{30}{3100}\,(75) \\[3mm] -\dfrac{600}{3100}\,(0) - \dfrac{20}{3100}\,(40) + \dfrac{50}{3100}\,(75) \end{bmatrix}$$

$$= \begin{bmatrix} \dfrac{2800}{3100} & - & \dfrac{1500}{3100} \\[2mm] \dfrac{2000}{3100} & + & \dfrac{2250}{3100} \\[2mm] \dfrac{800}{3100} & + & \dfrac{3750}{3100} \end{bmatrix} = \begin{bmatrix} \dfrac{1300}{3100} \\[2mm] \dfrac{4250}{3100} \\[2mm] \dfrac{2950}{3100} \end{bmatrix}$$

Therefore:

$$I_1 = \frac{1300}{3100} = 0.42\text{A}$$

$$I_2 = \frac{4250}{3100} = 1.375\text{A}$$

$$I_3 = \frac{2950}{3100} = 0.955\text{A}$$

Alternate Solution *(ref. 2, p. 23-10)*

A solution of this problem is also possible using loop equations. Although not the method of solution requested, this alternate solution would gain some credit.

Refer to Fig. 5-3. The equation for loop ABEF is

$$-E_1 + I_1 R_1 + I_1 R_2 - E_2 + I_2 R_2 = 0$$
$$I_1 (R_1 + R_2) + I_2 R_2 = E_1 + E_2$$
$$50I_1 + 20I_2 = 40$$

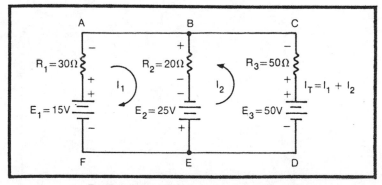

Fig. 5-3. Alternate solution to Problem 5-3.

from which we obtain:

$$I_1 = \frac{40 - I_2 \,(20)}{50}$$

The equation for loop BEDC is

$$-E_3 + I_2 \,(R_2 + R_3) + I_1 \,(R_2) - E_2 = 0$$
$$70I_2 + 20I_1 = 75$$

Substitute for I_1 and solve for I_2 :

$$70I_2 + 20 \left[\frac{40 - I_2 \,(20)}{50} \right] = 75$$

$$70I_2 + \frac{800}{50} - \frac{400I_2}{50} = 75$$

$$I_2 \,(70 - 8) = 75 - 16$$

$$I_2 = 0.95A$$

Go back to the equation for loop ABEF and substitute the value just obtained for I_2 :

$$50I_1 + 20(0.95) = 40$$
$$I_1 = 0.42A$$

The current flowing in the third branch is the total current:

$$I_3 = I_T$$
$$= 0.95 + 0.42$$
$$= 1.37A$$

PROBLEM 5-4

A certain series circuit requires 50V DC to cause 10A to flow. When the DC supply is removed and replaced by a 60 Hz AC supply, 100V is required to produce a current of 10A at 600W to the same series circuit. Calculate:

    (A) The DC resistance in ohms.
    (B) The AC resistance at 60 Hz.
    (C) The power factor at 60 Hz.
    (D) The reactance at 60 Hz.

Answer

    (A) Figure 5-4A shows the equivalent circuit for DC. The resistance is:

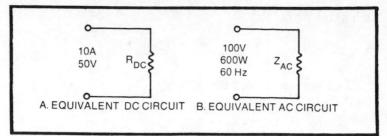

Fig. 5-4. Equivalent circuits.

$$R_{DC} = E_{DC} / I_{DC}$$
$$= 50/10$$
$$= 5\Omega$$

(B) Figure 5-4B shows the circuit for AC, from which

$$Z_{AC} = E_{AC} / I_{AC}$$
$$= 100/10$$
$$= 10\Omega$$

(C) The power factor is

$$P_F = W/VA$$
$$= 600/1000$$
$$= 0.6$$

(D) The reactance is

$$X = \sqrt{Z^2 - R^2}$$
$$= \sqrt{100 - 25}$$
$$= 8.66\Omega$$

PROBLEM 5-5

When 5V DC is applied across a certain circuit, the current flowing in the circuit is 0.1A. When the DC supply is replaced by an AC supply at 80 Hz, it is found that 10V is necessary to cause the same 0.1A current flow and that the power absorbed under the AC condition is 0.6W. Give the reasons for the difference and calculate:

(A) The DC resistance of the circuit.
(B) The power absorbed under DC conditions.
(C) The impedance of the circuit at 80 Hz.
(D) The power factor of the circuit.
(E) The AC circuit resistance.

How do you account for the fact that (E) is greater than (A)?

Answer

(A) From Fig. 5-5, the DC resistance is

$$R_{DC} = E/I$$
$$= 5/0.1$$
$$= 50\Omega$$

(B) The DC power is

$$P_{DC} = E \times I$$
$$= 5 \times 0.1$$
$$= 0.5W$$

(C) The impedance at 80 Hz is

$$Z = E/I$$
$$= 10/0.1$$
$$= 100\Omega$$

(D) The power factor is

$$P_F = \cos(\theta) = P/(EI)$$
$$= 0.6/(10 \times 0.1)$$
$$= 0.6$$

(E) The circuit resistance at 80 Hz is

$$R_{AC} = Z\cos(\theta)$$
$$= 100 \times 0.6$$
$$= 60\Omega$$

Answer (E) is greater than (A) because

1—The circuit contains inductance or capacitance losses.
2—The circuit configuration may be such that an inductance is in series with the resistance, a capacitance shunts the resistance, etc.
3—The phasor components shift from DC to AC conditions.

PROBLEM 5-6

It is required to match a 100Ω resistive load to an electronic generator having an internal impedance that is

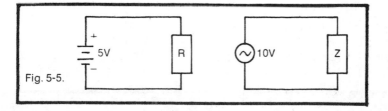

Fig. 5-5.

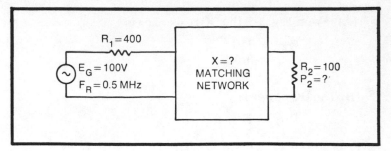

Fig. 5-6.

resistive and equal to 400Ω. A purely reactive network is to be used. The frequency of operation is to be 0.5 MHz.

(A) Design a suitable network yielding maximum power transfer.

(B) If the generator has an EMF of 100V, how much power does the circuit in (A) deliver to the load?

Answer *(ref. 2, pp. 211–213)*

The given information is summarized in Fig. 5-6.

(A) Maximum transfer of energy occurs when input and output impedances are matched. Using the network from reference 2, we have the circuit shown in Fig. 5-7A, for which

$$R_1 / R_2 = 400/100 = 4$$

If a phase angle is selected that would allow c to go to ∞, the π network would be converted into an L-section. The losses in an L-section are least for intermediate values of phase angle. Therefore, an angle of 50° is selected.

From reference 2, $R_1 / R_2 = 4$ at 50° and $c = \infty$. Then $b = 1.5$ (Fig. 76 of reference) and where $\beta < 90°$, $\alpha = 2$ (Fig. 75). The type-3 filter now becomes an L-section as in Fig. 5-7B with parameters

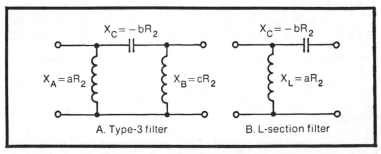

A. Type-3 filter   B. L-section filter

Fig. 5-7.

$$X_C = 1.5(100) = 150\Omega$$
$$X_L = 2(400) = 800\Omega$$

Capacitor C and inductor L become:

$$C = (\pi \times 0.5 \times 10^6 \times 150)^{-1} = 0.00213\,\mu F$$
$$L = (800)(\pi \times 0.5 \times 10^6)^{-1} = 256\,\mu H$$

(B) See ref. 2 (Fig. 79, p. 214). The ratio of the power lost to the network to the power delivered to the network is δ/Q. According to the figure, δ must be about 1.7 for the proper impedance match.

For minimum power loss, $Q = X_L/R$ must be large. At $L = 0.5$ mH, a Q of 80 is easily achieved. Therefore

$$R = X_L/Q$$
$$= 800/80$$
$$= 10\Omega$$

which means that losses here can be confined almost entirely to the l and R. Capacitor losses will thus be neglible. The equivalent circuit now becomes as in Fig. 5-8.

From ref. 2 (p. 213), the power delivered to network is P_{DN} and the power lost to network is P_{LN}, where

$$P_{DN} = P_{LN}\,Q/\delta$$

The generator power is

$$P_G = P_{DN}$$
$$= E_G^{\,2}/R_G$$
$$= 100^2/400$$
$$= 25W$$

Since Q is 80, it follows that

$$P_{LN} = P_{DN}\,\delta/Q$$
$$= 25 \times 1.7/80$$
$$= 0.53W$$

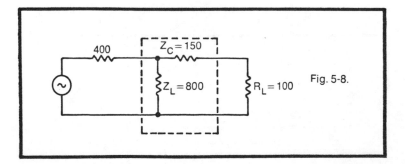

Fig. 5-8.

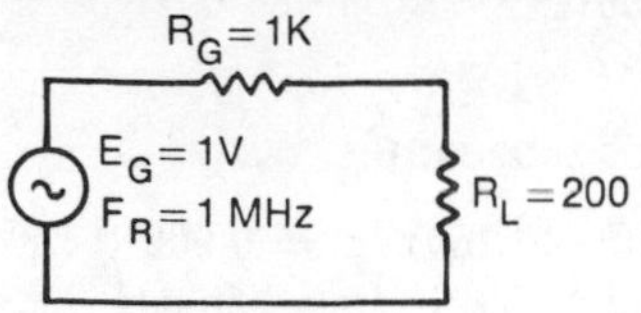

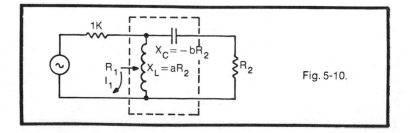

Fig. 5-9.

The power delivered to the load is then

$$P_2 = P_{DN} - P_{LN}$$
$$= 25 - 0.53$$
$$= 24.47W$$

PROBLEM 5-7

You have a generator that is rated at 1.0V output with no load current present. The generator operates at 1 MHz and has an internal resistance of 1000Ω. The generator is to work into a load impedance of 200Ω.

(A) Determine the power delivered to the load.

(B) Design a purely reactive L-section network to couple the generator to the load so that maximum power transfer is obtained. Draw a diagram and indicate values.

(C) Determine the maximum power that can be transferred to the load.

Answer

(Refer also to Problem 5-6, which is similar).

(A) The generator circuit is shown in Fig. 5-9. The total circuit resistance is 1200Ω, so the generator current is

$$I_G = E_G / 1200$$
$$= 0.833 \, mA$$

The power delivered to the load is then

$$P_L = I_G{}^2 \, R_L$$
$$= (8.33 \times 10^{-4})^2 \, (200)$$
$$= 139 \, \mu W$$

(B) For the L-section in Fig. 5-10, $R_1 / R_2 = 1000/200 = 5$

Fig. 5-10.

From ref. 2 (Figs. 75 and 76), $a = 2.5$ and $b = 1.7$, so that

$$X_L = a R_2 = 2.5 \times 200 = 500\Omega$$
$$X_C = - b R_2 = 1.7 \times 200 = 340\Omega$$

The component values of the L-section network in Fig. 5-11 are then:

$$L = X_L /(2\pi F_R) = 500/(6.28 \times 10^6) = 79.6\,\mu H$$
$$C = (2\pi F_R X_C)^{-1} = (6.28 \times 10^6 \times 340)^{-1} = 468\,pF$$

(C) At maximum power, the generator voltage will divide equally across R_G and Z_{IN}, producing 0.5V. This will produce a load current of

$$I_2 = 0.5/400 = 1.25\,mA$$

The maximum power delivered to the load is tnen

$$
\begin{aligned}
P_L &= I_2{}^2 R_L \\
&= (1.25 \times 10^{-3})^2 \ (200) \\
&= 313\,\mu W
\end{aligned}
$$

(Note: As a check on the solution, you should be able to short-circuit the generator end of the L-section and verify that the network impedance seen by the load on the output is also 200Ω, being equal to the load impedance as required for maximum power transfer.)

PROBLEM 5-8

Design a resistance pad having a loss of 10 dB and matching a 200Ω circuit to one having a resistance of 600Ω.
Answer   *ref. 1, p. 17-2)*
The required resistance pad would be an unbalanced T-network as in Fig. 5-12. The loss of the network is to be 10 dB, or

$$K_{MIN} = P_{IN} /P_{OUT} = 10$$

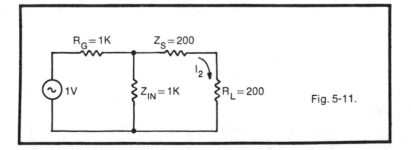

Fig. 5-11.

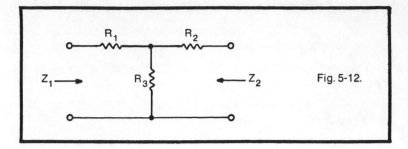

Fig. 5-12.

For the general T-network,

$$K_{MIN} = \frac{2Z_1}{Z_2} - 1 + 2\sqrt{\frac{Z_1}{Z_2}\left(\frac{Z_1}{Z_2} - 1\right)}$$

where $Z_1 = 600\Omega$ is the input impedance and $Z_2 = 200\Omega$ is the output impedance. Since Z_1 is greater than Z_2, we have:

$$R_1 = \frac{Z_1(K+1) - 2\sqrt{KZ_1\,Z_2}}{K-1}$$

$$= \frac{600(11) - 2(1100)}{9} = 488\Omega$$

and

$$R_2 = \frac{Z_2(K+1) - 2\sqrt{KZ_1\,Z_2}}{K-1} = \frac{200(11) - 2(1100)}{9} = 0$$

and

$$R_3 = \frac{2\sqrt{KZ_1\,Z_2}}{K-1}$$
$$= \frac{2(1100)}{9}$$
$$= 233\Omega$$

PROBLEM 5-9

It is necessary to couple a power amplifier to a transmitting antenna working at 1 MHz. The impedances of the

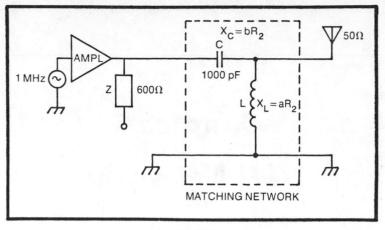

Fig. 5-13.

amplifier and of the antenna are resistances of 600Ω and 50Ω, respectively. The coupling network must isolate the antenna from the source of plate voltage in the amplifier as well as deliver maximum power to the antenna. Only one high-voltage capacitor of 1000 pF capacitance and suitable voltage rating is available. Design the matching network using this capacitor.

Answer

In Fig. 5-13, capacitor C is fixed, so that

$$X_C = \frac{1}{2\pi f C} = \frac{1}{2\pi\,(10^6)\,(10^{-9})} = 160\Omega$$

Now $X_C = bR_2$, and from ref. 2 (Fig. 76, p. 213) it is seen that $b = 3.4$. Therefore

$$X_C = (3.4)\,(50) = 170\Omega$$

which is within 10% of 160Ω and is acceptable.

From Fig. 75, it is also seen that $a = 3.6$, so

$$X_L = (3.6)\,(50) = 180\Omega$$

Solving for L:

$$L = \frac{1}{2\pi f X_L} = 28.6\,\mu H$$

Chapter 6
RLC Networks

Circuits comprised of resistive, inductive, and capacitive components make up the basic ingredients for all electrical networks. Therefore, this review of RLC networks serves to cover the basic equations of simple networks, leaving the more complex networks for later chapters. For study, consult references 1, 2, 3, 5, 6, 47, and 48.

RESONANCE CHARACTERISTICS

In a series-resonant circuit, resonant frequency f_0 is given by the formula

$$f_0 = \frac{1}{2\pi\sqrt{LC}}$$

The condition for resonance in a series circuit is that the capacitive and inductive reactances be equal; ie., $X_L = X_C$. The voltage across either L or C is then $E_L = E_C = QE_{IN}$ (see Fig. 6-1). The impedance of a series circuit is highest at resonance and is equal to

$$Z_0 = (2\pi f_0 L)^2 / R_S = L/RC$$

Series resonance features low tank Z and high line currents.

Parallel resonance occurs under approximately the same conditions as for series resonance, but the nature of the circuit leads to three different definitions of resonance: (1) $X_L = X_C$; (2) zero phase shift across the tank, which requires that the admittances in each parallel branch have

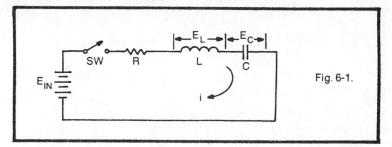

Fig. 6-1.

equal reactive components, $B_L = B_C$; and (3) maximum input impedance. Of the three definitions, (1) and (2) are most commonly used, with (1) generally being referred to as the *natural* resonant frequency. Problems involving parallel resonance are given in Chapter 7, which covers filters. Parallel resonance features high tank Z, low line current, and high tank current.

COUPLING CIRCUITS AND OTHER EQUATIONS

The phasor diagram of an RLC circuit is depicted in Fig. 6-2, showing the relationships between impedance, reactance, and resistance. The impedance of a series RLC circuit is given by the formula

$$Z = R^2 + (X_L - X_C)^2$$

Mutual coupling (Fig. 6-3) between two coils produces a mutual inductance

$$M = k\sqrt{L_1 L_2}$$

where k is the coupling coefficient. At resonance,

$$2\pi f_O M = \sqrt{R_P R_S}$$

The response of a series RLC circuit as shown in Fig. 6-1 is given by the formula

$$E_{IN} = e_R + e_L + e_C$$
$$= Ri + L di/dt + (1/c)\int_0^t i\, dt$$

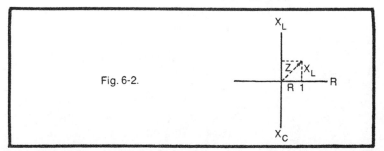

Fig. 6-2.

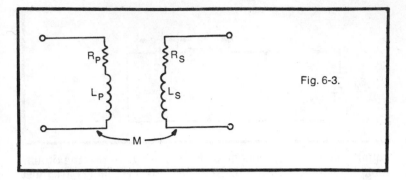

Fig. 6-3.

The response to a step input is of the form $i = A\epsilon^{st}$, which is explained in more detail in Chapters 15 and 16 on transients and Laplace transforms.

PROBLEM 6-1

A series RLC circuit has a resistance of 100 ohms, an inductance of 0.01 henry, and a capacitance of 1 microfarad.

(A) What is the natural frequency in hertz?

(B) What is the decrement factor, usually designated as *alpha*?

(C) What is the logarithmic decrement, usually designated as *delta*?

(D) What is the Q of the circuit?

Answer

(A) The natural resonant frequency of a series circuit is

$$f_0 = \frac{1}{2\pi\sqrt{LC}} = \frac{1}{6.28\sqrt{(0.01)(10^{-6})}} = 1590 \text{ Hz ANS}$$

(B) The decrement factor deals with the off-resonance to resonant frequency ratio. If f is the actual frequency and f_0 the resonant frequency, then

$\alpha = f/f_0$
$\quad = 1$ at resonance **ANS**

(C) The logarithmic decrement deals with the sharpness of a series-resonant curve. For this circuit

$$\delta = R/2f_0 L = \frac{100}{(2)(1590)(0.01)} = 3.145 \text{ ANS}$$

(D) For the circuit values given, the Q will indicate the figure of merit of the tank circuit and affect its response shape.

$$Q = 2\pi f_0 \, L/R$$
$$= (6.28)\,(1590)\,(0.01)/(100)$$
$$= 0.998 \text{ ANS}$$

PROBLEM 6-2

A series circuit containing resistance, reactance, and capacitance has the following values: $R = 25\Omega$, $L = 0.08$ H, and $C = 3.2 \, \mu$F. The input voltage is maintained at 100 volts, but the frequency is varied. At what frequencies will the power in the circuit be 150 watts? What frequency will produce the maximum voltage across L, and what will be this voltage?

Answer

The problem requires three answers: (A) frequencies where circuit power is 150W, (B) frequency where power is maximum, and (C) maximum voltage across L.

(A) Because the series circuit (Fig. 6-4) exhibits a resonance characteristic, there will be two frequencies at which the power dissipation will be the same, unless that power level is achieved only at the resonant frequency. Since $P = I^2 R$, it follows that for $P = 150$ watts and $R = 25$ ohms,

$$I = \sqrt{P/R}$$
$$= 150/25$$
$$= 2.45\text{A}$$

The circuit impedance must then be

$$Z = E/I$$
$$= 100/2.45$$
$$= 40.8\Omega$$

Also

$$Z^2 = R^2 + (X_L - X_C)^2$$
$$= 1665\Omega$$

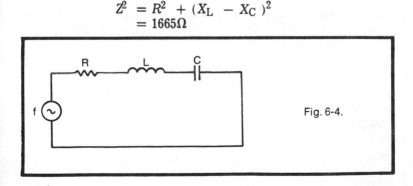

Fig. 6-4.

97

where

$$X_L = 2\pi f$$
$$= 2\pi(0.08)f$$
$$= 0.503f$$

and

$$X_C = 1/2\pi f$$
$$= 1/2\pi(3.2 \times 10^{-6})f$$
$$= 4974/f$$

Therefore

$$1665 = (25)^2 + (0.503f - 4974/f)^2$$

Solving for f yields the quadratic equation

$$0.503f^2 - 32.5f - 4974 = 0$$

The roots of this equation are $f = 72$ Hz and $f = 136$ Hz. **ANS**

PROBLEM 6-3

The equation of current in a resonant circuit is

$$I = \frac{800}{400 + (f - 1000)^2}$$

(A) Find the decrement of the circuit.
(B) If in the circuit $R = 15\Omega$, what is the value of L and C?

Answer

The response equation is similar to that of a series RLC circuit (Fig. 6-4).

(A) The logarithmic decrement reflects the sharpness of the resonant curve and is given as

$$\delta = R/(2\pi f_0 L)$$
$$= \pi R/X_L$$

Since R is given, f_0, L, and C must be found from the given equation by comparing it with the resonant response of a series circuit as given by:

$$I = \frac{E}{\sqrt{R^2 + (X_L - X_C)^2}}$$

Resonance occurs when $X_L - X_C = 0$. Therefore, by comparison,

$$f - 1000 = 0$$
$$f = 1000 \text{ Hz}$$

Rewriting the resonance equation in impedance form,

$$Z = E/I = \sqrt{R^2 + (X_L - X_C)^2}$$

so that

$$\sqrt{800/I} = \sqrt{400 + (f - 1000)^2}$$

Since R is said to be 15Ω, then $15^2 = 225\Omega$. Scaling the equation by $225/400$, we obtain

$$\sqrt{450/I} = \sqrt{225 + (0.75f - 750)^2}$$

Comparing terms, it is apparent that $X_C = 750$. At resonance, $X_L = X_C$ so $X_L = 750$. Therefore

$$\delta = 15\pi/750 = 0.0628 \text{ ANS}$$

(B) Since $X_L = X_C = 750\Omega$ and $f_0 = 1000$,

$$L = X_L / (2\pi f_0)$$
$$= 750/[2\pi(1000)]$$
$$= 119.4 \text{ mH ANS}$$

$$C = 1/(2\pi f_0 X_C)$$
$$= 1/[2\pi(1000)(750)]$$
$$= 0.2122 \ \mu\text{F ANS}$$

PROBLEM 6-4

A circuit is made up of a pure resistance of 50 ohms, an inductance of 0.50 henry, and a capacitance of 8 microfarads, all connected in series.

(A) What impedance does this circuit have when a 500 rad/sec voltage is applied?

(B) What is the power factor?

(C) What is the impedance and power factor at twice the frequency?

Answer

(A) For $\omega = 500$:

$$X_L = (500)(0.5) = 250\Omega$$
$$X_C = 1/(500)(8 \times 10^{-6}) = 250\Omega$$
$$Z = \sqrt{R^2 + (X_L - X_C)^2}$$
$$= \sqrt{50^2 + 0}$$
$$= 50\Omega \text{ ANS}$$

(B) Since $X_L = X_C$ the reactances cancel, so

$$P_f = R/Z = 50/50 = 1.00 \text{ ANS}$$

(C) At twice the frequency:

$$X_L = (2)\,(250) = 500\Omega$$
$$X_C = (0.5)\,(250) = 125\Omega$$
$$Z = \sqrt{50^2 + (500 - 125)^2} = 378.3\Omega \text{ ANS}$$
$$P_f = 50/378.3 = 0.132 \text{ ANS}$$

PROBLEM 6-5

A series circuit consists of a resistance, inductance, and capacitance. The circuit is connected to a variable-frequency oscillator whose voltage is maintained at 15 volts. At 1000 Hz the current is a maximum at 10 mA, but at 2000 Hz the current decreases to 1.0 mA. Find the values of the components in the circuit.

Answer

In a series circuit (Fig. 6-4) the current is maximum at resonance, at which time the circuit behaves as a simple resistance. Therefore,

$$R = E/I$$
$$= 15/0.01$$
$$= 1500\Omega \text{ ANS}$$

Off resonance, the circuit impedance follows the relationship

$$Z = \sqrt{R^2 + (X_L - X_C)^2}$$

Solving for the reactance term

$$X_L - X_C = \sqrt{Z^2 - R^2}$$
$$= \sqrt{(15/0.001)^2 - (1500)^2}$$
$$= 14.92\,\text{k}\Omega$$

Now 2000 Hz is twice the resonant frequency of 1000 Hz, so

$$X_L - X_C = 2X_0 - 0.5X_0$$
$$= 1.5X_0$$

where X_0 is the element impedance at resonance. Therefore,

$$1.5X_0 = 14920$$
$$X_0 = 9947\Omega$$

$$L = X_0/(2\pi f_0)$$
$$= 9947/[2\pi(1000)]$$
$$= 1.583\,\text{H ANS}$$

$$C = 1/(2\pi f_0 X_0)$$
$$= 1/[2\pi(1000)(9947)]$$
$$= 0.016\,\mu\text{F ANS}$$

PROBLEM 6-6

A series circuit has component values of $L = 1.5$ mH, $C = 40$ pF, and $R = 100\Omega$. What voltage will appear across each component if 2 volts is applied to the circuit at its resonant frequency?

Answer

First determine the resonant frequency:

$$f = \frac{1}{2\pi\sqrt{LC}}$$

$$= \frac{1}{2\pi\sqrt{(1.5 \times 10^{-3})(40 \times 10^{-12})}} = 649.7\,\text{kHz}$$

$$
\begin{aligned}
X_L &= 2\pi fL \\
&= 2\pi(649.7 \times 10^3)(1.5 \times 10^{-3}) \\
&= 6124\Omega
\end{aligned}
$$

Since $X_L = L_C$ at resonance,

$$X_C = 6124\Omega$$

The current flowing through the circuit at resonance is

$$
\begin{aligned}
I &= E/R \\
&= 2/100 \\
&= 0.02\text{A}
\end{aligned}
$$

Therefore:

$$
\begin{aligned}
E_R &= IR \\
&= (0.02)(100) \\
&= 2\text{V ANS}
\end{aligned}
$$

$$
\begin{aligned}
E_L &= IX_L \\
&= (0.02)(6124) \\
&= 122.5\text{V ANS}
\end{aligned}
$$

$$
\begin{aligned}
E_C &= IX_C \\
&= (0.02)(6124) \\
&= 122.5\text{V ANS}
\end{aligned}
$$

PROBLEM 6-7

A new electronic device has a nonlinear volt-ampere characteristic of:

$$I_N = -10^{-4}\,E_N{}^2$$

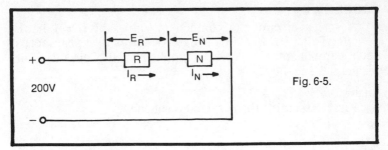

Fig. 6-5.

The linear device is connected in series with a resistive device having a linear characteristic of:

$$I_R = (100 - E_R)10^{-3}$$

If 200 volts is connected to the series circuit thus formed, find I_N, E_R, and E_N.

Answer

From Fig. 6-5 it is clear that $E_N + E_R = 200$ and that $I_N = I_R$ since it is a series circuit. First, solve for E_R, noting that $E_N = 200 - E_R$ and $I_N = I_R$:

$$-10^{-4} E_N{}^2 = (100 - E_R)10^{-3}$$
$$(200 - E_R)^2 = -(100 - E_R)10^1$$
$$200^2 - 400E_R + E_R{}^2 = -1000 + 10E_R$$
$$E_R{}^2 - 410E_R + 41000 = 0$$

By the quadratic equation,

$$E_R = \frac{-(-410) \pm (-410)^2 - 4(1)(41000)}{2(1)}$$

$$= \frac{410 \pm 4100}{2}$$

$$= 205 \pm 32$$

$$= 237 \text{ or } 173$$

Since the supply voltage is only 200V, the answer cannot be 237V, so $E_R = 173V$ (**ANS**). As a result, $E_N = 2000 - E_R = 27V$ (**ANS**).

The current flow through the series circuit is then

$$I_N = I_R = (100 - E_R)10^{-3}$$
$$= (100 - 173)10^{-3}$$
$$= -0.073A \text{ ANS}$$

PROBLEM 6-8

Two meshes in a certain network are connected by mutual inductance coupling. The secondary self-inductance coupling.

The secondary self-inductance L_2 is 0.62 mH and the resistance in the secondary mesh R_2 is 10Ω. The resistance R_1 of the primary mesh is 10 kΩ and the capacitance C_1 of the primary capacitor is 0.005 μF. Assume that the coefficient of coupling is unity. The supply generator produces an emf of $E = 2 \sin (2\pi 10^5 \; t)$ into mesh number 1. Primary inductance L_1 is then adjusted for maximum current in the secondary.

(A) Determine the value of capacitance C_2 needed in the secondary mesh to allow maximum current to flow in the secondary, and give the value of this current.

(B) If the value of L_2 is now changed to 0.20 mH, what would be the best adjustment of C_1 and C_2 for maximum current to exist in the secondary? The coefficient of coupling remains the same as before.

Answer   *(ref. 2, p. 150)*

The problem circuit would appear as in Fig. 6-6. The operating frequency of the generator is 100 kHz since the argument of the sine term is $2\pi 10^5 \; t = 2\pi ft$, making $f = 100$ kHz.

The mutual coupling is $M = k\sqrt{L_1 \; L_2}$. Given that coupling coefficient $k = 1$, then $M = \sqrt{L_1 \; L_2}$. For maximum possible secondary current to occur, M must satisfy the condition $\omega M \sqrt{R_1 \; R_2}$. The minimum possible M is then

$$M = \sqrt{R_1 \; R_2} /\omega = \frac{\sqrt{(10^4 \;) (10)}}{2\pi 10^5} = 5.033 \times 10^{-4}$$

The voltage across C_2 will peak when $R_1 >> \omega L_1$. But before this condition can be tested, the value of L_1 must be found. Knowing M, we can solve for L_1 :

$$M = \sqrt{L_1 \; L_2}$$

$$
\begin{aligned}
L_1 &= M^2 /L_2 \\
&= (5.033 \times 10^{-4} \;)^2 /(0.62 \times 10^{-3} \;) \\
&= 4.086 \times 10^{-4} \\
&= 0.4086 \text{ mH}
\end{aligned}
$$

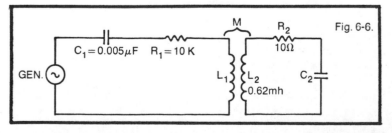

Fig. 6-6.

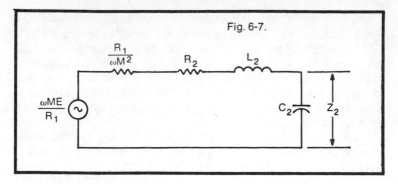

Fig. 6-7.

Now, test the condition for R_1 :

$$R_1 >> \omega L_1$$
$$10^4 >> 2\pi(10^5)(4.086 \times 10^{-4})$$
$$10^4 >> 257$$

The test condition is true, so the equivalent circuit shown in Fig. 6-7 can be drawn.

(A) Secondary current I_2 will be maximum when $X_{L2} = X_{C2}$ at resonance. That is,

$$I_2 = E_2 / Z_2$$

$$= \frac{\omega ME/R_1}{(\omega M)^2 /R_1 + R_2 + (X_{L2} - X_{C2})}$$

$$= \frac{2\pi 10^5 (5.033 \times 10^{-4})E/10^4}{[(2\pi 10^5 \times 5.033 \times 10^{-4})^2 /10^4] + 10}$$

$$= \frac{3.162 \times 10^{-2} E}{20}$$

$$= 1.581 \times 10^{-3} E$$

The excitation voltage is given as $2 \sin(2\pi ft)$, so $E = 2$. Thus,

$$I_2 = (1.581 \times 10^{-3})(2)$$
$$= 3.162 \text{ mA ANS}$$

Since $X_{L2} = X_{C2}$ at resonance,

$$C_2 = 1/(2\pi f)^2 L_2$$
$$= 1/[(2\pi \times 10^5)^2 (0.62 \times 10^{-3})]$$
$$= 4.086 \times 10^{-9}$$
$$= 4.086 \text{ nF ANS}$$

(B) If inductor L_2 now becomes 0.2 mH, R_1 will still be much greater than ωL_1. Therfore, the condition for M and L_2 will remain unchanged. The inductance is then

$$
\begin{aligned}
L_1 &= M^2 / L_2 \\
&= (5.033 \times 10^{-4})^2 / (0.2 \times 10^{-3}) \\
&= 1.267 \times 10^{-3} \\
&= 1.267 \text{ mH \textbf{ANS}}
\end{aligned}
$$

Tuning capacitance C_1 is then

$$
\begin{aligned}
C_1 &= 1 / [(2\pi f)^2 L_1] \\
&= 1 / [(2 \times 10^5)^2 / (1.267 \times 10^{-3})] \\
&= 2.00 \times 10^{-9} \\
&= 2 \text{ nF \textbf{ANS}}
\end{aligned}
$$

And capacitance C_2 becomes

$$
\begin{aligned}
C_2 &= 1 / [(2\pi f)^2 L_2] \\
&= 1 / [(2\pi \times 10^5)^2 (0.2 \times 10^{-3})] \\
&= 1.267 \times 10^{-8} \\
&= 12.67 \text{ nF \textbf{ANS}}
\end{aligned}
$$

PROBLEM 6-9

In the magnetic network shown, find the number of uniformly distributed turns needed to produce a total flux of 1.5 webers.

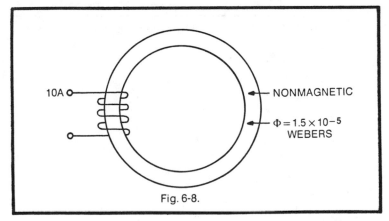

10A   NONMAGNETIC

$\Phi = 1.5 \times 10^{-5}$ WEBERS

Fig. 6-8.

Answer   *(ref. 1, p. 14-4; ref. 51)*

For this problem, use the conversion factors in Table 6-1. Since

$$
\Phi = N_0 (NI) A / l
$$

Table 6-1. Magnetic Units and Conversion Factors.

MULTIPLY	BY	TO OBTAIN
F in ampere-turns	$0.4\pi = 1.257$	F in gilberts
F in gilberts	$1/(0.4\pi) = 0.796$	F in ampere-turns
F in pragilberts*	0.1	F in gilberts
F in gilberts	10	F in pragilberts*
R in ampere-turns	$4\pi = 12.57$	F in pragilberts*
F in pragilberts*	$1/4(\pi) = 0.0796$	F in ampere-turns
H in ampere-turns/in.	$0.4\pi/2.54 = 0.495$	H in oersteds
H in oersteds	$2.54/(0.4\pi) = 2.02$	H in ampere-turns/in.
H in praoersteds*	10^{-3}	H in oersteds
H in oersteds	10^3	H in praoersteds*
H in ampere-turns/in.	495	H in praoersteds*
H in praoersteds*	0.00202	H in ampere turns/in.
B in maxwells/sq in.	$1/6.45 = 0.1555$	B in gauss
B in gauss B in mazwells/sq cm. (lines/in²) }	6.45	B in maxwells/sq in.
B in webers/sq meter*	10^4	B in gauss
B in gauss	10^{-4}	B in webers/sq meter*
B in mazwell/sq in.	$10^{-4}/6.45 = 0.155 \times 10^{-4}$	B in webers/sq meter*
B in webers/sq meter*	6.45×10^4	B in maxwells/sq in.
Φ in maxwells	10^{-8}	Φ in webers*
Φ in lines of flux	10^{-8}	Φ in webers*
Φ in webers*	10^8	Φ in mazwells
1 gauss	1	maxwell/sq cm

*MKS system.

then

$$N = \Phi l/N_0\ IA$$

$$= \frac{(1.5 \times 10^{-5}\ \text{w})\ (26\pi\ \text{cm})\ (10^{-2}\ \text{m/cm})}{(4\pi \times 10^{-7}\ \text{w/m-A})\ (10\ \text{A})\ (\pi)\ (2\ \text{cm})^2\ (10^{-4}\ \text{m}^2/\text{cm}^2)}$$

$$= 776\ \text{turns} \textbf{ ANS}$$

Chapter 7
Filters
and Attenuators

The following topics present a review of various filter types and their responses. Among the textbooks that provide more complete design information are references 1, 6, and 47.

FILTER DEFINITIONS

Figure 7.1 shows the basic response shapes and serves to illustrate the following definitions.

Passband—That portion of the frequency spectrum that shows the least attenuation.

Stopband—That portion of the frequency spectrum that possesses the greatest amount of attenuation.

Slope (or transition region)—That portion of the curve that slopes from the passband to the stopband, or in reverse. This is the transition rate and is usually expressed in decibels per octave. It is sometimes specified as a specific attenuation at a given frequency.

Insertion Loss—This is loss in signal strength. It is a comparison of input power to output power, expressed in decibels.

Rejection—Same as insertion loss, except it refers to the stopband region only.

Cutoff Frequency—This is the point at which the passband and the slope intersect.

Crossover Frequency—Where more than one filter is used, this is the frequency common to adjacent channels.

Crossover Insertion Loss—Crossover of the multiplexer type response will appear to possess an additional 3 dB insertion loss.

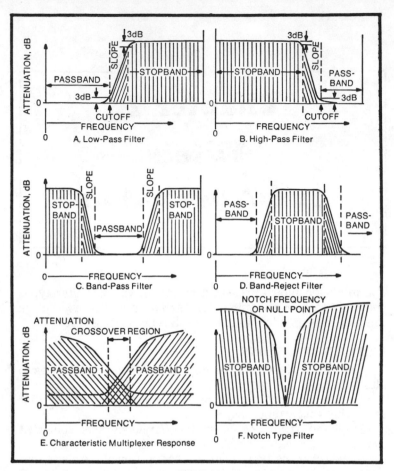

Fig. 7-1.

TYPES OF FILTERS AND THEIR CHARACTERISTICS

M-Derived (ladder)—These are four-terminal devices consisting of symmetrical T or pi (π) sections composed of reactive elements. The reactive elements can be either series or parallel LC legs. When a large difference exists between input and output impedances the filters may break down into unsymmetrical half T's or half pi's.

Constant-k—In a constant-k filter, the product of the shunt or series impedance is a constant and is independent of frequency. For example,

$$k = \sqrt{Z_1 \, Z_2} = R$$

where R is the terminating resistance. Also, if nondissipative elements were used,

$$k = \sqrt{L_1/C_2} = \sqrt{L_2/C_1} = \ldots = R$$

In a constant-k filter, the frequency of maximum attenuation exists at zero and at infinite frequencies. Where greater attenuation is desired in the immediate vicinity of the passband, m-derived filters are normally used.

Bandpass—These are of the single- or double-tuned LC tank filters used in IF stages.

Lattice—This filter consists of T and pi symmetrically arranged into the configurations shown in Fig. 7-2. It features steep attenuation slopes.

Bridged-T, Parallel-T (or notch)—This is a form of the lattice type configuration shown in Fig. 7-2. The bridged-T usually consists of R and C components and are therefore a form of attenuator. Bridged-T's are most popular in "notching" out one specific frequency; hence the reason for its name, notch filter. Parallel-T attenuators possess characteristics similar to the bridged-T. The circuits are shown in Fig. 7-3.

Butterworth Filters—This filter is renowned for its "flat top." It is sometimes referred to as the *maximally flat* (within

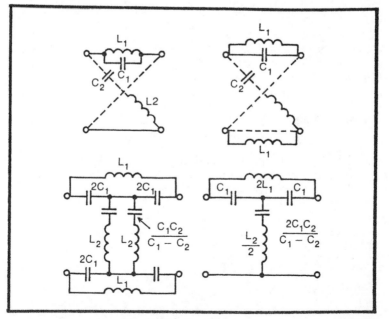

Fig. 7-2. Lattice filters.

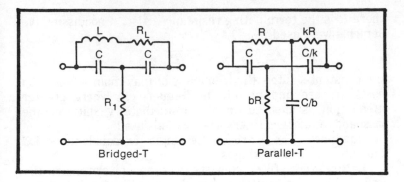

Fig. 7-3. Bridged-T and parallel-T filters.

the passband) filter. Its slope can be governed by the number of poles used.

Bessel (or Bessel-Thompson)—The Bessel filter is known as the *linear phase* filter. It receives its excellent phase-shift linearity at the expense of possessing a more gradual cutoff then, say, Butterworth and most other types. This filter is ideal for passing square pulses with very little overshoot and no time delay at wide frequency ranges.

Tchebyscheff—This filter is characterized by its "equal ripple," although its feature is sharp cutoff. The ripple is usually 2 dB or less. A lower ripple can be obtained, but at the expense of the sharp cutoff.

Transitional Butterworth Thompson—The TBT filter is a trade-off of the last three filters described. It provides greater attenuation than the Bessel in the passband but at the expense of slightly more overshoot and at the sacrifice of some of the flatness of the Butterworth.

PROBLEM 7-1

For the following symmetrical π section, find at 400 Hz:

(A) The open-circuit impedance.
(B) The short-circuit impedance.
(C) The characteristic impedance.

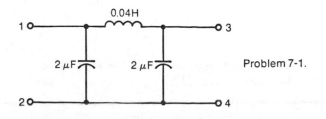

Problem 7-1.

Answer *(ref. 1, pp. 23-31 and 23-50)*

(A) The open-circuit impedance Z_{OC} is given by the formula

$$Z_{OC} = \frac{2Z_2 \, (Z_1 \, + \, 2Z_2 \,)}{Z_1 \, + \, 4Z_2}$$

From the network in the problem,

$$Z_1 \, = 0.04\omega \angle 90 = 0 + 0.04\omega j$$
$$Z_2 \, = 10^6 \, /(2\omega \angle -90) = 0 - 10^6 \, j/(2\omega)$$

Substituting for Z_1 and Z_2 :

$$Z_{OC} \, = \frac{2(-10^6 \, j/2\omega) \, [(0.04\omega j) \, + \, 2(-10^6 \, j/2\omega)]}{(0.04\omega j) \, + \, 4(-10^6 \, j/2\omega)}$$

$$= \frac{-10^6 \, j(0.04\omega^2 \, j \, - \, 10^6 \, j)}{(0.04\omega^2 \, j \, - \, 2 \times 10^6 \, j)\omega}$$

$$Z_{OC} = \frac{-10^6 \, j[0.04(2\pi400)^2 \, j \, - \, 10^6 \, j]}{[0.04(2\pi400)^2 \, j \, - \, 2 \times 10^6 \, j] \, (2\pi400)}$$

$$= \frac{-10^6 \, j \, (0.2527 \times 10^6 \, - \, 10^6 \,)}{(0.2527 \times 10^6 \, - \, 2 \times 10^6 \,) \, (0.2527 \times 10^6 \,)}$$

$$= -1.692j$$
$$= 1.692 \angle -90 \text{ ANS}$$

(B) The short-circuit impedance is given by the formula

$$Z_{SC} \, = \frac{2Z_1 \, Z_2}{Z_1 \, + \, 2Z_2}$$

Substituting values,

$$Z_{SC} \, = \frac{2(0.04\omega \angle 90) \, (10^6 \, /2\omega \angle -90)}{(0.04\omega j) \, + \, 2(-10^6 \, j/2\omega)}$$

$$= \frac{0.04\omega \, (-10^6 \, j)}{0.04\omega^2 \, - 10^6}$$

$$= \frac{-0.04 \, (2\pi400) \, (10^6 \, j)}{0.04 \, (2\pi400)^2 \, - 10^6}$$

$$= 134.5j, \text{ or } 134.5 \angle 90 \text{ ANS}$$

111

(C) The characteristic impedance is given by the formula

$$Z_{0\Pi} = \sqrt{\frac{4Z_1 Z_2^2}{Z_1 + 4Z_2}}$$

Substituting for Z_1 and Z_2 :

$$Z_{0\Pi} = \sqrt{\frac{4(0.04\omega \angle 90)(10^6 /2\omega \angle -90)^2}{(0.04\omega j) + 4(-10^6 j/2\omega)}}$$

$$= \frac{0.04j(-10^6 j)^2}{(0.04\omega^2 j) + 2(-10^6 j)}$$

$$= \sqrt{\frac{-0.04 \times 10^{12}}{0.04(2\pi400)^2 - 2 \times 10^6}}$$

$$= \sqrt{2.289 \times 10^4}$$
$$= 151.3 \text{ ANS}$$

PROBLEM 7-2

In the parallel-T (notch type) filter shown, a null is desired at 60 Hz that has the narrowest attenuation band possible. Find the parameter values for R and C.

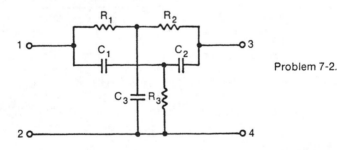

Problem 7-2.

Answer *(ref. 1, p. 16-24)*

A complete null occurs at ω_0 , for which the values of R and C must satisfy the equation

$$RC = 1/\omega_0$$
$$= 1/2\pi(60)$$
$$= 2.65 \times 10^{-3}$$

Since R and C are independent, assign a convenient value for R and solve for C. Let $R = 10\text{k}\Omega$ making

112

$$C = 2.65 \times 10^{-3} / (10 \times 10^3)$$
$$= 0.265 \times 10^{-6}$$
$$= 0.265 \mu F$$

For the narrowest attenuation band possible, $k = 100$ (from ref. 1) in the formula

$$B = k/(k + 1)$$
$$= 100/(100 + 1)$$
$$= 0.99$$

Thus for the parallel-T circuit, the component values would be:

$$R_1 = R = 10 \text{ k}\Omega$$
$$R_2 = kR = 1 \text{ M}\Omega$$
$$R_3 = BR = 9900\Omega$$
$$C_1 = C = 0.265 \mu F$$
$$C_2 = C/k = 2650 \text{ pF}$$
$$C_3 = C/B = 0.268 \mu F$$

PROBLEM 7-3

For a pi filter in which the series arm consists of a 75 mH inductance coil, the resistance of which is 25Ω, and the two arms consist each of a $0.2\mu F$ capacitor, the resistance of which is negligible, find the following:

(A) The open-circuit and short-circuit impedances at 400 Hz.

(B) The characteristic impedance at 400 Hz and at 1000 Hz.

Answer

From the circuit in Fig. 7-4A, it is seen that

$$X_{C1} = X_{C2} = 1/2\pi fC$$

$$= \frac{1}{2\pi(400)(0.2 \times 10^{-6})}$$

$$= 1989\Omega$$

so that in Fig. 7-4B, $Z_{C1} = Z_{C2} = 0 - 1989j = 1989 \angle -90$. Similarly,

$$X_L = 2\pi fL$$
$$= 2\pi(400)(0.075)$$
$$= 188.5\Omega$$

so that $Z_L = 25 + 180.5j = 188.5 \angle 13$.

113

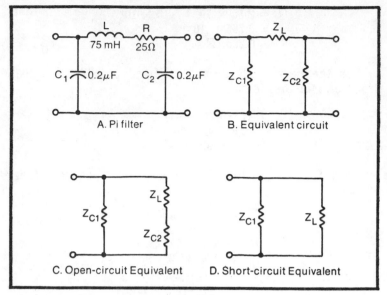

Fig. 7-4.

The circuit can be redrawn as in Fig. 7-4C, making it possible to combine Z_{C2} and Z_L as an equivalent impedance, Z_{EQ} :

$$Z_{EQ} = 25 + 180.5j - 1989j$$
$$= 251808j$$

The resultant is essentially capacitive in nature, so the open-circuit impedance is approximately

$$Z_{OC} = \frac{Z_{C1} + Z_{EQ}}{Z_{C1} \ Z_{EQ}} = \frac{(1989) \ (1808)}{1989 + 1808} = 947\Omega \text{ ANS}$$

The short-circuit impedance is derived from the circuit in Fig. 7-4D, for which Z_L may be assumed equal to 188.5Ω.

$$Z_{SC} = \frac{Z_{C1} \ Z_L}{Z_{C1} + Z_L} = \frac{(-1989) \ (188.5)}{(-1989) + (188.5)} = 208\Omega \text{ ANS}$$

(B) The characteristic impedance at 400 Hz will be equal to the open-circuit impedance, as calculated in part (A):

$$Z_0 = Z_{OC} = 947\Omega \text{ ANS}$$

At 1000 Hz the element impedances will change from that at 400 Hz, making

$$Z_{C1} = Z_{C2} = (400/1000)(1989j) - 796j$$
$$Z_L = 25 + (1000/400)(180.5j) = 25 + 451j$$

Thus, the characteristic impedance is equal to the new Z_{OC}, which is computed approximately as follows:

$$Z_{EQ} = 25 + 451j - 796j$$
$$= 25 - 345j$$
$$= 346 \angle 4$$

$$Z_{OC} = Z_0 = \frac{(796)(346)}{(790) + (346)} = 242\Omega \text{ ANS}$$

PROBLEM 7-4

For a symmetrical pi type filter with full series arm of 0.1 H inductance and shunt-arm capacitors of $0.02\mu F$ each, calculate the following:

(A) Neglecting the resistance of the circuit elements, find the propagation constant if the frequency is 500 Hz and 2000 Hz.

(B) Find the characteristic impedance at each of the preceding frequencies.

Answer *(ref. 1, chs. 7 and 8; ref. 6, p. 16-2)*

(A) When a filter section is terminated on an image impedance basis, the transmission (or propagation) constant is

$$\gamma = \alpha + j\beta$$
$$= 2 \ln (\sqrt{1 + (Z_1/4Z_2)} + \sqrt{Z_1/4Z_2})$$

where Z_1 and Z_2 are defined in Fig. 7-5. Solving:

$$Z_1 = 0.1 j\omega$$
$$2Z_2 = 2/X_C = (5 \times 10^7)/j\omega$$
$$Z_2 = (2.5 \times 10^7)/j\omega$$
$$\frac{Z_1}{4Z_2} = \frac{0.1(j\omega)^2}{4(2.5 \times 10^7)} = -1 \times 10^{-9} \times \omega^2$$

At 500 Hz, $\omega = 3141$ and $\omega^2 = 9.87 \times 10^6$, so that

$$\gamma = 2 \ln[\sqrt{1 - (1 \times 10^{-9})(9.87 \times 10^6)}$$
$$+ \sqrt{-(1 \times 10^{-9})(9.87 \times 10^6)}]$$
$$= 2 \ln [0.995 + 0.0993j]$$

115

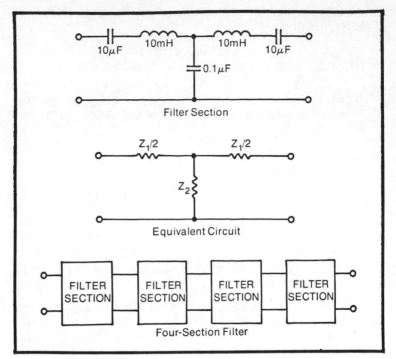

Fig. 7-5. Four-section filter network.

Converting to polar form in radians,

$$\gamma = 2 \ln [1 \exp(0.0995j)]$$
$$= 0 + 2(0.0995j)$$
$$= 0 + 0.199j \text{ ANS}$$

The answer indicates that the attenuation constant (α) is zero, so the transmission loss is zero nepers, and also zero decibels. The phase shift (β) is 0.199 radians.

At 2000 Hz, $\omega = 1.256 \times 10^4$ and $\omega^2 = 1.579 \times 10^8$, so

$$\gamma = 2 \ln[\sqrt{1 - (1 \times 10^{-9})(1.579 \times 10^8)}$$
$$+ \sqrt{-(1 \times 10^{-9})(1.579 \times 10^8)}]$$
$$= 2 \ln[0.918 + 0.397j]$$
$$= 2 \ln[1 \exp(0.408j)]$$
$$= 0 + 0.816j \text{ ANS}$$

(B) The characteristic impedance is given by the formula

$$Z_0 = \sqrt{\frac{Z_1 Z_2}{1 + Z_1 / 4Z_2}}$$

where $Z_1 Z_2 = 2.5 \times 10^6$
At 500 Hz,

$$Z_1 / Z_2 = (-4 \times 10^{-9})(9.87 \times 10^6)$$
$$= -3.948 \times 10^{-2}$$

and at 2000 Hz,

$$Z_1 / Z_2 = (-4 \times 10^{-9})(1.579 \times 10^8)$$
$$= -6.316 \times 10^{-1}$$

Therefore, at 500 Hz the characteristic impedance is

$$Z_0 = \sqrt{(2.5 \times 10^6)/[1 + (-3.948 \times 10^{-2})/4]}$$
$$= 1.589 \times 10^3$$
$$= 1.589 \text{ k}\Omega \text{ ANS}$$

At 2000 Hz the characteristic impedance is

$$Z_0 = \sqrt{(2.5 \times 10^6)/[1 + (-6.316 \times 10^{-1})/4]}$$
$$= 1.723 \times 10^3$$
$$= 1.723 \text{ k}\Omega \text{ ANS}$$

PROBLEM 7-5

For the circuit shown in the figure, find:

(A) L_2 for parallel resonance at 15 kHz.
(B) Z_{BC} at 45 khz.
(C) L_1 or C_1 (as the case may be) to provide the lowest impedance at 45 kHz.
(D) Z_{AB} at 45 kHz.
(E) Z_{AB} at 15 kHz.

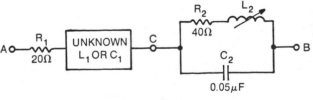

Problem 7-5.

Answer

Since L_1 or C_1 is chosen for minimum impedance at 45 kHz, let this be the pass frequency, f_1. And since L_2 is chosen for maximum tank impedance at 15 kHz, let this be the stop frequency, f_2.

(A) To block signals at f_2, impedance Z_{BC} is adjusted for parallel resonance at f_2. This means that inductive branch $R_2 - L_2$ and capacitive branch C_2 must have equal susceptances: $B_L = B_C$. Using the relationship, $Y = G - jB$:

$$Y_C = 1/Z_C = 1/(-jX_{C2}) = j/X_{C2}$$

$$Y_L = 1/Z_L = 1/R_2 + jX_{L2})$$

$$= \left[\frac{1}{R_2 + jX_{L2}} \right] \left[\frac{R_2 - jX_{L2}}{R_2 - jX_{L2}} \right]$$

$$= \left[\frac{R_2}{R_2{}^2 + X_{L2}{}^2} \right] - \left[\frac{jX_{L2}}{R_2{}^2 + X_{L2}{}^2} \right]$$

For the capacitive branch,

$$B_C = 1/X_{C2}$$

For the inductive branch,

$$B_L = \frac{X_{L2}}{R_2{}^2 + X_{L2}{}^2}$$

Setting B_C equal to B_L:

$$\frac{X_{L2}}{R_2{}^2 + X_{L2}{}^2} = \frac{1}{X_{C2}}$$

$$X_{L2} X_{C2} = R_2{}^2 + X_{L2}{}^2$$

$$X_{L2}{}^2 - X_{C2} X_{L2} + R_2{}^2 = 0$$

$$X_{L2} = \frac{X_{C2}}{2} \pm \sqrt{\frac{X_{C2}{}^2 - 4R_2{}^2}{2}}$$

Solving for L_2 at $f_2 = 15$ kHz:

$$2\pi f_2 L_2 = \frac{1}{4\pi f_2 C_2} \pm \tfrac{1}{2}\sqrt{(1/2\pi f_2 C_2)^2 - 4(40)^2}$$

$$L_2 = 1/8\pi^2 f_2{}^2 C_2 \pm 1/4\pi f_2 \sqrt{(1/2\pi f_2 C_2)^2 - 6400}$$

$$= 1/8\pi^2 (15 \times 10^3)^2 (0.05 \times 10^{-6})$$

$$\pm 1/4\pi(15 \times 10^3)\sqrt{1/2\pi(15 \times 10^3)(0.05 \times 10^{-6})]^2 - 6400}$$

$$= 1.126 \times 10^{-3} \pm 1.043 \times 10^{-3}$$

which yields answers of 0.083 mH and 2.169 mH. The larger value of inductance is the answer since it will produce the lowest conductance, and hence the highest total current. Therefore, $L_2 = 2.169$ mH **ANS**.

(B) To compute impedance Z_{BC} at 45 kHz, it is necessary to find the parallel impedance of tuned circuit $R_2 - C_2 - L_2$; for which we use the value of L_2 found in part (A) of the problem:

$$Z_{L2} = R_2 + X_{L2} j$$
$$= 40 + 2\pi(45 \times 10^3)(2.169 \times 10^{-3})j$$
$$= 40 + 613j$$

$$Z_{C2} = -jX_{C2}$$
$$= -j/[2\pi(45 \times 10^3)(0.05 \times 10^{-6})]$$
$$= -70.7j$$

$$Z_{BC} = \frac{Z_{L2}\, Z_{C2}}{Z_{L2} + Z_{C2}}$$

$$= \frac{(40 \times 613j)(-70.7j)}{40 + 613j - 70.7j}$$

$$= \left(\frac{4.33 \times 10^4 - 2828j}{40 + 542.3j} \right) \times \left(\frac{40 - 542.3j}{40 - 542.3j} \right)$$

$$= \frac{1.984 \times 10^5 - 2.359 \times 10^7\, j}{2.96 \times 10^5}$$

$$= 0.670 - 79.7j \text{ ANS}$$

(C) Since impedance Z_{BC} in the preceding step was negative, to give the lowest impedance at 45 kHz, it is necessary to use an inductance in the series arm. Thus,

$$X_L = 79.7 = 2\pi f_2\, L_1$$
$$L_1 = 79.7/[2\pi(45 \times 10^3)]$$
$$= 0.282 \text{ mH ANS}$$

(D) Impedance Z_{AB} at 45 kHz is the total series impedance offered by the network, and this is equal to $Z_{Ac} + Z_{BC}$:

$$Z_{AB} = (20 + 79.7j) + (0.670 - 79.7j)$$
$$= 20.670\Omega \text{ ANS}$$

119

(E) Impedance Z_{AB} at 15 kHz is also equal to $Z_{AC} + Z_{BC}$ but must be recomputed for the different frequency:

$$Z_{AB} = 20 + \frac{R_2{}^2 + X_{LZ}{}^2}{R_2} + \frac{79.7j}{3}$$

$$= 20 + \frac{(40)^2 + [2\pi(15000)(2.169 \times 10^{-3})]^2}{40} + 26.6j$$

$$= 1.105 \times 10^3 + 26.6j$$

$$|Z|_{AB} = 1105\Omega \text{ ANS}$$

PROBLEM 7-6

If at input terminals AB of the filter system shown, a 120 Hz voltage of amplitude 200V is applied, to what value will this be reduced at load terminals DC?

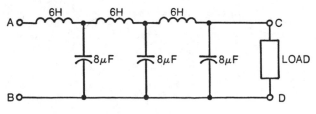

Problem 7-6.

Answer *(ref. 47, p. 313)*

This is a 3-section LC filter. For one section, the output voltage is reduced by the factor:

$$\frac{X_C}{X_L + X_C} = \frac{1/\omega Cj}{\omega Lj + 1/\omega Cj} = \frac{1}{1 - \omega^2 LC}$$

In most cases, the term $\omega^2 LC$ is much greater than unity, so this factor reduces to $1/\omega^2 LC$ for one section. For two sections, this factor becomes $1/\omega^4 L^2 C^2$, and for three sections, $1/\omega^6 L^3 C^3$. Inserting component values, the reduction factor for the filter system is:

$$E_0 = \frac{1}{(2\pi 120)^6 (6)^3 (8 \times 10^{-6})^3} = 49.2 \times 10^{-6}$$

120

If the input voltage was a pure sign wave, the output voltage across the load would be equal to the reduction factor times the input voltage, or about 18.1 mV **ANS**.

Though not so stated in the problem, the input signal is commonly a full-wave rectified signal, derived from the 60 Hz power line. In this case, the second harmonic (120 Hz) would dominate the higher-order harmonics, since the amount of output ripple decreases inversely as sixth power of the frequency in this system. Thus, the output voltage would consist mostly of this ripple, and its peak value would be:

$$E_{PK} = \left(\frac{4E_M}{3\pi} \right) \left(\frac{1}{\omega^6 \, L^3 \, C^3} \right)$$

$$= \frac{4\,(200)\,(49.2 \times 10^{-6})}{3\pi}$$

$$= 4.18 \, \text{mV}$$

PROBLEM 7-7

The wave trap shown has a 60-henry inductor with $Q = 5$ at 120 Hz. The capacitive losses are negligible.

(A) Find the value of C to produce resonance at 120 Hz.

(B) Compare the solution in part (A) with that for a solution in which R_L is negligible.

(C) Find the impedance ratio of the wave trap at 120 Hz to that at 60 Hz.

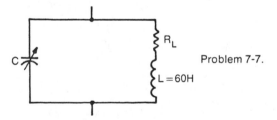

Problem 7-7.

Answer

Since L and Q are known, the value of R_L is then

$$\begin{aligned} R_L &= X_L \,/Q = 2\pi f L/Q \\ &= 2\pi (120)\,(60)/(5) \\ &= 9050 \end{aligned}$$

(A) Parallel resonance occurs when $B_L = B_C$. The admittances of the capacitive and inductive branches are derived as follows:

$$Y_L = \frac{1}{R_L + X_L j} \times \frac{R_L - X_L j}{R_L - X_L j}$$

$$= \frac{R_L}{R_L{}^2 + X_L{}^2} - \frac{X_L j}{R_L{}^2 + X_L{}^2}$$

$$Y_C = (1/X_C j)(-j/-j)$$

$$= j/X_C$$

For the inductive branch, $Y_L = G_L - jB_L$, so

$$B_L = \frac{X_L}{R_L{}^2 + X_L{}^2}$$

For the capacitive branch, $Y_C = G_L + jB_C$, so

$$B_C = 1/X_C$$

Setting B_L equal to B_C :

$$\frac{X_L}{R_L{}^2 + X_L{}^2} = \frac{1}{X_C}$$

$$X_C = \frac{R_L{}^2 + X_L{}^2}{X_L}$$

$$= \frac{(9050)^2 + [2\pi(120)(60)]^2}{2\pi(120)(60)}$$

$$= 47.05 \times 10^3$$

$$\frac{1}{2\pi(120)C} = 47.05 \times 10^3$$

$$C = 0.0282\mu\text{F ANS}$$

(B) When $R_L = 0$, X_C simple equals X_L for resonance.
Thus,

$$X_C = X_L = 2\pi fL$$
$$= 2\pi(120)(60)$$
$$= 45.2\,\text{k}\Omega$$

$$C = 1/[2\pi fX_C]$$
$$= 1/[2\pi(120)(45.2) \times 10^3)]$$
$$= 0.0293\,\mu\text{F ANS}$$

(C) At 120 Hz, $X_L = 45.2\,k\Omega$ and $X_C = 45.4\,k\Omega$. At 60 Hz, $X_L = 22.6\,k\Omega$ and $X_C = 90.8\,k\Omega$. In both cases, R_L remains 9050Ω. The impedance at 120 Hz is then:

$$Z_{120} = \frac{(9050 + 45.2 \times 10^3\, j)\,(-45.4 \times 10^3\, j)}{9050 + 45.2 \times 10^3\, j - 45.4 \times 10^3\, j}$$

$$= \frac{2.05 \times 10^9 - 4.11 \times 10^8\, j}{9050 - 200j}$$

$$|Z_{120}| = 231\,k\Omega$$

The impedance at 60 Hz is:

$$Z_{60} = \frac{(9050 + 22.6 \times 10^3\, j)\,(-90.8 \times 10^3\, j)}{9050 \times 22.6 \times 10^3\, j - 90.8 \times 10^3\, j}$$

$$= \frac{2.05 \times 10^9 - 8.22 \times 10^8\, j}{9050 - 6.82 \times 10^4\, j}$$

$$|Z_{60}| = 32.1\,k\Omega$$

Therefore:

$$\frac{|Z_{120}|}{|Z_{60}|} = \frac{231}{32.1} = 7.20 \text{ ANS}$$

PROBLEM 7-8

A filter network is required for a control system application. One of the requirements of the network is that the current in the output circuit cannot be removed more than 60 degrees from the current in the input terminals. The filter is made up of four identical T-sections connected in series, each T having two series arms, each arm composed of a 10 mH coil of a 0.1 μF capacitor. The receiving end, or terminating impedance, is equal to the characteristic impedance at 1000 Hz. There is a 1000 Hz sine-wave voltage impressed across the sending-end terminals, and the current at the sending end is 0.01A RMS.

(A) Does the network satisfy the imposed requirement?

(B) What is the RMS value of the required input voltage?

(C) What is the RMS value of the current through, and the voltage across, the terminating impedance?

Answer (ref. 1, p. 16-3)

(A) The filter network would appear as in Fig. 7-5. For each T-section,

$$\alpha + \beta j = 2 \ln[\sqrt{1 + (Z_1/4Z_2)} + \sqrt{Z_1/4Z_2}\,]$$

Solving for Z_1 only:

$$Z_1 = 2[-j/\omega(10^{-5}) + j\omega(0.01)]$$

$$= \frac{-2j + 2(2\pi \times 10^3)^2 (0.01) (10^{-5})j}{(2\pi \times 10^3) (10^{-5})}$$

$$= 93.8j$$

$$4Z_2 = 4[-j/\omega(0.1 \times 10^{-6})]$$
$$= -4j/(2\pi \times 10^3) (0.1 \times 10^{-6})$$
$$= -6366j$$

$$Z_1/4Z_2 = 93.8j/(-6366j) = -0.01473$$

$$\alpha + \beta j = 2\ln(\sqrt{1 - 0.01473} + \sqrt{-0.01473})$$

$$= 2\ln(0.993 + 0.121j)$$

Converting to polar form in degrees,

$$\alpha + \beta j = 2\ln[1\exp(6.95°j)]$$
$$= 0 + 13.9°j$$

Therefore the phase shift (β) per section is 13.9°.

With a phase shift of 13.9° per filter section, a four-section filter will have a phase shift of 55.6°, which satisfies the phase requirement of being less than 60°.

(B) The characteristic impedance at 1000 Hz is

$$Z_{OT} = \sqrt{Z_1 Z_2 [1 + (Z_1/4Z_2)]}$$
$$= \sqrt{(93.8j) (-1592j)[1 + (93.8j)/(-6366j)]}$$
$$= \sqrt{(1.493 \times 10^5) (0.985)}$$
$$= 384\Omega$$

The required RMS input voltage is then

$$e_{IN} = i_{IN} Z_T$$
$$= 2i_{IN} Z_{OT}$$
$$= 2(0.01) (384)$$
$$= 7.68V \text{ ANS}$$

(C) Since the filter network is lossless ($\alpha = 0$), the input and output currents will be the same, or 0.01A. Also, since the characteristic impedance is stated to be equal to the terminating impedance, the voltage across the terminating load will be merely half of the input voltage, or 3.84V.

124

PROBLEM 7-9

(A) Design a low-pass filter having a cutoff frequency of 2000 Hz and a zero-frequency output resistance of 500 ohms. Sketch the filter.

(B) Design a high-pass filter with a cutoff frequency at 2000 Hz and a characteristic impedance at infinite frequency of 500 ohms. Sketch the filter.

Answer

(A) For the low-pass filter shown in Fig. 7-6A:

$$f_C = 1/\pi\sqrt{L_1 C_2}$$

$$\begin{aligned} L_1 C_2 &= (1/f_C \pi)^2 \\ &= (1/2000\pi)^2 \\ &= 2.53 \times 10^{-8} \end{aligned}$$

$$Z_1 Z_2 = (\omega L_1\, j)\,(-j/\omega C_2) = L_1/C_2$$

$$\frac{Z_1}{4Z_2} = \frac{\omega L_1\, j}{4(-j/\omega C_2)}$$

$$= -\omega^2 L_1 C_2 /4$$

which is zero at frequency of zero.

$$\begin{aligned} R_0 = Z_{OT} &= \sqrt{Z_1 Z_2 (1 + Z_1/4Z_2)} \\ &= \sqrt{Z_1 Z_2} = \sqrt{L_1/C_2} \\ &= 500\,\Omega \end{aligned}$$

$$L_1/C_2 = (500)^2$$

$$L_1 = 25 \times 10^4\, C_2$$

$$L_1 C_2 = 2.53 \times 10^{-8} = 25 \times 10^4\, C_2{}^2$$

$$C_2 = \sqrt{2.53 \times 10^{-8}/25 \times 10^4} = 0.318\ \mu\text{F ANS}$$

$$L_1 = 25 \times 10^4\,(0.318 \times 10^{-6}) = 79.5\ \text{mH ANS}$$

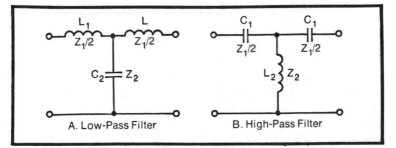

Fig. 7-6. Filter designs.

125

(B) For the high-pass filter in Fig. 7-6B:

$$f_C = 1/4\pi\sqrt{C_1 L_2}$$

$$\begin{aligned}
C_1 L_2 &= (1/4\pi f_C)^2 \\
&= [1/4\pi(2000)]^2 \\
&= 1.58 \times 10^{-9}
\end{aligned}$$

$$Z_1 Z_2 = (-j/\omega C_1)(\omega L_2 j) = L_2/C_1$$

$$Z_1/4Z_2 = 0 \text{ at frequency of zero.}$$

$$\begin{aligned}
R_0 &= Z_{OT} = \sqrt{Z_1 Z_2 (1 + Z_1/4Z_2)} \\
&= \sqrt{Z_1 Z_2} = \sqrt{L_2/C_1} = 500\Omega
\end{aligned}$$

$$L_2 = (500)^2 C_1 = 25 \times 10^4 C_1$$

$$C_1 L_2 = 1.58 \times 10^{-9} = 25 \times 10^4 C_1{}^2$$

$$C_1 = \sqrt{1.58 \times 10^{-9}/25 \times 10^4} = 0.0795\,\mu\text{F ANS}$$

$$L_2 = (25 \times 10^4)(0.0795 \times 10^{-6}) = 19.9\,\text{mH ANS}$$

PROBLEM 7-10

Find the magnitude of the first three harmonics at the output of the filter shown when the square wave signal is applied at the input.

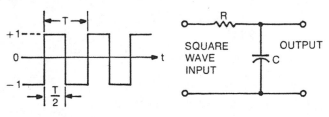

Problem 7-10.

Answer (ref. 47, p. 297)

The Fourier series for a square wave is

$$e_0 = 2/T\int_0^T f(t) \sin(n\omega t)\, dt$$

where $n = 1, 3, 5, \ldots$

For the integration, from $t = 0$ to $t = T/2$, $f(t)$ is a constant; set this equal to $+E$. From $T/2$ to T, $f(t)$ is also a constant; set this equal to $-E$. Therefore:

$$\begin{aligned}
e_0 &= 2/T\int_0^{T/2} E \sin(n\omega t)\, dt - 2/T\int_{T/2}^T E \sin(n\omega t)\, dt \\
&= -2E/n\omega T[\cos(n\omega t)]_0^{T/2} + 2E/n\omega T[\cos(n\omega t)]_{T/2}^T
\end{aligned}$$

so $\omega = 2\pi f$ and $f = 1/T$.

Therefore, $\omega = 2\pi/T$ or $T\omega = 2\pi$. Substituting,

$$e_0 = -2E/2\pi n[\cos(\omega T/2) - \cos(0)]$$

$$+ 2E/2\pi n[\cos(n\omega T) - \cos(n\omega T/2)]$$

$$= -E/\pi n[\cos(2n\pi/2) - 1]$$

$$+ E/\pi n[\cos(2n\pi) - \cos(2n\pi/2)]$$

$$= -E/\pi n[\cos(n\pi) - 1]$$

$$+ E_\pi /n[\cos(2n\pi) - \cos(n\pi)]$$

$$= E/\pi n[\cos(2n\pi) - \cos(n\pi) - \cos(n\pi) + 1]$$
$$= E/\pi n[\cos(2n\pi) - 2\cos(n\pi) + 1]$$

Let $E = 1$ volt and $n = 1$, then $\cos(2\pi) = +1$, $\cos(\pi) = -1$, and

$$e_0 = (1/\pi)(1 + 2 + 1) = 4/\pi$$
At $n = 2$, $e_0 = (1/2\pi)(1 - 2 + 1) = 0$.
At $n = 3$, $e_0 = (1/3\pi)(1 + 2 + 1) = 4/3\pi$.
Factoring out a 4 and writing terms for all odd numbers,

$$f(t) = 4/3\pi[\sin(\omega t) + \sin(3\omega t)/3$$

$$+ \sin(5\omega t)/5 + \sin(7\omega t)/7 + \ldots]$$

Alternate Answer *(ref. 52, p. 228)*

The Fourier series for a square wave as shown is

$$f(t) = 4V/\pi \sin(t) + 4V/3\pi \sin(3\omega t) + 4V/5\pi$$
$$\sin(5\omega t) + 4V/7\pi \sin(7\omega t) + \ldots$$

In the expression for the Fourier series, the first term is the fundamental frequency, the second term is the third harmonic, and so on for all odd harmonics.

For the low-pass filter in the problem, the network response is:

$$\frac{E_{\text{OUT}}}{E_{\text{IN}}} = \frac{1/\omega C}{\sqrt{R_2 + (1/\omega C)^2}} = \frac{1}{\sqrt{\omega^2 R^2 R^2 + 1}}$$

Since $\omega = 2\pi/T$ for the fundamental, the third harmonic occurs at $\omega = 3(2\pi/T) = 6\pi/T$. For which:

$$E_{IN} = 4/3\pi \sin(3\omega t)$$

$$E_{OUT} = \frac{(4/3\pi)\sin(3\omega t)}{\sqrt{\omega_3{}^2 R^2 C^2 + 1}}$$

$$= \left| \frac{4/3\pi}{\sqrt{(6\pi/T)^2 R^2 C^2 + 1}} \right|$$

Fifth harmonic:

$$E_{OUT} = \frac{(4/5\pi)\sin(5\omega t)}{\sqrt{\omega_5{}^2 R^2 C^2 + 1}}$$

$$= \frac{4/5\pi}{\sqrt{(10\pi/T)^2 R^2 C^2 + 1}}$$

Seventh harmonic:

$$E_{OUT} = \frac{(4/7\pi)\sin(7\omega t)}{\sqrt{\omega_7{}^2 R^2 C^2 + 1}}$$

$$= \left| \frac{4/7\pi}{\sqrt{(14\pi/T)^2 R^2 C^2 + 1}} \right|$$

Chapter 8
Solid-State and
Vacuum-Tube Diodes

Most of the diodes covered in this chapter involve problems dealing with basic fundamentals. Consequently, a deep review of equations and comparisons is not needed because the problems themselves contain this information. For detailed information, see references 1, 2, 8, and 12.

Rectifier type diodes are specifically covered in Chapter 9 on power supplies. For a rundown on various types of solid-state diodes, see Chapter 10 where all types of semiconductors are combined to show the differences in operating characteristics and their symbols. Chapter 13 on oscillators and modulation also contains material of interest.

As a source of reference it is strongly recommended that the P.E. candidate equip himself with one or more of the many semiconductor manuals readily available from manufacturers such as Westinghouse, General Electric, RCA, Motorola, Fairchild, National Semiconductor, Texas Instruments, Teledyne Crystalonics, and the like.

SOLID-STATE DIODE CHARACTERISTICS

The general expression for forward current to reverse, or leakage, current is

where
$$I_F = I_S \left[\exp(\pm qV/kT) - 1 \right]$$

I_F = forward current
I_S = reverse current (also I_R or I_{CO})
q = electron charge (1.6×10^{-19} coulomb)

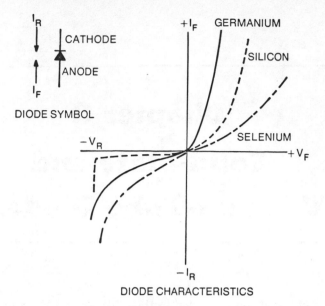

DIODE CHARACTERISTICS

Chart C-8-1.

V = applied bias voltage; positive if forward, negative if reverse

k = Boltzmann's constant (1.38×10^{-23} watt-sec/°K)

T = absolute temperature (K); about 300°K at room temperature

VACUUM-TUBE DIODE

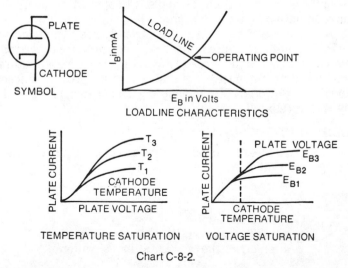

LOADLINE CHARACTERISTICS

TEMPERATURE SATURATION

VOLTAGE SATURATION

Chart C-8-2.

GAS-FILLED TUBES

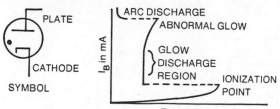

CHARACTERISTICS

Chart C-8-3.

USEFUL EQUATIONS
Definition of terms:

m = mass of the electron (see Table 8-1)
m_i = mass of the ion
n = ratio of ion charge to electron charge
e = electron charge (see Table 8-1)
F = voltage gradient in statvolts
s = distance in centimeters
t = time in seconds
ΔE_B = change in plate voltage
ΔI_B = change in plate current
T = cathode temperature

Electron acceleration:

$$a = eF/m \quad \text{cm/sec}^2$$

Electron velocity, from rest:

$$v = eFt/m = s\sqrt{ZeF/m} \quad \text{cm/sec}$$

Distance traveled by an electron:

$$S = eFt^2/2m$$

Velocity of electron falling through a voltage potential E:

$$v = (5.97 \times 10^7)\sqrt{E}$$

Velocity of an ion moving through an electric field:

$$v = (5.97 \times 10^7)Enm/m_i \quad \text{cm/sec}$$

Plate resistance of vacuum diode:

$$r_p = \Delta E_B/\Delta I_B \quad \bigg| \quad T = \text{constant}$$

131

Table 8-1. Physical Properties of Electrons and Ions

CHARGE

Electronic charge (e) .... 4.767×10^{-10} statcoulomb

1.590×10^{-20} abcoulomb

1.590×10^{-19} coulomb

MASS

Mass of an electron * .... 9.038×10^{-28} gram

Mass of a proton .... 1.6608×10^{-24} gram

Mass of a hydrogen atom .... 1.6617×10^{-24} gram

Mass of hypothetical atom of unit atomic weight (m_O) ...... 1.649×10^{-24} gram

RATIOS

Charge to mass of an electron * .... 1.769×10^{7} abcoulombs per gram

5.303×10^{17} statcoulombs per gram

1.769×10^{8} coulombs per gram

Mass of hydrogen atom to that of electron * 1848.0

* Measured at rest.

PROBLEM 8-1

From your knowledge of semiconductor diodes, make a comparison between silicon, germanium, selenium, and copper oxide diodes. List at least 15 characteristics and parameters for comparison. Which would you recommend for:

(A) High-frequency operation.

(B) Lowest forward voltage drop.

(C) Lowest reverse leakage.

(D) Surge suppression.

(E) Highest temperature operation.

Answer

Table 8-2 compares the four diode types.

(A) Silicon or germanium has the best high-frequency characteristics, with silicon being the superior of the two.

(B) Germanium is preferable because of its small size and ease of heat sinking.

(C) Silicon has the lowest leakage of all.

(D) Selenium diodes connected in reverse bias, or possibly silicon avalanche diodes if they may be considered.

(E) Silicon has the best high-temperature qualities.

PROBLEM 8-2

For what voltage will the current in a PN junction diode reach 90% of its saturation current at room temperature?

Answer *(ref. 8, p. 25)*

This problem has two solutions, depending on whether forward or reverse saturation is inferred. The governing equation is:

$$I_F = I_S [\exp(\pm qV/kT) - 1]$$

Table 8-2. Comparison of Semiconductor Diodes

	SILICON	GERMANIUM	SELENIUM	COPPER OXIDE
Size	very small	very small to small	large to very large	large
Weight	very light	very light	light	heaviest
Cost (low power)	low	low – med	med	med – high
Cooling	natural, heat sink, forced air	natural, heat sink, forced air	natural, forced air	natural, forced air
Aging (R_F)	none	none	increases. (reforms)	stabilizes
Foward loss (at same current density)	good	excellent	fair (highest)	excellent
DC forward drop (approx. per cell)	0.9V	0.65V	1.0V	0.5V
Leakage (reverse current)	lowest	very low	low	fair
Recovery from voltage transient	no	no	best	good
Recovery from current surges	yes	yes	very good	very good
Loss of rectif characteristics	no	no	yes (some)	no
Series operation	good	good	excellent	excellent
Parallel operation	good (with care)	good (with care)	excellent	excellent
Operating temperature limit	200°C	100°C	130°C	75°C
Thermal capacity	poor	poor	fair	best
Low-Voltage efficiency	good	excellent	fair	good
Effect of humidy	hermetic seal	hermetic seal	negligible	negligible
High frequency use	good	good	poor	fair
Life	over 100,000 hours	over 100,000 hours	60,000 to 100,000 hrs. (depends on temperature)	fair— better than vacuum tube

At room temperature, $q/kT = 38.5$, so

$$I_F = I_S [\exp(\pm 38.5V) - 1]$$

or

$$\ln \left[\frac{I_F}{I_S} + 1 \right] = \pm 38.5V$$

(A) For 90% forward saturation, $I_F/I_S = 0.9$, making the left-hand side of the equation equal $\ln(1.9)$, making $V = -16.65$ mV.

133

(B) For 90% reverse saturation, $I_S/I_F = 0.9$, for which the solution becomes $V = -59.3$ mV.

PROBLEM 8-3

Two PN junction diodes are connected in series opposing as indicated in Fig. 8-3. A 5-volt battery is connected to this series arrangement. Find the voltage across each diode at room temperature. Assume the zener avalanche voltage is greater than 5 volts.

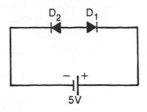

Problem 8-3.

Answer

The figure indicates that diode D1 is reverse biased, and that D2 is forward biased. In the equation

$$I_F = I_S [\exp(qV/kT) - 1]$$

if the diode's normal bias is reversed, both V and I_S become negative quantities. The equation for each diode is then:

$$I_{D1} = -I_{S1} [\exp(-qV/kT) - 1]$$

and

$$I_{D2} = I_{S2} [\exp(qV/kT) - 1]$$

Being a series circuit, $I_{D1} = I_{D2}$. Assuming identical diodes where $I_{S1} = I_{S2}$,

$$-\exp(-qV/kT) + 1 = \exp(qV/kT) - 1$$

Knowing that k is Boltzmann's constant, $k = 1.28 \times 10^{-23}$ W−sec/°K. T is temperature in degrees Kelvin; $T = 300°K$ at room temperature. So at room temperature $KT/q = 0.026$.

$$\exp(-v_1/0.026) + \exp(V_2/0.026) = 2$$

The voltage drops around the loop will be
$5 - V_1 - V_2 = 0$, or $-V_1 = V_2 - 5$. Thus,

$$\exp[(V_2 - 5)/0.026] + \exp(V_2/0.026) = 2$$

$$\exp(V_2/0.026)[1 + \exp(-5/0.026)] = 2$$

The term $\exp(-5/0.026)$ is extremely small, so it can be ignored. Therefore,

$$\exp(V_2/0.026) = 2$$
$$V_2/0.026 = \ln(2)$$
$$V_2 = 0.693(0.026) = 18 \text{ mV ANS}$$
$$V_1 = -(V_2 - 5) = 5 - 0.018 = 4.982V \text{ ANS}$$

PROBLEM 8-4

An ideal silicon PN junction diode has a reverse saturation current of 30 microamperes. At a temperature of 125°C, find the dynamic resistance for a bias of 0.2 volt:

(A) In the forward direction.
(B) In the reverse direction.

Answer

$$I = I_S \left[\exp(qV/kT) - 1\right]$$

For the given conditions,

$$\frac{q}{kT} = \frac{1.602 \times 10^{-19}}{(1.38 \times 10^{-23})(273 + 125)}$$

$$= 29.17$$

(A) For forward bias:

$$1/r_{DYN} = (30 \times 10^{-6})(29.17)\exp(29.17V)$$
$$= 8.75 \times 10^{-4} \quad \exp(29.17V)$$
$$r_{DYN} = 1143 \quad \exp[(29.17)(-0.2)]$$
$$= 3.34\Omega \text{ ANS}$$

(B) For reverse bias:

$$1/r_{DYN} = (30 \times 10^{-6})(29.17)\exp(-29.17V)$$
$$r_{DYN} = 1143 \quad \exp[(29.17)(0.2)]$$
$$= 390 \text{ k}\Omega \text{ ANS}$$

PROBLEM 8-5

In the zener circuit shown, determine the values of R_1 and R_L if the load voltage is to be held at 8 volts at a zener current of 20 mA. Assume an unregulated input of 12V and that zener resistance Z_R equals R_L. What wattage zener would you use?

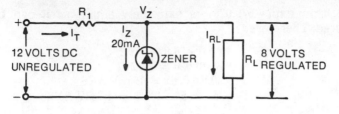

Problem 8-5.

Answer

In the circuit the total input current is

$$I_T = I_Z + I_L$$
$$= 20 + 20$$
$$= 40 \text{ mA}$$

Furthermore, the voltage drop across R_1 is

$$E_{R1} = 12 - 8 = 4V$$

(A) The input dropping resistance would then be

$$R_1 = E_{R1} / I_T$$
$$= 4/0.04$$
$$= 100\Omega \text{ ANS}$$

(B) The proper load resistance would be

$$R_L = V_Z / I_{RL}$$
$$= 8/0.02$$
$$= 400\Omega \text{ ANS}$$

(C) The worst-case condition would occur if the load resistance was removed, making R_L infinite. Then I_Z would become 40 mA when regulating the output at 8V. The power dissipated by the zener at this time would be:

$$P_Z = E_Z I_Z$$
$$= (8) (0.04)$$
$$= 0.32W$$

Therefore, a wattage rating of 0.5W is recommended.

PROBLEM 8-6

A diode with two flat plates, one of them the emitter, is spaced 3/16 inch apart. With this spacing, a current of 120 mA flows when 40V is applied across the plates. If the spacing is now reduced to 1/8 inch and the voltage raised to 50V, what will the new current be?

Answer *(ref. 2, p. 281)*

The Child-Langmuir law states that

$$I = \sqrt{2}/9\pi \, (e/m)^{1/2} \, E^{3/2} / d^2$$

where e

$$= 4.767 \times 10^{-10} \text{ statcoulomb}$$
$$m = 9.038 \times 10^{-28} \text{ gm}$$
$$E = \text{plate potential in volts}$$
$$d = \text{distance between plates in centimeters}$$

Substituting values in the expression, it simplifies to:

$$I = (2.33 \times 10^{-6}) E^{3/2} / d^2$$

And since only a ratio is involved in this problem, it can be expressed as:

$$I_2 = E_2^{\,3/2} \, d_1^{\,2}$$
$$I_1 = E_1^{\,3/2} \, d_2^{\,2}$$

or

$$I_2 = I_1 \, (E_2 / E_1)^{3/2} \, (d_1 / d_2)^2$$

Solving for I_2:

$$I_2 = (0.120) \, (1.25)^{3/2} \, (1.5)^2$$
$$= (0.120) \, (1.398) \, (2.25)$$
$$= 0.3775$$
$$= 377.5 \text{ mA ANS}$$

PROBLEM 8-7

How do two diodes, one of them operating under temperature saturation, the other under space-charge saturation, react when:

(A) The voltage across the diode is changed.
(B) The temperature of the cathode is changed.

Answer *(ref. 1, p. 2-13; ref. 2, p. 281)*

A diode that is space-charge limited will follow the equation:

$$I_B = K(E_B)^{3/2} / d^2$$

As shown in Fig. 8-1, there are two regions of operation defined, above and below temperature T_1. Below T_1 a change in E_B (assuming K and d remain constant) will produce a change in I_B. Beyond T_1 the current is limited due to the space charge following the relationship:

$$\frac{I_{B2}}{I_{B1}} = K \left(\frac{E_{B1}}{E_{B2}} \right)^{3/2}$$

A diode that is lmited by temperature saturation follows the equation

$$I_B = AT^2 \exp(-\phi/T)$$

where A and ϕ are constant. If now the temperatures are varied, the plate current will change by the ratio

$$\frac{I_{B1}}{I_{B2}} = K \quad \frac{T_1{}^2 \quad \exp(-\phi/T_1)}{T_2 \quad \exp(-\phi/T_2)}$$

Beyond E_{B1} the plate current continues to increase, but at a reduced rate because of the Shottky effect; the electrostatic field tends to reduce the work function of the emitter.

(A) For the space-charge limited diode, when voltage is changed in the T_0 to T_1 region, space current increases. For the temperature-saturated diode, when voltage is changed between E_{BO} and E_{B1}, space current increases, and between E_{B1} and E_{B2}, current remains relatively constant.

(B) For space-charge limiting, when the cathode temperature is changed in the T_1 to T_2 region, space current remains somewhat constant. For temperature saturation, when the temperature of the cathode is changed, the plate current between E_{B0} and E_{B1} is the same; beyond E_{B1}, space current increases, but becomes constant after E_{B2} for a given temperature.

PROBLEM 8-8

Two parallel plates, one of them an emitter of electrons, are spaced 1/2 inch apart in an evacuated vessel. If 20 volts is

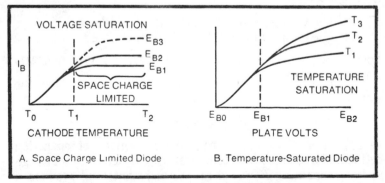

Fig. 8-1. Comparison of diode limitations.

applied between the plates, with the emitter negative, how long will it take for an electron to fly from the emitter to the other plate? Assume a uniform electric field exists between the plates.

Answer

(A) With space-charge limiting, the transit time is

$$t = 3d/\sqrt{2eE_B/m}$$

where $e = 1.6 \times 10^{-19}$ coulomb
$m = 9.1 \times 10^{-13}$ kg

For the voltage and spacing given in the problem:

$$t = \frac{3\,(0.5)\,(0.0254 \text{ m/in.})}{\sqrt{\dfrac{2(1.6 \times 10^{-19}\,)\,(20)}{9.1 \times 10^{-31}}}}$$

$$= 0.0381/(2.65 \times 10^6\,)$$
$$= 14.37 \times 10^{-9} \text{ sec, or } 14.37 \text{ nsec ANS}$$

(B) Without space charge, the transit time is

$$t = 2d/\sqrt{2eE_B/m}$$
$$= 0.0254/(2.65 \times 10^6\,)$$
$$= 9.58 \times 10^{-9} \text{ sec, or } 9.58 \text{ nsec ANS}$$

PROBLEM 8-9

A gas type rectifier tube with a 15-volt drop across the tube is used as a full-wave rectifier with a transformer supplying 350 volts RMS, 60 Hz, secondary voltage. The load is a resistor shunted by a 16 microfarad capacitor. It is noted that conduction begins 18 degrees before the transformer voltage reaches its peak value. Determine this peak charging current.

Answer

Assuming that the filter capacitor is completely charged, less 15 volts diode drop, at the peak of the previous half-cycle, the capacitor will commence discharging through the load resistance until 18 degrees before the peak of the next half-cycle. (Fig. 8-2). At this point in the charging cycle, the charging current is given by $i_C = C dv_C/dt$. Since the transformer secondary voltage is sine wave, the capacitor voltage is then

$$v_C = 350\sqrt{2}\sin(2\pi ft) - 15$$

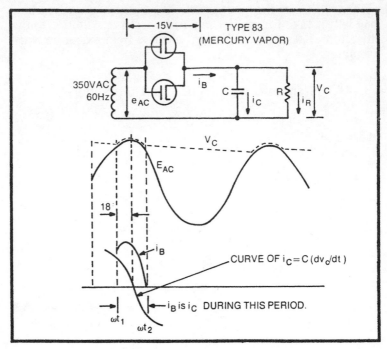

Fig. 8-2. Determining peak rectifier current.

where the -15 volts represents the drop of the rectifier. The charging current flowing into the capacitor is then:

$$i_C = C(2\pi f)\,(350\sqrt{2})\,\cos\,(2\pi f t)$$

Setting the cosine argument equal to $90-18$ degrees, at which point the surge current should be greatest, the equation becomes

$$\begin{aligned}
i_C &= (16 \times 10^{-6})\,(377)\,(350\sqrt{2})\,\cos(72)\\
&= (2.986)\,(0.309)\\
&= 0.923\text{A ANS}
\end{aligned}$$

Alternate Answer $\hspace{4cm}$ (ref.1)

From Termin's handbook, assuming that the capacitor voltage at $\omega t_1 \times 2\pi$ is the same as the voltage at $\omega t = 2\pi$ (no appreciable error is introduced), the following equation can be used to find ωRC. For the given rectifier, $\omega t = 125$, so

$$R = \frac{125}{(377)\,(16 \times 10^{-6})} = 20.8\,\text{k}\Omega$$

140

However, if ωRC is greater than 3.5, the peak tube current occurs at the beginning of conduction; that is, at $\omega t = \omega t_1$. The peak current is then

$$I_{PEAK} = (E_M/R)\sqrt{1 + \omega^2\,R^2\,C^2}\,\sin(\omega t_2 - \omega t_1)$$

At $\omega t_2 = 90°$, $\omega RC = 125$. Solving for E_M in the equation for I_{PEAK}, we get

$$\begin{aligned} E_M &= 350\sqrt{2} - 15 \\ &= 495 - 15 \\ &= 480\text{V} \end{aligned}$$

Therefore,

$$\begin{aligned} I_{PEAK} &= (480/20{,}800)\sqrt{1 + (125)^2}\,\sin(90° - 72°) \\ &= (2.885)\sin(18°) \\ &= 0.891\text{A ANS} \end{aligned}$$

Chapter 9
Power Supplies
and Regulators

Solid-state and vacuum diodes are reviewed in Chapter 8. This chapter provides problems and solutions of single-phase and multiphase power supplies, using various loads and filters. The final problem involves the design of a transistorized regulator.

Figure 9-1 shows the typical half-wave rectifier circuit with associated waveforms and normalized values. Included in the figure is the familiar Fourier equation that contains the DC, fundamental, and harmonic values inherent in the half-wave rectified waveform, while Table 9-1 lists the basic equations relating thereto. Figure 9-2 and Table 9-2 similarly show the full-wave characteristics of rectification, the Fourier expansion, and associated equations.

As a matter of convenience, Table 9-3 is included to provide a quick reference to the division of a sine wave into commonly referenced angles and equivalent radians.

Table 9-4 is a matrix of common single-phase and multiphase rectifier circuits. It enables the conversion of average, peak, and RMS values of current and voltage for all the rectifier types shown. To convert from the extreme right-hand column values to the extreme left-hand values, simply multiply the right-hand values by the conversion factor under the specific circuit of concern. This table also enables the comparison between the various rectifier circuits and related parameters.

To use Table 9-4, assume a problem, say, a full-wave, centertap, single-phase arrangement to provide an average

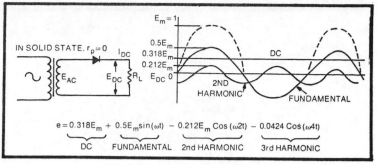

$$e = 0.318E_m + 0.5E_m \sin(\omega t) - 0.212E_m \cos(\omega 2t) - 0.0424 \cos(\omega 4t)$$

DC FUNDAMENTAL 2nd HARMONIC 3rd HARMONIC

Fig. 9-1. Half-wave rectification.

DC output of 250 volts at 100 mA, using a choke input filter. Find: (1) the peak current per leg, (2) transformer secondary RMS volts per leg, and (3) primary volt-amperes.

(1) Under the full-wave centertapped column, at the intersection of "peak current per rectifier leg" (inductive load), find the multiplier, 1.00. The average DC output thus becomes the peak current per leg. (Notice, with a resistive load, the peak would have become 157 mA.)

(2) The intersection for RMS secondary volts per leg is 1.11 (to centertap). Thus, $250 \times 1.11 = 277.5$ volts.

(3). At the intersection of "transformer primary volt-amperes per leg," find 1.11. Before multiplying DC watts by 1.11, the DC watts output must be established, which is determined by multiplying the initially given DC volts and DC amperes. Thus, $250 \times 0.1 = 25$ watts. Now, $25 \times 1.11 = 27.75$ voltamperes per leg.

Table 9-5 is added as an additional reference, and it deals with single-phase and multiphase gas type rectifiers. The equations to the right of each circuit refer to the parameters labelled therein.

Table 9-1. Half-Wave Equations.

DC OUTPUT VOLTS	$E_{DC} = E_O$	$= E_M/\pi = \sqrt{2}E_{RMS}/\pi$
DC OUTPUT CURRENT	$I_{DC} = I_O$	$= E_{DC}/R_L = E_M/\pi R_L = \sqrt{2}E_{RMS}/\pi R_L$
FORM FACTOR	F.F.	$E_{RMS}/E_{AVE} = 1.57$
INPUT POWER	$P_{DISS} = P_{IN}$	$[E_M/2(R_P + r_P)]^2(R_L + r_P) = E_{DC}I_{DC}(F.F.)^2\cos\theta$
OUTPUT POWER*	P_O	$I_{DC}^2 R_L = [E_M/\pi(R_L + r_P)]^2 R_L$
RIPPLE	γ	$\sqrt{(F.F.)^2 - 1} = [(I_{RMS}/I_{DC})^2 - 1]^{1/2}$
EFFICIENCY*	η	$(P_O/P_{IN}) \times 100 = 40.6/(1 + r_P/R_2)$
PEAK INVERSE V.	PIV	$2E_M$

* In solid-state diodes $R_p = 0$.

143

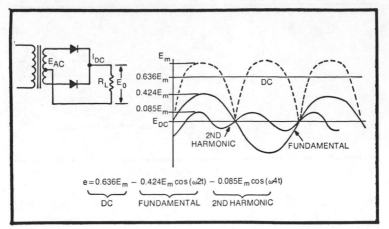

Fig. 9-2. Full-wave.

Sometimes a P.E. exam problem will specify the length of lead from a high-current power supply to the work space with a given wire size. *Do not ignore the reasons behind such statements.* Be certain to check the voltage drop across the lead, and if excessive, be sure to state it in your solution. The nomograph in Fig. 9-3 will provide a quick check by merely placing a straightedge at the intersection of lines (1) and (2) and reading the voltage drop (E) per distance (d).

Note: when using a nomograph or a table of any kind during the P.E. exam, be sure to indicate the referenced source or attach a copy of same.

PROBLEM 9-1

A sinusoidally varying alternating voltage with an RMS value of 115 volts is connected to a series combination of a perfect rectifier and a 500-ohm resistor.

Table 9-2. Full-Wave Equations.

DC OUTPUT VOLTS	$E_{DC} = E_O$	$2E_{rm}/\pi = 2\sqrt{2}E_{RMS}/\pi$
DC OUTPUT CURRENT	$I_{DC} = I_O$	$E_{DC}/R_L = 2E_M/\pi R_L = (2/\pi)I_M$
FORM FACTOR	F.F.	RMS/AVE = 1.11
INPUT POWER	$P_{DISS} = P_{IN}$	$E_{DC} I_{DC} (F.F.)^2 \cos\theta$
OUTPUT POWER*	$P_{OUT} = P_O$	$I_{DC}^2 R_L = 4(I_M/\pi)^2 R_L$
PEAK INVERSE VOLTS	PIV	$2E_m = 2\sqrt{2}E_{RMS}$
EFFICIENCY*	η	$(P_O/P_{IN}) \times 100 = 8/\pi^2 (R_L/r_p + R_L) +$ $= 81.2/(1 + r_p/R_L)^*$

*In solid-state diodes $r_p = 0$.

144

Table 9-3. Sine-Wave Divisions.

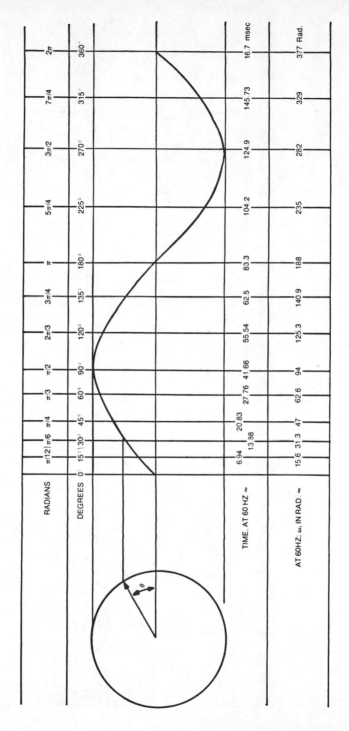

RADIANS	$\pi/12$	$\pi/6$		$\pi/4$	$\pi/3$	$\pi/2$	$2\pi/3$	$3\pi/4$	π	$5\pi/4$	$3\pi/2$	$7\pi/4$	2π
DEGREES	0°	15°	30°	45°	60°	90°	120°	135°	180°	225°	270°	315°	360°
TIME, AT 60 HZ =		6.94	13.88	20.83	27.76	41.66	55.54	62.5	83.3	104.2	124.9	145.73	16.7 msec
AT 60HZ; ω, IN RAD =		15.6	31.3	47	62.6	94	125.3	140.9	188	235	282	329	377 Rad.

Table 9-4. Characteristics of Common Rectifier Circuits.

TYPE OF CIRCUIT	SINGLE PHASE HALF WAVE	SINGLE PHASE FULL WAVE CENTER-TAP	SINGLE PHASE FULL WAVE BRIDGE	THREE PHASE STAR (HALF WAVE)	THREE PHASE FULL WAVE BRIDGE	SIX PHASE STAR (THREE PHASE DIAMETRIC)	THREE PHASE DOUBLE WYE WITH INTERPHASE TRANSFORMER	*TO DETERMINE ACTUAL VALUE OF PARAMETER, MULTIPLY FACTOR SHOWN BY VALUE OF:
PRIMARY								*Assumes zero forward drop and zero reverse current in rectifiers and no AC line or source reactance.
SECONDARY								
ONE CYCLE WAVE OF RECTIFIER OUTPUT VOLTAGE (NO OVERLAP)								
NUMBER OF RECTIFIER LEGS IN CIRCUIT	1	2	4	3	6	6	6	
AVERAGE D-C VOLTS OUTPUT	1.00	1.00	1.00	1.00	1.00	1.00	1.00	AVERAGE D.C. VOLTAGE OUTPUT
RMS D-C VOLTS OUTPUT	1.57	1.11	1.11	1.02	1.00	1.00	1.00	AVERAGE D.C. VOLTAGE OUTPUT
PEAK D-C VOLTS OUTPUT	3.14	1.57	1.57	1.21	1.05	1.05	1.05	AVERAGE D.C. VOLTAGE OUTPUT
PEAK INVERSE VOLTS PER RECTIFIER LEG	3.14	3.14	1.41	2.09	1.05	2.09	2.42	AVERAGE D.C. VOLTAGE OUTPUT
RMS SECONDARY VOLTS PER TRANSFORMER LEG	1.41	2.82	1.00	2.45	2.45	2.83	2.83	RMS SECONDARY VOLTS PER TRANSFORMER LEG
AVERAGE D-C OUTPUT CURRENT	1.00	1.00	1.00	1.00	1.00	1.00	1.00	AVERAGE D.C. OUTPUT CURRENT
AVERAGE D-C OUTPUT CURRENT PER RECTIFIER LEG	1.00	0.500	0.500	0.333	0.333	0.167	0.167	AVERAGE D.C. OUTPUT CURRENT
RMS CURRENT PER RECTIFIER LEG — RESISTIVE LOAD	1.57	0.785	0.785	0.587	0.579	0.409	0.293	AVERAGE D.C. OUTPUT CURRENT
RMS CURRENT PER RECTIFIER LEG — INDUCTIVE LOAD	----	0.707	0.707	0.577	0.577	0.408	0.289	

Table 9-4. (continued).

Parameter									Reference
		RESISTIVE LOAD				**INDUCTIVE LOAD OR LARGE CHOKE INPUT FILTER**			
PEAK CURRENT PER RECTIFIER LEG	RESISTIVE LOAD	3.14	1.57	1.57	1.21	1.05	1.05	0.525	AVERAGE D.C. OUTPUT CURRENT
	INDUCTIVE LOAD	----	1.00	1.00	1.00	1.00	1.00	0.500	AVERAGE D.C. OUTPUT CURRENT
PEAK TO AVERAGE RATIO	RESISTIVE LOAD	3.14	3.14	3.14	3.63	3.15	6.30	3.15	
	INDUCTIVE LOAD CURRENT PER LEG	----	2.00	2.00	3.00	3.00	6.00	3.00	
% RIPPLE $\left(\dfrac{\text{RMS OF RIPPLE}}{\text{AVERAGE OUTPUT VOLTAGE}}\right)$		121%	48%	48%	18.3%	4.2%	4.2%	4.2%	
TRANSFORMER SECONDARY RMS VOLTS PER LEG		2.22	1.11 (TO CENTER TAP)	1.11 (TO‑AL)	0.855 (TO NEUTRAL)	0.428 (TO NEUTRAL)	0.740 (TO NEUTRAL)	0.855 (TO NEUTRAL)	AVERAGE D.C. VOLTAGE OUTPUT
TRANSFORMER SECONDARY VOLT‑AMPERES PER LEG		3.49	1.57	1.11	1.48	1.05	1.81	1.48	D.C. WATTS OUTPUT
TRANSFORMER PRIMARY RMS AMPERES PER LEG		1.57	1.00	1.00	0.471	0.816	0.577	0.408	AVERAGE D.C. OUTPUT CURRENT
TRANSFORMER PRIMARY VOLT‑AMPERES PER LEG		3.49	1.11	1.11	1.21	1.05	1.28	1.05	D.C. WATTS OUTPUT
AVERAGE OF PRIMARY AND SECONDARY VOLT‑AMPERES		3.49	1.34	1.11	1.35	1.05	1.55	1.26	D.C. WATTS OUTPUT
PRIMARY LINE CURRENT		----	1.00	1.00	0.817	1.41	0.817	0.707	AVERAGE D.C. OUTPUT CURRENT
LINE POWER FACTOR		----	0.900	0.900	0.826	0.955	0.955	0.955	

Table 9-5. Gas Rectifiers.

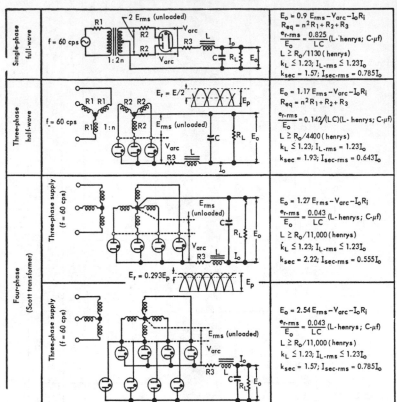

(A) What is the peak value of the current flowing in the resistor?

(B) What current would a DC meter connected in series with the resistor show?

(C) If the resistor is rated at 10 watts, will it be overloaded?

Answer

The peak voltage (see Fig. 9-4) is

$$E_P = E_{RMS} \sqrt{2}$$
$$= (115) (1.414)$$
$$= 162.6V$$

The average DC voltage is

$$E_{DC} = E_P / \pi$$
$$= 162.6/3.141$$
$$= 51.8V$$

Table 9-5. (continued).

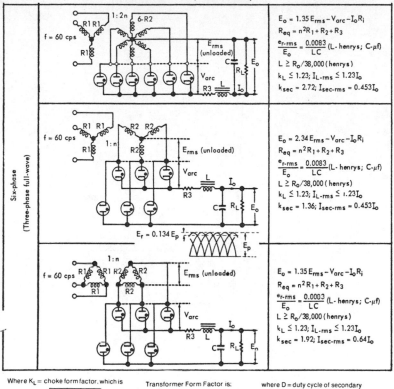

Where K_L = choke form factor, which is

$$k_L = \sqrt{1 + 0.5\left(\frac{R_L}{3\omega L}\right)^2}$$

Transformer Form Factor is:

$$k_{sec} = 1.2\sqrt{2D}$$

where D = duty cycle of secondary

$I_{sec\text{-}rms}$ = RMS current of transformer secondary.

(A) The peak current through R_L is

$$
\begin{aligned}
I_P &= E_P / R_L \\
&= 162.6/500 \\
&= 0.3523\text{A } \textbf{ANS}
\end{aligned}
$$

(B) The DC current through R_L is

$$
\begin{aligned}
I_{DC} &= E_{DC} / R_L \\
&= 51.8/500 \\
&= 0.1036\text{A } \textbf{ANS}
\end{aligned}
$$

(C) The power dissipated in the resistor is

$$
\begin{aligned}
P &= I_{DC}^2 \, R_L \\
&= (0.1036)^2 \ (500) \\
&= 5.36\text{W}
\end{aligned}
$$

The resistor will not be overloaded because it is operating at about half of its rated capacity.

149

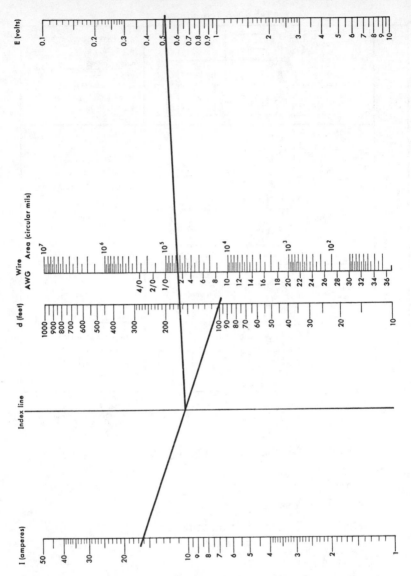

Fig. 9-3. Voltage drop versus wire size. To use nomograph, connect wire current to d (feet), which is the length of wire. From index line draw a line through wire size, if known, and read the voltage drop, or if desired voltage drop is known, read wire size. The problem can also be worked in reverse. To find the length of wire or the maximum current per given wire size and desired voltage drop, draw a line through E (volts) and wire size to establish crossing point at index line. If length of line is known draw a line through it and the index line and read I (amps) or if I(amps) is known, draw a line through it and the index line and read length of line to give referenced voltage drop.

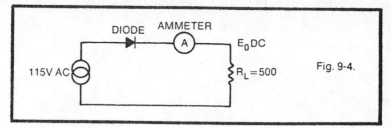

Fig. 9-4.

PROBLEM 9-2

A single-phase half-wave rectifier consists of a 115 to 300 volt stepup transformer, an ideal diode, and a load resistor of 7250 ohms. At the load, determine:

(A) The DC voltage.
(B) The DC amperage.
(C) The load power.
(D) The percent efficiency.

Answer

The problem circuit is shown in Fig. 9-5. The peak voltage across the load resistor is

$$E_P = 300\sqrt{2}$$
$$= 424V$$

(A) The DC voltage is

$$E_{DC} = E_P/\pi$$
$$= 424/\pi$$
$$= 135V \text{ ANS}$$

(B) The DC amperage is

$$I_{DC} = E_{DC}/R_L$$
$$= 135/7250$$
$$= 18.62 \text{ mA ANS}$$

(C) The load dissipation for half-wave rectifiers is

$$P_D = [E_P/2(R_L + R_D)]^2 R_L$$
$$= (424/14500)^2 (7250)$$
$$= 6.21W \text{ ANS}$$

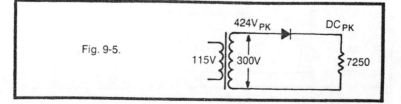

Fig. 9-5.

(D) The actual DC power delivered to the load is

$$P_{DC} = E_{DC}\,I_{DC}$$
$$= [E_P\,/\pi(R_L\,+\,R_D\,)]^2\,(R_L\,+\,R_D\,)$$
$$= 2.52\text{W}$$

Therefore, the efficiency is

$$\eta = P_{DC}\,/P_D$$
$$= 6.21/2.52$$
$$= 40.6\%\ \textbf{ANS}$$

PROBLEM 9-3

A full-wave rectifier has a transformer secondary whose voltage measures 300 volts from centertap to each end terminal. The rectifier load resistance is 7250 ohms. At the load, determine:

(A) The DC voltage.
(B) The DC amperage.
(C) The power.
(D) The percent efficiency.

Answer

The circuit is drawn in Fig. 9-6. The peak output voltage is

$$E_P = 300\sqrt{2}$$
$$= 424\text{V}$$

(A) The DC output voltage is

$$E_{DC} = 2E_P\,/\pi$$
$$= 270\text{V}\ \textbf{ANS}$$

(B) The DC output current is

$$I_{DC} = E_{DC}\,/R_L$$
$$= 37.2\ \text{mA}\ \textbf{ANS}$$

(C) The output power dissipation is

$$P_D = E_P^{\,2}\,/2R_L$$
$$= 12.4\text{W}\ \textbf{ANS}$$

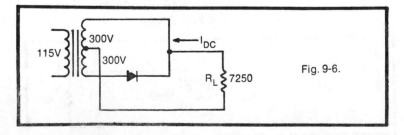

Fig. 9-6.

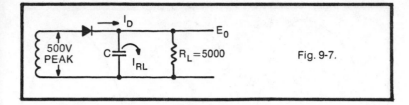

Fig. 9-7.

(D) The DC output power is

$$P_{DC} = \left(\frac{2E_p}{\pi R_L} \right)^2 \ R_L = 10W$$

So the power efficiency is

$$\eta = P_{DC}/P_D$$
$$= 80.6\% \ \textbf{ANS}$$

PROBLEM 9-4

A 60 Hz half-wave rectifier is energized by a transformer having a peak output of 500V. The load consists of a 5000Ω resistance shunted by a 10 μF capacitor. Assume an ideal transformer and diode. Calculate:

(A) The maximum current through the rectifier.
(B) The DC voltage across the load.

Answer

As shown in Fig. 9-7, the input drive to the rectifier is

$$E_S = E_P \ \sin(\omega t) = 500 \sin(377t)$$

(A) During conduction, $I_D = I_C + I_{RL}$. During non-conduction, $I_C = -I_{RL}$; and during conduction,

$$I_C = C \ dE_S \ /dt$$
$$= \omega C E_P \ \cos(\omega t)$$
$$= 1.89 \cos(\omega t)$$

Letting $\omega t = \theta_2$;

$$\theta_2 = \tan^{-1}(-\omega R_L \ C)$$
$$= \tan^{-1}(-18.85)$$
$$= 86.96°$$

Referring to Fig. 9-8.

$$\sin(\theta_1) = \sin(\theta_2) \quad \exp\left[\frac{-2\pi(\theta_2 - \theta_1)}{\omega C R_L} \right]$$

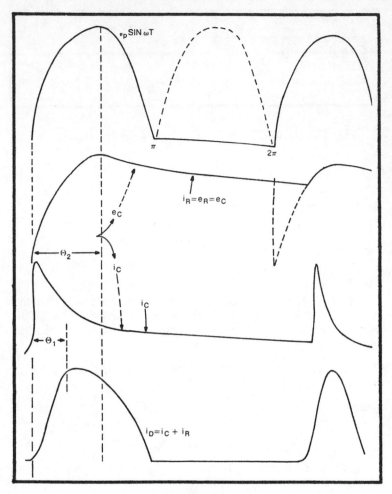

Fig. 9-8.

For which the solution is $\theta_1 = 50°$. Thus:

$$I_C = 1.89 \cos (\theta_1) = 1.215A$$
$$I_{RL} = (E_p / R_L) \sin(\theta_1) = 0.0766A$$
$$I_D = 1.215 + 0.077 = 1.292A \text{ ANS}$$

(B) The DC voltage across the load is given by the formula:

$$E_{DC} = (E_p / 2\pi)\sqrt{1 + (\omega C R_L)^2} [1 - \cos(\theta_2 - \theta_1)]$$
$$= (79.57)(18.88)(1 - 0.731)$$
$$= 403.6V \text{ ANS}$$

154

PROBLEM 9-5

A 5 μF capacitor and a perfect rectifier are connected as shown to a sinusoidally varying alternating voltage with an RMS value of 200V at the instant when the voltage passes through zero. Plot the current flowing through the rectifier. The frequency is 60 Hz.

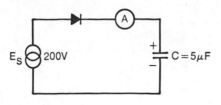

Problem. 9-5.

Answer

The voltage is given by $E_S = 200\sqrt{2} \sin(377t)$. When the rectifier is conducting, the current through the capacitor, and hence the rectifier, will lead the voltage by 90°. The value of this current is determined by the size of the capacitor and the frequency of the power source:

$$\begin{aligned} I_C &= C\,dE_S/dt \\ &= C(200\sqrt{2})\,(377)\cos(377t) \\ &= (5 \times 10^{-6})\,(1.066 \times 10^5)\sin(377t) \\ &= 0.533A\,(peak) \end{aligned}$$

This information is plotted in Fig. 9-9.

PROBLEM 9-6

Design a choke-input power supply to give 300 volts DC at load currents ranging from 1.5A to 0.2A, using a single-phase

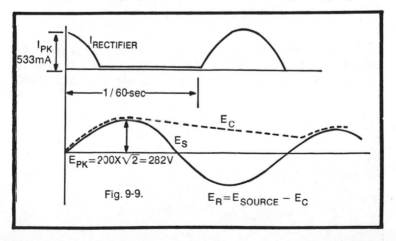

Fig. 9-9.

155

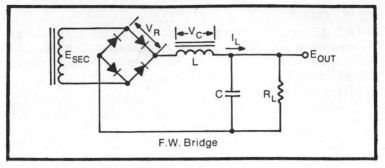

F.W. Bridge

Fig. 9-10.

bridge rectifier operating at 60 Hz. The maximum ripple is
0.4% and the maximum voltage across the choke is 10 volts.
Assume that the voltage drop across each rectifier is 1 volt.

Answer   *(ref. 1, p. 15-4)*

For a full-wave bridge (Fig. 9-10) the RMS output voltage
is

$$E = 1.11(E_0 + V_D)$$

where V_D is the voltage drop across the choke and rectifiers,
which in this case is 12 volts because there are two bridge
diodes at 1V each and 10V across the choke. This leads to an
RMS output of 346V.

The optimum value of choke inductance is first found by
calculating the minimum load resistance:

$$\begin{aligned} R_{L(MIN)} &= E_0/I_{L(MAX)} \\ &= 300/1.5 \\ &= 200\Omega \end{aligned}$$

Referring to Fig. 15-3 (ref. 1), the value of "A" is found by
entering the graph at 60 Hz and moving up to the 1ϕ line. At
this intersection, move left to the ordinate and read 1100V,
which is "A."

$$\begin{aligned} L_{CRIT} &= R_{L(MAX)}/A \\ &= 1500/1100 \\ &= 0.1.36H \end{aligned}$$

The optimum inductance is then

$$\begin{aligned} L_{OPT} &= 2L_{CRIT} \\ &= 2.72H \end{aligned}$$

The minimum filter capacitance is

$$C_{MIN} = 5/[2\pi f R_{L(MIN)}]$$
$$= 5/[2\pi(60)(200)]$$
$$= 6.63 \times 10^{-5}$$
$$= 66\ \mu F$$

(This calculation of capacitance is based on the assumption that the capacitive reactance is much less than one-fifth R_{LMIN}.)

Using the ripple attenuation chart (ref. 1), for a 0.4% ripple:

$$f^2\ LC = 8 \times 10^5$$

$$LC = (8 \times 10^5)/(60^2)$$
$$= 222\ (\mu F - H)$$

For an optimum L of 2.72H,

$$C = 222/2.72$$
$$= 81.7\ \mu F$$

Since this value is greater than C_{MIN}, you can use a 100 μF unit.

The peak inverse voltage of the rectifiers is

$$PIV = 1.57 E_0$$
$$= (1.57)(300)$$
$$= 471V$$

So 500—PIV rectifiers would be advised.

PROBLEM 9-7

A 5Y3 rectifier tube is to be used in a full-wave rectifier circuit to furnish 60 mA at 250V DC. How high must the AC voltage per plate be for:

(A) A choke-input filter (choke is 5 henrys or larger)?
(B) A capacitor-input filter (capacitor is 4 μF or larger)?

Answer   *(ref. 1; 53; 54)*

(A) For a choke-input filter as in Fig. 9-11A, a tube manual shows that a 5Y3 will provide the necessary voltage and current with 320V RMS per plate, assuming that the filter loss is negligible.

(B) For a capacitor-input filter as in Fig. 9-11B, the same tube manual shows that 240V RMS per plate is required.

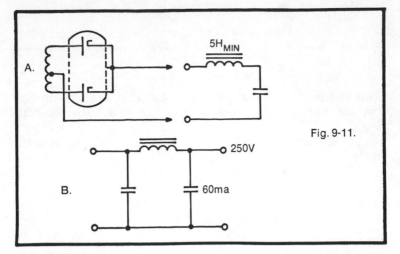

Fig. 9-11.

PROBLEM 9-8

A 3-phase rectifier has a 2-section capacitor-input filter connected to its output. The primaries of the three rectifier transformers are connected in delta and the secondaries in distributed wye. In the filter L_1 = 1.5H, L_2 = 3H, C_1 = 10 μF, and C_2 = 5 μF. Load resistance R_L = 500Ω.

(A) Sketch the circuit.

(B) If the RMS value of the line voltage is 230V, 3-phase, 60 Hz, what is the turns ratio of each of the transformers if the rectifier is to supply 1000V DC at no load?

Answer   *(refs. 6 and 54)*

(A) The circuit appears in Fig. 9-12.

(B) The required secondary voltage in each leg is

$$\begin{aligned}
E_{\text{SEC}} &= 0.855\, E_{\text{DC}} \\
&= (0.855)\,(1000) \\
&= 855\text{V RMS}
\end{aligned}$$

as taken from Table 9-4 in this chapter. Assuming that the diode and choke voltage drops are negligible as compared to the 855V, this voltage will suffice for the remaining computations.

The leg-to-leg voltage of the secondary circuit is

$$\begin{aligned}
E_{\text{L-L}} &= \sqrt{3}\, E_{\text{SEC}} \\
&= (1.732)\,(855) \\
&= 1481\text{V RMS}
\end{aligned}$$

158

With a primary voltage of 230V RMS and a secondary voltage of 855V, the turns ratio of an ideal transformer would be

$$N = N_{SEC} / N_{PRI}$$
$$= E_{SEC} / E_{PRI}$$
$$= 855/230$$
$$= 3.717 : 1 \text{ ANS}$$

PROBLEM 9-9

A stack of selenium rectifier disks is rated at 2 amperes RMS and 20 volts inverse peak per disc. A rectifier is desired to operate at 208 volts, 3-phase, 4-wire without transformers. Specify (1) the number of discs per stack, (2) the number of single stacks required, (3) the voltage capacity, and (4) the current capacity, assuming 15 percent voltage regulation from no load to full load, for the following cases:

(A) The rectifier that will give the largest output.

(B) The rectifier that will give the smallest voltage ripple.

(C) The rectifier that will require the smallest number of discs and still draw a balanced load. Compare the ripple in this case with that in part (B).

Answer   *(ref. 14, pp. 5-6 to 5-11; ref. 42)*

(A) From the rectifier handbook (or Table 9-4 in this chapter), choose a half-wave or 3-phase bridge circuit (Fig. 9-13). Both are rated at 2.45 PIV per leg, but the bridge requires a PIV rating of $1.05E_{DC}$ whereas the half-wave requires $2.09E_{DC}$. Thus the half-wave configuration will require about twice as many discs per stack but only about half as many stacks. However, the half-wave circuit can provide only half as much DC output current as the bridge

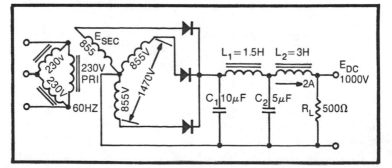

Fig. 9-12.

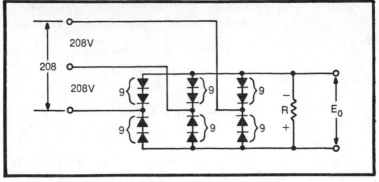

Fig. 9-13.

circuit, so the bridge will provide the greatest amount of output power.

(1) For the bridge circuit:

$$E_{RMS} = 208/\sqrt{3} = 120V$$
$$E_{PEAK} = E_{RMS} \sqrt{2} = 170V$$
$$E_{PIV} = E_{PEAK} \sqrt{3} = 294V$$

At 15% regulation,

$$E_{RMS} = (120)(1.15) = 138V \text{ maximum}$$
$$E_{PIV} = (294)(1.15) = 338V \text{ maximum}$$

So the number of discs per stack is

$$338/20 = 17 \text{ discs ANS}$$

(2) The bridge rectifier requires six stacks.

(3) The voltage capacity is based on the PIV rating of the stack, which is $20 \times 17 = 240V$ peak. Working backwards, this yields a maximum output capacity of 139V RMS, which is within the regulation range.

(4) The output current capacity will remain at 2 amperes per stack, providing a total bridge output of $2/0.579 = 3.45A$ RMS.

(B) The bridge circuit also provides the lowest amount of ripple at 4.2%, so the values remain as in part (A) of the problem. (See Table 9-4 in this chapter.)

(C) Since the PIV requirements in each leg of the half-wave circuit are double that of each bridge leg, but only half as many legs are required, the same total number of discs (6×7 or 3×14) would be needed for either circuit. Both the half-wave and bridge rectifier draw a balanced load, so there is no apparent advantage for either circuit based on this criterion.

The ripple of the half-wave rectifier is 18.3% as compared to the 4.2% for the bridge, and this would make the bridge preferable.

PROBLEM 9-10

Design a single-phase, bridge, power supply to provide 500 mA at 400V DC. The power transformer operates from a 230V, 60 Hz line and provides an open-circuit secondary of 340V. The resistance of the primary winding is 4Ω, and of the secondary, 20Ω. The rectifiers each have an equivalent forward resistance of 2Ω. The power supply is to have a maximum ripple of 5%,

Answer

(1) This power supply will require a series resistor as shown in Fig. 9-14 to help minimize output ripple voltage. The value of this resistor is

$$R_S = R_{SEC} + R_{PRI} N^2 + 2R_F$$

where R_{SEC} = secondary resistance of transformer
R_{PRI} = primary resistance of transformer
N = turns ratio of transformer (sec/pri)
R_F = forward resistance of rectifier

Inserting values:

$$R_S = 20 + 4(340/230)^2 + 2(2)$$
$$= 20 + 8.74 + 4$$
$$= 32.74\Omega$$

The load resistance is

$$R_L = 400/0.5A$$
$$= 800\Omega$$

(2) The ratio $R_S/R_L = 4.1\%$ corresponds to the parameter shown in Fig. 9-15. Using a 5% ripple, the graph yields a design value of 12 for $\omega C R_L$. So:

$$C = 12/2\pi(60)(800)$$
$$= 3.98 \times 10^{-5}$$
$$= 39.8 \,\mu F$$

For a convenient value, use 40 μF.

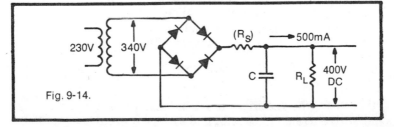

Fig. 9-14.

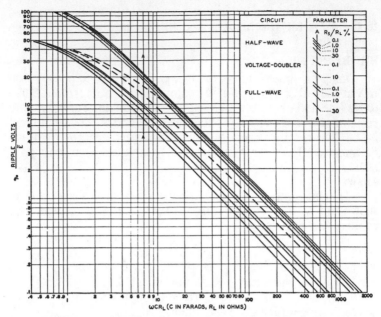

Fig. 9-15. Root-mean-square ripple voltage for capacitor-input circuits.

(3) Check to see now that $E_0 = 400$ volts, or approximately so, as required. In Fig. 9-16 for $R_S/R_L = 4\%$ and $\omega CR_L = 12$, the graph indicates that $E_0/E_P = 0.84$. Therefore,

$$\begin{aligned} E_0 &= 0.84E_P \\ &= (0.84)(340\sqrt{2}) \\ &= 404\text{V} \end{aligned}$$

(4) Determine now the rectifier current ratings:

$$\begin{aligned} I_{SURGE} &= E_P/R_S \\ &= 340\sqrt{2}/32.7 \\ &= 14.7\text{A} \end{aligned}$$

From Fig. 9-17, find the peak forward current, I_P. Since we are using a full-wave rectifier, $n = 2$. Therefore, $R_S/nR_L = 2\%$ and $n\omega CR_L = 24$, yielding an I_P/I_0 value of about 7.8. Since the rectifiers conduct alternately on each half cycle, output current I_0 for each rectifier is half of 500 mA, or 250 mA. The peak rectifier current is then:

$$\begin{aligned} I_P &= 7.8I_0 \\ &= 1.95\text{A} \end{aligned}$$

Therefore, the rectifiers must be able to handle an average forward current of about 2A and a surge current of about 15A.

(4) The final factor to determine is the peak inverse voltage (PIV) of the rectifiers, which in this bridge circuit is equal to the peak secondary voltage of the transformer. Thus, the PIV rating of the rectifiers must be at least $340\sqrt{2} = 481V$. A 500V rating is commonly available, but a 600V unit would be desirable to handle powerline fluctuations.

PROBLEM 9-11

Design a full-wave centertapped power supply to provide 200 mA at 150V DC. Use a $30\mu F$ capacitor as the input of the filter. Assume 60 Hz power and specify all necessary components.

Answer

(1) Start by finding load resistance R_L in Fig. 9-18:

$$R_L = 150V/0.2A$$
$$= 750\Omega$$

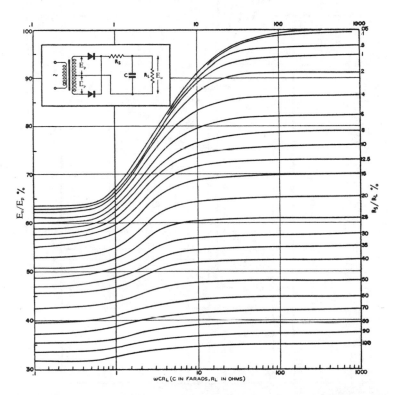

Fig. 9-16. Relation of applied alternating peak voltage to direct output voltage in full-wave capacitor-input circuits.

163

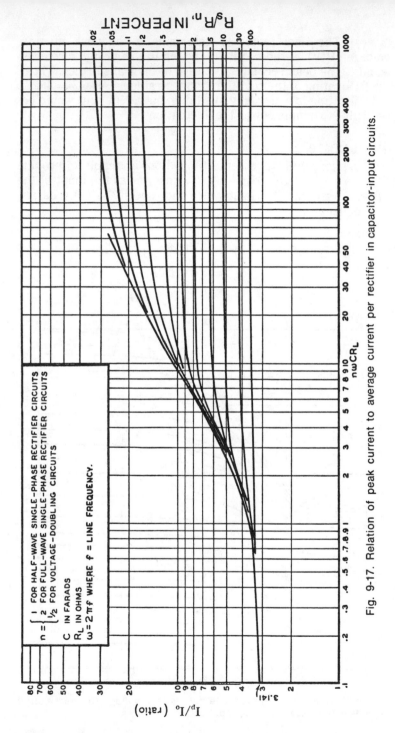

Fig. 9-17. Relation of peak current to average current per rectifier in capacitor-input circuits.

164

(2) Assume series resistance R_S in the rectifier circuit to be about 10% of the load resistance, or 75Ω.

(3) Find the required secondary voltage of the transformer. For this, use the universal chart shown in Fig. 9-16. To enter the graph, find

$$\omega C R_L = 2 (60) (30 \times 10^{-6}) (750)$$
$$= 8.48$$

Use 8.5 to enter the graph at the bottom. For $R_S / R_L = 10\%$ the intersection occurs at $E_O / E_P = 74\%$. The peak output voltage of each half of the transformer secondary is then

$$E_P = 150/0.74$$
$$= 203V$$

The transformer secondary must then produce

$$E_S = 203/\sqrt{2}$$
$$= 143V \text{ RMS}$$

for each half, or 286V across the entire winding.

(4) The surge current through the rectifiers is

$$I_{SURGE} = E_p / R_S$$
$$= 203/75$$
$$= 2.7A$$

(5) The peak rectifier current is found from Fig. 9-17 (as in the preceeding problem). For a full-wave rectifier, $n = 2$, so $R_S / 2R_L = 5\%$. Enter the graph at 5% and note the point at which it intersects $2\omega C R_L = 17$, which yields the value $I_p / I_O = 6.2$. The average current (I_O) through each rectifier is half of the 200 mA, or 100 mA, so

$$I_p = (6.2) (0.1)$$
$$= 0.62A$$

Since a typical 1A rectifier will have a 30A surge rating, there is little trouble in satisfying these current ratings.

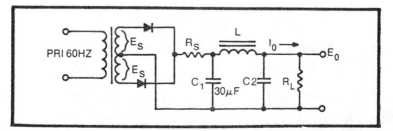

Fig. 9-18.

165

(6) The PIV rating of the rectifier in this full-wave circuit must be at least twice the peak voltage across one-half of the transformer secondary, or 406V minimum. Therefore, use 750 to 1000V PIV rating to allow for surges.

PROBLEM 9-12

A zener regulator is shown in the accompanying figure.
(A) What are I_{RL} , I_Z , and I_{R1} when $R_L = 450\Omega$?
(B) What are the currents when R_L doubles?
(C) What are the currents when R_L drops to 250Ω?
(D) What zener wattage would you recommend?

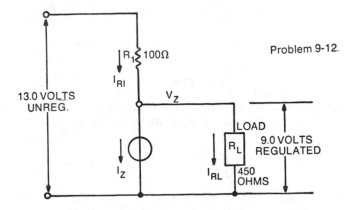

Problem 9-12.

Answer

(A) When $R_L = 450\Omega$:

$$I_{RL} = V_Z /R_L$$
$$= 9/450$$
$$= 20 \text{ mA ANS}$$

$$I_{R1} = (E_{IN} - V_Z)/R_1$$
$$= (13 - 9)/100$$
$$= 40 \text{ mA ANS}$$

$$I_Z = I_{R1} - I_{RL}$$
$$= 40 - 20$$
$$= 20 \text{ mA ANS}$$

(B) When R_L doubles, becoming 900Ω:

$$I_{RL} = 9/900 = 10\text{mA ANS}$$

$$I_{R1} = 40 \text{ mA (unchanged) ANS}$$

$$I_Z = 40 - 10 = 30 \text{ mA ANS}$$

166

(C) At $R_L = 250\Omega$:

$$I_{RL} = 9/250 = 36\,mA \text{ ANS}$$
$$I_{R1} = 40\,mA \text{ (unchanged) ANS}$$
$$I_Z = 40 - 36 = 4\,mA \text{ ANS}$$

(D) For the given circuit configuration, the worst-case condition occurs when R_L is infinite (open circuit). At this point I_Z will go to 40 mA, corresponding to a dissipation of

$$P_Z = V_Z\,I_Z$$
$$= (9)(0.04)$$
$$= 0.36W$$

Therefore, a wattage rating of 1/2 watt (ANS) would be needed.

PROBLEM 9-13

Design a transistorized series voltage regulator with the following requirements:

Output voltage: 28V DC, ±0.1V	Nominal load current: 0.67A
Input voltage: 45V DC, ±5V	Minimum load current: 0.5A
Source resistance: 8Ω	Maximum load current: 1.0A
Load resistance: 42Ω, ±14Ω	Chassis temperature: 55°C

Answer

The basic circuit is illustrated by the block diagram in Fig. 9-19A, with the completed circuit in Fig. 9-19B. The design steps are as follows:

(1) Select a zener reference diode to supply approximately half the required output voltage: In this case a 12V zener is an easily obtained commercial unit rated at 400 mW at 25°C and 300 mW at 55°C. The maximum diode current is then limited to

$$I_Z = (400\,mW)/12V$$
$$= 33\,mA \text{ at } 25°C \text{ (25 mA at 55°C)}$$

(2) *Calculate* V_2 *and* I_2 *in Fig. 9-19A*: For a single amplifier stage, $V_2 = V_R$; for additional stages, $V_2 = V_R + 0.5(V_0 - V_R)$. Thus, V_2 is 12V for a one-stage amplifier and 20V for a two-stage. Assuming a two-stage amplifier design, I_2 becomes equal to $I_{B1} + I_{C2}$, where

$$I_2 \leqslant I_L /V_2$$
$$\leqslant 0.67/20$$
$$\leqslant 33.5\,mA$$

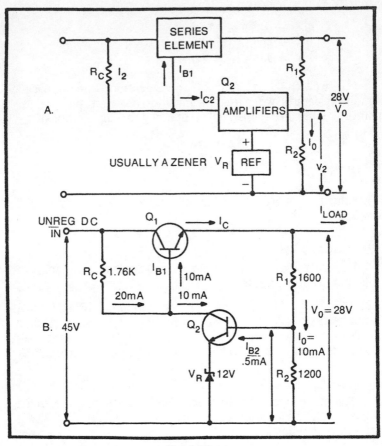

Fig. 9-19.

So, choose $I_2 = 20$ mA for convenience.

(3) Series pass transistor Q_1 must have certain parameters satisfied to work in this circuit:

$$V_{CE} \geq V_S + \Delta V_S - V_0 \left(1 + \frac{R_S}{R_L + \Delta R_L} \right)$$

$$\geq 45 + 5 - 28(1 + 8/56)$$
$$\geq 18V$$

$$V_{CE(SAT)} \leq V_S - \Delta V_S - V_0 \left(1 + \frac{R_S}{R_L - \Delta R_L} \right)$$

$$\leq 45 - 5 - 28(1 + 8/28)$$
$$\leq 4V$$

$$I_{C(MAX)} \geq I_{LOAD} = 1.0A$$

Also, collector resistor R_C will be

$$R_C = (V_S - V_S - V_L - V_{BE})/I_2$$
$$= (45 - 5 - 28 - 0.8)/0.02$$
$$= 560\Omega \ (1/2 \ watt)$$

(4) The parameters for Q_2 are as follows:

$$V_{CE(MAX)} \geq V_L + V_{BE1}$$
$$\geq 28.1 + 0.8$$
$$\geq 28.9V$$

$$I_{C(MAX)} = I_2 - I_{B1} = 1 \ mA$$

Assume $h_{fe} \geq 20$. Then

$$I_{B2} = I_{C2}/h_{fe}$$
$$= (10 \ mA)/20$$
$$= 0.5 \ mA$$

(5) Finally, determine R_2 and R_1 : Because I_0 should be much greater than I_{B2} to prevent loading, let $I_0 = 10 \ mA$. Feedback voltage V_2 will be only slightly higher than zener voltage V_R, about 0.6V greater if Q_2 is a silicon transistor. Therefore, V_2 will be about 12.6V when using a 12V zener.

$$R_2 = V_2/I_0$$
$$= 12.6/0.01$$
$$= 1260\Omega$$

$$R_1 = (V_0 - V_2)/I_0$$
$$= (28 - 12.6)/0.01$$
$$= 1540\Omega$$

Chapter 10
Transistors, Semiconductors, and Digital Circuits

The solid-state diode and the transistor are the forerunners of all semiconductors. Basic advantages of semiconductors over vacuum tubes are the size, weight, life, reliability, and efficiency, owing to the absence of filaments. Discrete semiconductors are giving way in many areas to integrated circuits (ICs) in which complete circuits are fabricated on one "chip." However, the understanding of discrete semiconductors and their behavior makes the understanding of ICs easier because of the similarity of input and output parameters. This chapter provides a brief review of transistor amplifiers and their characteristics, a rundown of all types of semiconductors, and an abbreviated peek at digital logic fundamentals.

Semiconductor problems are not restricted to this chapter, however. Solid-state diodes are used in Chapter 8 as well as Chapter 9 (power supplies). Chapter 12 contains problems on amplifiers, which are more general in nature since they can be either semiconductor or vacuum tube amplifiers. Also, Chapter 13 on oscillators and modulation contains problems involving solid-state oscillators.

DIODE CHARACTERISTICS

More complete information on the semiconductor diode can be found in Chapter 8 and by consulting reference 8, page 24.

The current-voltage characteristic of the diode is described by the formula

$$I_F = I_S \left[\exp(\pm qV/kT) - 1 \right]$$

where I_F = forward current
I_S = reverse current
q = electron charge (1.6×10^{-19} coulombs)
v = applied bias (+ if forward, − if reverse)
k = Boltzmann's constant (1.38×10^{-23} watt-sec/°K)
T = absolute temperature (°K), which is about 300°K at room temperature

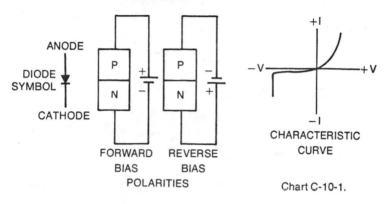

Chart C-10-1.

TRANSISTOR CHARACTERISTICS

This section contains basic transistor information. For further information consult references 6, 8, and 16.

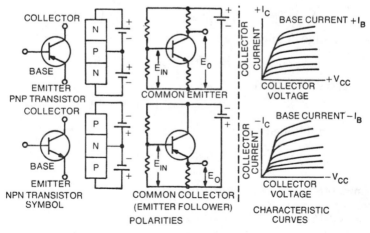

Chart C-10-2.

Black Box

A transistor amplifier can be treated as a "black box" having the usual four-terminal amplification and impedance parameters. Depending upon the connection scheme employed, there are three common transistor configurations with differing parameters.

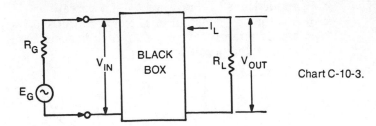

Chart C-10-3.

Common-Base Configuration

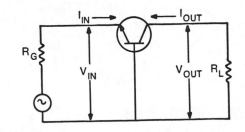

Chart C-10-4.

Current amplification:

$$A_I = h_{fb} = \alpha = I_C / I_E = -h_{fe} / (1 + h_{fe})$$

Voltage amplification:

$$A_V = h_{fb} R_L / h_{ib} = -h_{fe} R_L / h_{ie}$$

Input Impedance:

$$Z_{IN} = h_{ib} = h_{ie} / (1 + h_{fe})$$

Output impedance:

$$Z_{OUT} = 1/h_{ob} = (1 + h_{fe}) h_{oe}$$

Common-emitter Configuration

Current amplification:

$$A_I = h_{fe} = \beta = I_C / I_B = -h_{fb} / (1 + h_{fb})$$

172

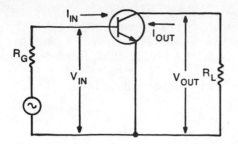

Chart C-10-5.

Voltage amplification:

$$A_V = -h_{fe} R_L / h_{ib} = h_{fb} R_L / h_{ib}$$

Input impedance:

$$Z_{IN} = h_{ie} = h_{ib} / (1 + h_{fb})$$

Output impedance:

$$Z_{OUT} = 1/h_{oe} = (1 + h_{fb})/h_{ob}$$

Common-Collector Configuration

This configuration is also commonly known as an emitter follower.

Chart C-10-6.

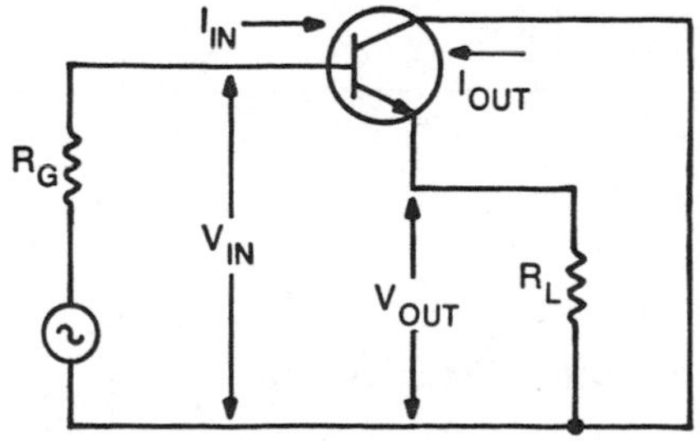

Current amplification:

$$A_I = h_{fc} = -(1 + h_{fe}) = -1/(1 + h_{fb})$$

Voltage amplification:

$$A_V = h_{fe} / (1 + h_{fe}) = \beta/(1 + \beta) \leqslant 1$$

Input impedance:

$$Z_{IN} = (1 + h_{fe})R_L = R_L / (1 + h_{fb})$$

Output impedance:

$$Z_{OUT} = R_G / (1 + h_{fe}) = (1 + h_{fb})R_G$$

173

JUNCTION FIELD-EFFECT TRANSISTORS

Junction FETs are one of two basic groups within the FET family tree, shown in Fig. 10-1. The gate electrode of a junction FET, or JEFT, is actually a PN-junction diode formed over the conducting channel, which may be either P or N type semiconductor material. For further information, consult references 35 and 36.

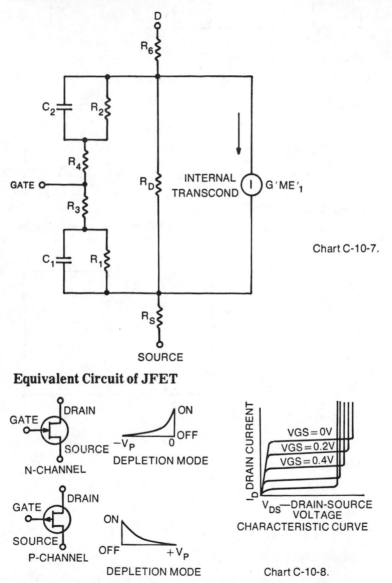

Chart C-10-7.

Equivalent Circuit of JFET

Chart C-10-8.

174

Basic Parameters

Relative magnitudes of equivalent circuit parameters:

$$R_1, R_2 >> r_D >> R_3, R_4, R_5, R_6$$

External transconductance:

$$g_m = \frac{\Delta I_D}{\Delta V_{GS}}\bigg|_{V DS = constant} = \frac{g'm}{1 + g'mR_S}$$

where g_m = transconductance of JFET circuit
I_D = drain current
V_{GS} = gate-source voltage
$g'm$ = internal transconductance, also known as y_{fs}, or forward transadmittance
R_S = source resistance or load

Voltage gain:

$$A_V = g_m R_L$$

where R_L is the load resistance connected to the drain terminal.

Noise figure:

$$F = 10 \log[E_N^2/(4E_G^2 kTBR_G)]$$

where F = noise figure
E_N = output noise
E_G = open-circuit generator voltage
k = Boltzmann's constant
T = temperature (°K)
B = bandwidth
R_G = generator resistance

The noise generated (E_N) is equal to:

$$E_N^2 = 4kTBR_G$$

MOS FIELD-EFFECT TRANSISTORS

The MOSFET is the second family of FETs shown in Fig. 10-1. Originally this family was known as IGFETs, for *insulated-gate* FETs, but they are most generally referred to now as MOSFETs since the fabrication process generally used is described as *metal oxide semiconductor*, or MOS.

While the JFET is limited to depletion-mode operation, in which the gate voltage is applied to *reduce* channel con-

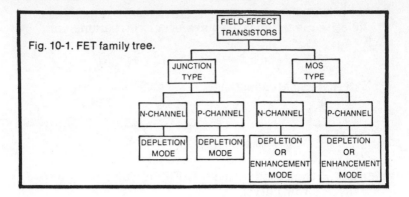

Fig. 10-1. FET family tree.

ductivity, the MOSFET device is also capable of enhancement-mode operation if constructed accordingly. An intermediate mode is sometimes fabricated, in which the device may be operated in either the enhancement or depletion mode, depending upon the gate bias chosen. The enhancement mode is desirable in selected circuit applications because the device is off, or nonconducting, unless a gate signal is deliberately applied. But the depletion mode is most often used in discrete circuits because it is easily made into a self-biasing circuit configuration, much as with the cathode or emitter bias arrangement of vacuum tubes and bipolar transistors.

For further information see references 36, 37, 38, and 43.

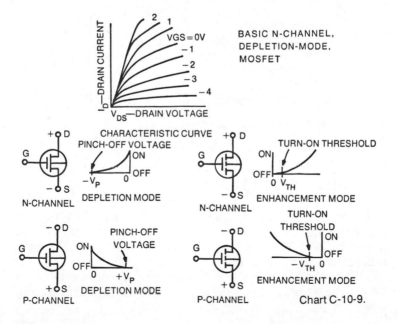

Chart C-10-9.

Amplifier Configurations

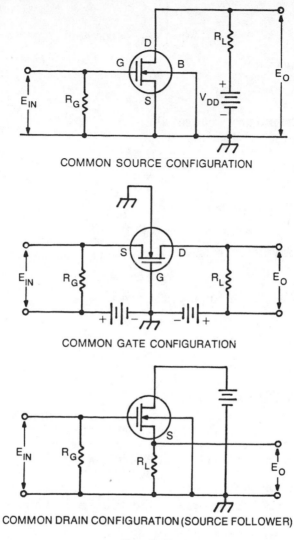

COMMON SOURCE CONFIGURATION

COMMON GATE CONFIGURATION

COMMON DRAIN CONFIGURATION (SOURCE FOLLOWER)

Chart C-10-10.

Basic Parameters

Transconductance:

$$g_m = \frac{\Delta I_D}{\Delta V_{GS}} \bigg|_{V_{DS}} = \text{constant}$$

Channel resistance:

$$r_D = \left. \frac{\Delta V_D}{\Delta I_D} \right|_{V_{GS}} = \text{constant}$$

Voltage gain:

$$A_V = g_m \, r_D$$

Common-source voltage gain:

$$A_V = \frac{g_m \, r_D \, R_L}{r_D + R_L}$$

Output impedance:

$$Z_{OUT} = r_D + (g_m \, r_D + 1)R_S$$

Effective input resistance

$$R_{IN} = R_G / (1 - A_V)$$

Source-follower output resistance:

$$R_{OUT} = \frac{r_D \, R_S}{r_D + (g_m \, r_D + 1)R_S}$$

UNIJUNCTION TRANSISTORS

Unijunction transistors (UJTs) and programmable UJTs (PUTs) operate in much the same manner when used as relaxation oscillators and timers. Further information can be found in reference 8 and also in Problem 13-1.

Basic Parameters

Compensating resistance: For firing-point stability with variations in temperature and supply voltage, resistor R_2 is chosen to be approximately

$$R_2 = 0.70 \, R_{BB} / \eta V_1$$

in which case the peak firing voltage will be about equal to

$$V_P = \eta V_1$$

where $R_{BB} = R_{B1} + R_{B2}$ = interbase resistance

η = intrinsic standoff ratio

178

Peak point voltage:

$$V_P = \eta V_{BB} + V_D$$

where V_D is the forward voltage drop across the emitter diode and is typically on the order of 0.5 to 0.7 volts.

Equivalent Circuit

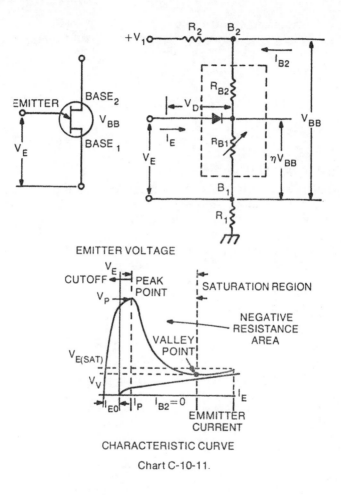

CHARACTERISTIC CURVE

Chart C-10-11.

DIAC

Diacs are used mostly to control triacs. As the characteristic curve illustrates, diacs are bidirectional breakdown devices. Consult reference 39 for further information.

179

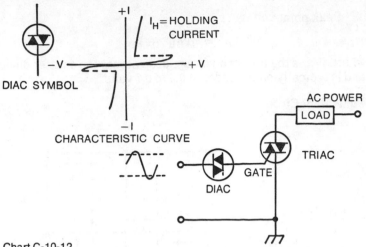

DIAC SYMBOL

I_H = HOLDING CURRENT

CHARACTERISTIC CURVE

AC POWER

LOAD

TRIAC

GATE

DIAC

Chart C-10-12.

TYPICAL APPLICATION

FOUR-LAYER DIODE

Four-layer diodes, also known as Schockley diodes, are made by diffusing four semiconductor layers to form a PNPN device, which is similar to an SCR but without the gate terminal. It features low leakage, submicrosecond switching times, and a conducting voltage drop of less than one volt. Consult references 33 and 34 for further information.

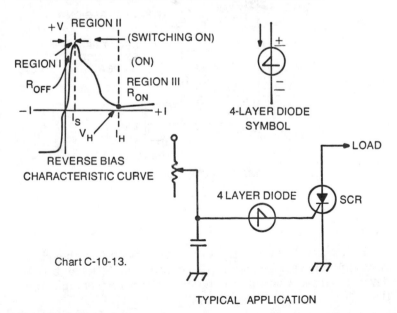

REGION II

(SWITCHING ON)

REGION I

(ON)

R_{OFF}

REGION III
R_{ON}

I_S

V_H I_H

REVERSE BIAS
CHARACTERISTIC CURVE

4-LAYER DIODE
SYMBOL

LOAD

4 LAYER DIODE

SCR

Chart C-10-13.

TYPICAL APPLICATION

SILICON CONTROLLED RECTIFIER

SCRs are a type of thyristor and are quite useful in switching applications involving both AC and DC power. These devices are discussed at length in references 16, 39, 40, 41, and 42.

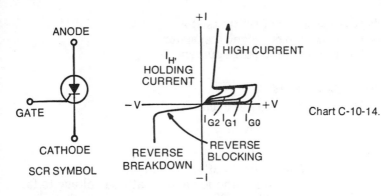

Chart C-10-14.

Equivalent Circuit

An SCR resembles two transistors connected with base and collector terminals tied together as shown. This arrangement, once triggered, remains turned on and conducting until the conduction current is reduced below the device's holding current, at which point the gain of the transistor-like amplifier becomes insufficient to maintain the circuit in saturation, at which point the device reverts to its nonconducting state.

Chart C-10-15.

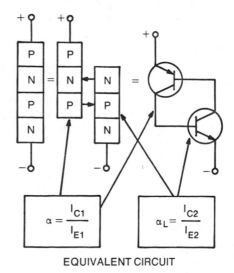

EQUIVALENT CIRCUIT

181

The trigger current depends upon both the gain of the individual transistors and upon the leakage current:

$$I_T = \frac{I_{CO}}{1 - \alpha_1 - \alpha_2}$$

where α_1 = gain of first transistor
     α_2 = gain of second transistor
     I_{CO} = cutoff leakage current

Circuit Parameters

Power dissipation per cycle:

$$W = V_A \, I_A \, t_r \, / 4.6$$

where V_A = anode voltage prior to switching
     I_A = anode current after switching
     t_r = switching time for anode-cathode voltage to fall from 90% to 10%

Average power dissipation:

$$W = f V_A \, I_A \, t_r \, / 4.6$$

where f is the switching frequency.

It should also be noted that rated power dissipation is normally given for a full 180° conduction angle into a resistive load. Shorter conduction angles and different load conditions require derating factors to be applied, such as the two preceding dissipation parameters that approximate the heating effect during the SCR's turn-on times.

TRIAC

Triacs perform like two SCRs connected in a manner that allows them to pass alternating currents. The following information applies to triacs, but dissipation factors and other pertinent information can be obtained by consulting the references given earlier under SCRs.

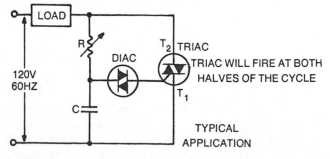

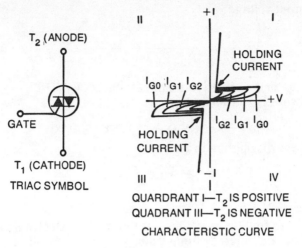

Chart C-10-16.

ZENER DIODE

This workhorse of modern electronics has seen much use in voltage regulators and in surge and voltage-protection circuits. Reference 14 is just one of many sources of further information.

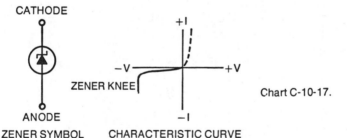

Chart C-10-17.

ZENER SYMBOL   CHARACTERISTIC CURVE

Simple Voltage Regulator

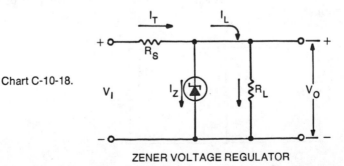

Chart C-10-18.

ZENER VOLTAGE REGULATOR

Output voltage:

$$V_{\text{OUT}} = \frac{V_{\text{IN}}}{R_S/R_L + R_S/R_Z + 1}$$

where R_Z is the effective zener resistance, V_O/I_Z.

Power dissipation in R_S:

$$P_D = (V_{\text{IN(MAX)}} - V_{\text{OUT}})I_{\text{L(MAX)}}$$

Power dissipation in zener:

$$P_Z = V_{\text{OUT}} I_{\text{Z(MAX)}}$$

TUNNEL DIODE

Though finding much less use in modern electronics than originally anticipated, tunnel diodes are still found on many tests and are still studied as classic examples of negative-resistance devices. See reference 15 for more detailed information.

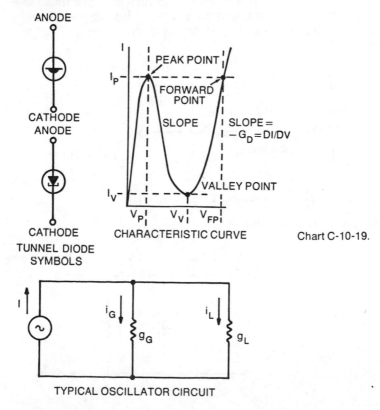

CHART C-10-19.

TUNNEL DIODE SYMBOLS

CHARACTERISTIC CURVE

TYPICAL OSCILLATOR CIRCUIT

Circuit Parameters

Maximum power delivered to load, assuming $g_G = g_L$:

$$P_{L(MAX)} = i^2 / (4g_L)$$

REVIEW OF BASIC DIGITAL LOGIC

The following is a summary of logic rules and basic information. See also references 44, 45, and 46.

Fundamental Elements

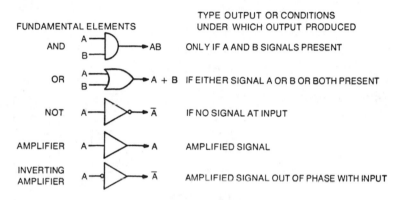

		TYPE OUTPUT OR CONDITIONS UNDER WHICH OUTPUT PRODUCED
AND	AB	ONLY IF A AND B SIGNALS PRESENT
OR	A + B	IF EITHER SIGNAL A OR B OR BOTH PRESENT
NOT	$\overline{A}$	IF NO SIGNAL AT INPUT
AMPLIFIER	A	AMPLIFIED SIGNAL
INVERTING AMPLIFIER	$\overline{A}$	AMPLIFIED SIGNAL OUT OF PHASE WITH INPUT

Chart C-10-20.

Combinatorial Elements

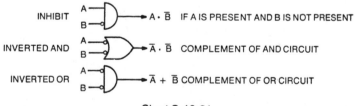

INHIBIT	$A \cdot \overline{B}$	IF A IS PRESENT AND B IS NOT PRESENT
INVERTED AND	$\overline{A} \cdot \overline{B}$	COMPLEMENT OF AND CIRCUIT
INVERTED OR	$\overline{A} + \overline{B}$	COMPLEMENT OF OR CIRCUIT

Chart C-10-21.

Laws

Commutative: $A + B = B + A$
$B \cdot A \qquad A \cdot B =$
Associative: $A + (B + C) = (A + B) + C$
$\qquad A \cdot (B \cdot C) = (A \cdot B) \cdot C$
Distributive: $A \cdot (B + C) = (A \cdot B) + (A \cdot C)$
$\qquad A + (B \cdot C) = (A + B) \cdot (A + C)$

185

Identities

$$A \cdot 0 = 0$$
$$U \cdot A = A$$
$$A \cdot \overline{A} = 0$$
$$A + A = A$$
$$0 + A = A$$
$$U + A = U$$
$$A + \overline{A} = U$$
$$\overline{\overline{A}} = A$$

Important Relations

$$(\overline{A + B}) = \overline{A} \cdot \overline{B}$$
$$(\overline{A \cdot B}) = \overline{A} + \overline{B}$$
$$(\overline{A} \cdot B) + (A \cdot \overline{B}) = (\overline{A} + B) \cdot (A + B)$$
$$(\overline{A} \cdot \overline{B}) + (A \cdot B) = (A + \overline{B}) \cdot (\overline{A} + B)$$

Other Elements

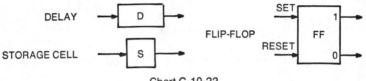

Chart C-10-22.

Positional Number Systems

DECIMAL	OCTAL	TERNARY	BINARY
10	8	3	2
0	0	0	0
1	1	1	1
2	2	2	10
3	3	10	11
4	4	11	100
5	5	12	101
6	6	20	110
7	7	21	111
8	10	22	1000
9	11	100	1001
10	12	101	1010
11	13	102	1011
12	14	110	1100
13	15	111	1101
14	16	112	1110
15	17	120	1111
16	20	121	10000
17	21	122	10001
18	22	200	10010
19	23	201	10011
20	24	202	10100
40	50	1111	101000
100	144	10201	1100100

Binary Addition

$A + B = \text{Sum}$
$0 + 0 = 0$
$0 + 1 = 1$
$1 + 0 = 1$
$1 + 1 = 0$ and 1 to carry

Binary Subtraction

$A - B = \text{Difference}$
$0 - 0 = 0$
$0 - 1 = 1$ and 1 to borrow
$1 - 0 = 1$
$1 - 1 = 0$

Binary Multiplication

$A \times B = \text{Product}$
$0 \times 0 = 0$
$0 \times 1 = 0$
$1 \times 0 = 0$
$1 \times 1 = 1$

Binary Division

$A/B = \text{Quotient}$
$0/0 = ?$
$0/1 = 0$
$1/0 = ?$
$1/1 = 1$

Binary to Decimal Conversion

32 16 8 4 2 1

1 0 1 1 0 1 $= 32 + 8 + 4 + 1 = 45$
      1 0 1 0 $= 8 + 2 = 10$
1 1 0 1 1 1 $= 32 + 16 + 4 + 2 + 1 = 55$

Decimal to Binary Conversion (Convert 1 0 1 1 0 1)

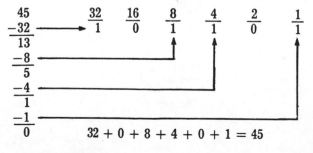

$$32 + 0 + 8 + 4 + 0 + 1 = 45$$

187

Truth Tables

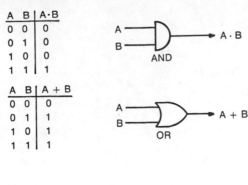

A	B	A·B
0	0	0
0	1	0
1	0	0
1	1	1

A	B	A + B
0	0	0
0	1	1
1	0	1
1	1	1

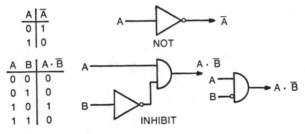

A	$\overline{A}$
0	1
1	0

A	B	A·$\overline{B}$
0	0	0
0	1	0
1	0	1
1	1	0

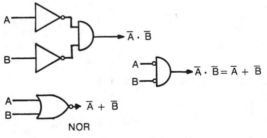

A	B	$\overline{A} + \overline{B}$
0	0	1
0	1	0
1	0	0
1	1	0

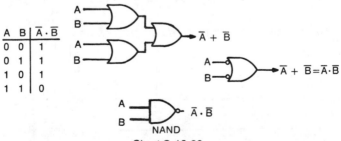

A	B	$\overline{A} \cdot \overline{B}$
0	0	1
0	1	1
1	0	1
1	1	0

Chart C-10-23.

188

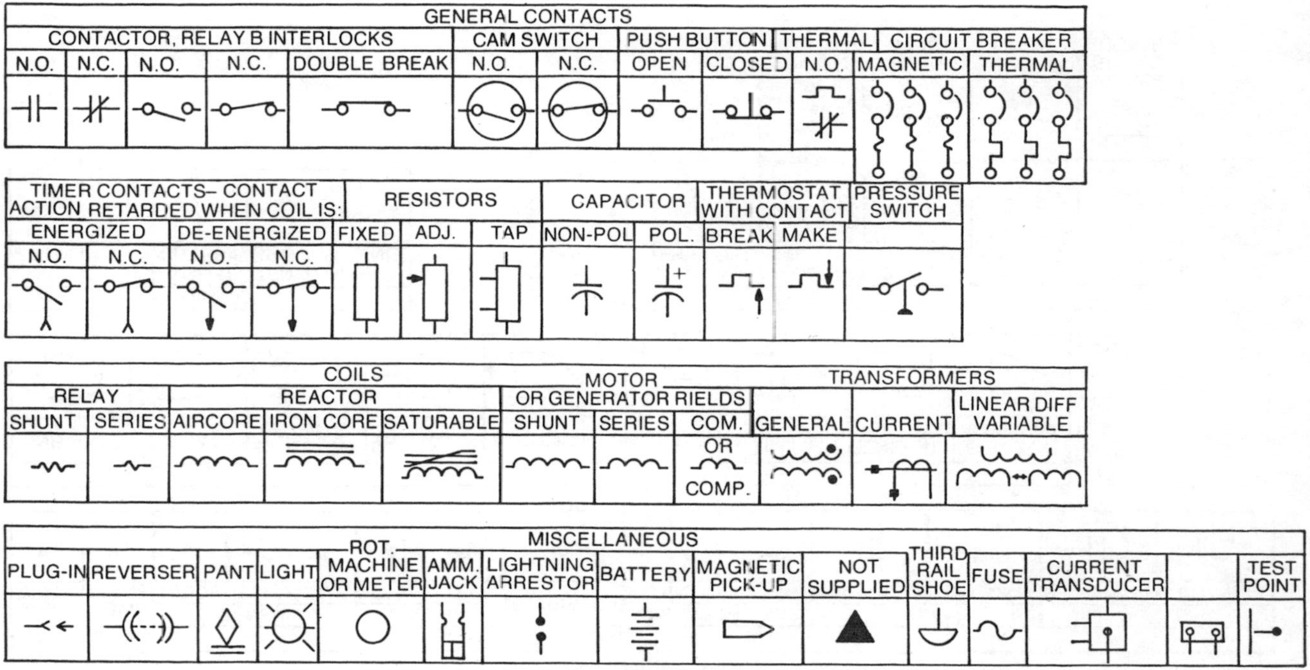

GENERAL CONTACTS

CONTACTOR, RELAY B INTERLOCKS				CAM SWITCH		PUSH BUTTON		THERMAL	CIRCUIT BREAKER		
N.O.	N.C.	N.O.	N.C.	DOUBLE BREAK	N.O.	N.C.	OPEN	CLOSED	N.C.	MAGNETIC	THERMAL

PRESSURE SWITCH

CAPACITOR				THERMOSTAT WITH CONTACT	
FIXED	ADJ.	TAP	NON-POL.	POL.	BREAK	MAKE

TIMER CONTACTS— CONTACT
ACTION RETARDED WHEN COIL IS:

ENERGIZED		DE-ENERGIZED	
N.O.	N.C.	N.O.	N.C.

RESISTORS

FIXED	ADJ.	TAP

COILS

RELAY		REACTOR	
SHUNT	SERIES	AIRCORE	IRON CORE	SATURABLE

MOTOR OR GENERATOR FIELDS

SHUNT	SERIES	COM.
		OR COMP.

TRANSFORMERS

GENERAL	CURRENT	LINEAR DIFF VARIABLE

MISCELLANEOUS

ROT. MACHINE OR METER	AMM. METER	JACK	LIGHTNING ARRESTOR	BATTERY	MAGNETIC PICK-UP	NOT SUPPLIED	THIRD RAIL SHOE	FUSE	CURRENT TRANSDUCER	TEST POINT

PLUG-IN	REVERSER	PANT	LIGHT

Chart C-10-24. Graphic Symbols for Electrical, Electronic, and Logic Diagrams.

189

Chart 10-24. (continued).

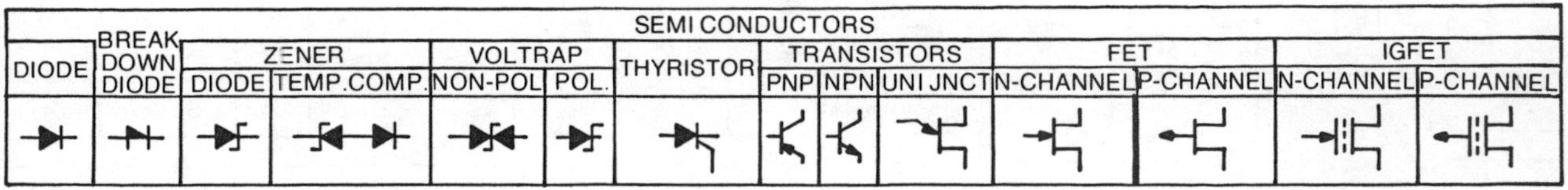

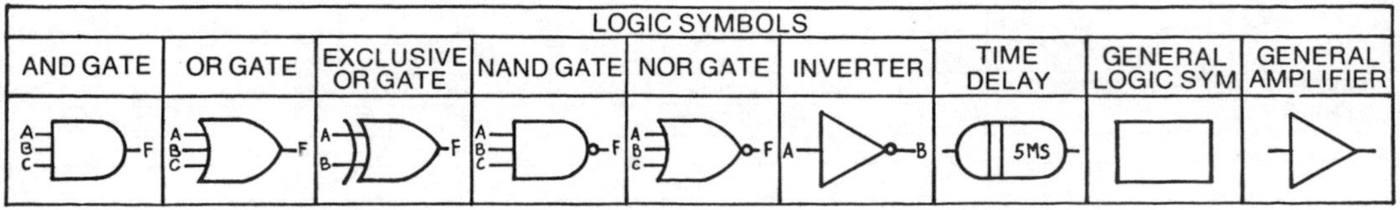

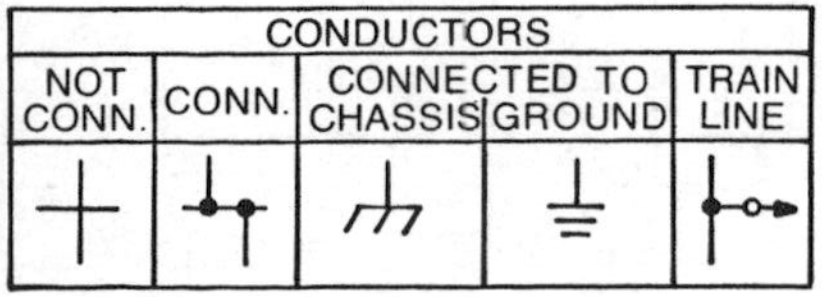

SEMI CONDUCTORS

	BREAK DOWN DIODE	ZENER		VOLTRAP		THYRISTOR	TRANSISTORS			FET			IGFET	
DIODE	DIODE	DIODE	TEMP.COMP.	NON-POL.	POL.		PNP	NPN	UNI JNCT	N-CHANNEL	P-CHANNEL	N-CHANNEL	N-CHANNEL	P-CHANNEL
⏚	⏚	⏚	⏚	⏚	⏚	⏚	⏚	⏚	⏚	⏚	⏚	⏚	⏚	⏚

LOGIC SYMBOLS

AND GATE	OR GATE	EXCLUSIVE OR GATE	NAND GATE	NOR GATE	INVERTER	TIME DELAY	GENERAL LOGIC SYM	GENERAL AMPLIFIER
						5MS		

CONDUCTORS

	NOT CONN.	CONN.	CONNECTED TO CHASSIS	GROUND	TRAIN LINE

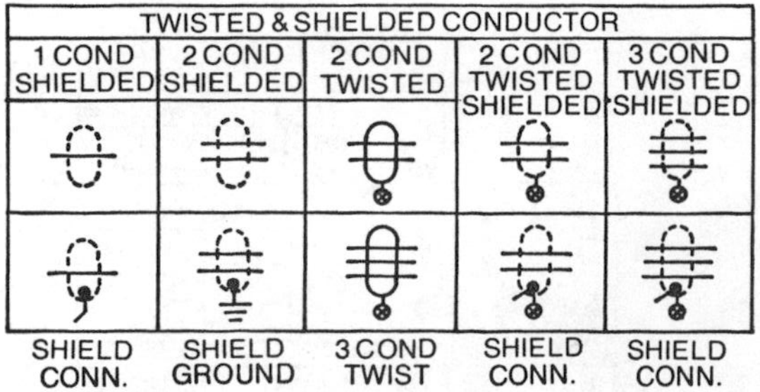

TWISTED & SHIELDED CONDUCTOR

1 COND SHIELDED	2 COND SHIELDED	2 COND TWISTED	2 COND TWISTED SHIELDED	3 COND TWIST	3 COND TWISTED SHIELDED
SHIELD CONN.	SHIELD GROUND	3 COND TWIST	SHIELD CONN.	SHIELD CONN.	SHIELD CONN.

190

PROBLEM 10-1

Draw the T-equivalent circuit of a transistor in terms of parameters r_e, r_b, r_c, and α.

Answer

The T-equivalent circuits are obtained from the solution of the diffusion equation and the addition of r_b'. Figure 10-2 shows the low-frequency equivalent T circuits.

(A) The emitter junction resistance at 300°K is

$$r_e = 0.026/I_e$$

which is in the order of 20 ohms and is an intrinsic parameter.

(B) The collector resistance at 300°K is

$$r_c = 0.026/I_{SAT}$$

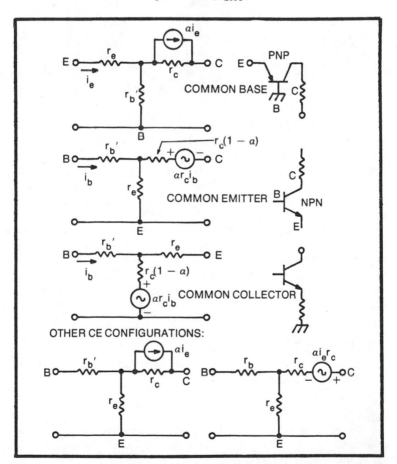

Fig. 10-2. Low-frequency equivalent circuits.

191

The collector resistance is also an intrinsic parameter, and it is high with respect to r_e because it develops from the reverse-biased junction between the base and collector. This resistance is in the order of 10 kΩ.

(C) The base spreading resistance is:

$$r_b{}' = \rho_B \, 1/(24\omega_0 \, d)$$

This resistance is caused by the majority carrier drift flow in the direction transverse to the flow between emitter and collector. It is an extrinsic parameter and is significant because of the low doping density of the base compared to the emitter and collector. This resistance should be low for good frequency response and has the effect of reducing gain.

(D) The parameter alpha is given by the various relationships:

$$\alpha = \left. \frac{dI_{EC}}{dI_E} \right|_{V_{BC} \,=\, \text{constant}} \qquad \alpha = \left. \frac{dI_C}{dI_E} \right|_{V_{BC} \,=\, \text{constant}} \qquad \alpha \approx \left. \frac{\Delta I_C}{\Delta I_E} \right|_{V_{EC} \,=\, \text{constant}}$$

(E) Diffusion capacitance C_{DE} pertains to the base-emitter junction. It is the capacitance associated with a forward-biased junction and varies in proportion to the emitter current. It is associated with the tailing off of majority carriers from the transition region. This capacitance determines the frequency response of the transistor and is an intrinsic parameter on the order of 100 pF.

(F) Transition capacitance C_{TC} pertains to the base-collector junction. It is associated with a reverse-biased junction and varies as the square root or cube root of the junction voltage. It is associated with minority carriers in the depletion region and is in the order of 10 pF. This is an extrinsic parameter.

PROBLEM 10-2

Find the voltage gain of a grounded-base amplifier having the following parameters: $r_c = 2$ MΩ, $r_b = 1$ kΩ, $r_e = 20Ω$, $\alpha = 0.98$. Use a load resistance of 100 kΩ.

Answer

The T-equivalent circuit would appear as in Fig. 10-3. The problem can be worked by solving the two loop equations, designated loop 1 and loop 2.

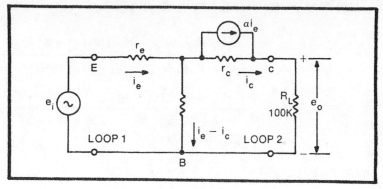

Fig. 10-3. Equivalent circuit.

For loop 1:

$$e_i - i_e r_e - r_b (i_e - i_c) = 0$$
$$e_i - i_e (r_e - r_b) + i_c r_b = 0$$

Thus,

$$i_e = \frac{e_i + i_c r_b}{r_e + r_b}$$

For loop Z:

$$-i_c r_c + \alpha i_e r_c - i_c R_L + r_b (i_e - i_c) = 0$$
$$-i_c (r_c + R_L + r_b) + i_e (\alpha r_c + r_b) = 0$$
$$i_c = \frac{i_e (\alpha r_c + r_b)}{r_c + R_L + r_b}$$

Substituting for i_e :

$$i_c = \left[\frac{e_i + i_c r_b}{r_e + r_b} \right] \left[\frac{\alpha r_c + r_b}{r_c + R_L + r_b} \right]$$

$$i_c (r_e + r_b)(r_c + R_L + r_b) =$$
$$e_i (\alpha r_c \times r_b) + i_c (\alpha r_c r_b + r_b)$$

$$i_c [(r_e + r_b)(r_c + R_L + r_b) -$$
$$\alpha r_c r_b - r_b{}^2] = e_i (\alpha r_c + r_b)$$

$$i_c / e_i = e_0 / e_i = \frac{(\alpha r_c + r_b) R_L}{(r_e + r_b)(r_c + R_L + r_b) - \alpha r_c r_b - r}$$

193

Inserting values:

$$e_0/e_i = \frac{[0.98(2 \times 10^6) + 10^3](10^5)}{(20 + 10^3)(2 \times 10^6 + 10^5 + 10^3) - (0.98)(2 \times 10^6)(10^3) - (10^3)^2}$$

$$= \frac{1.961 \times 10^{11}}{(1.02 \times 10^3)(2.101 \times 10^6) - 1.96 \times 10^9 - 10^6}$$

$$= 1077 \text{ ANS}$$

Alternate Answer (ref. 8)

From the transistor manual, the voltage gain is:

$$A_v = \frac{R_L(r_b + \alpha r_c)}{(R_S + r_e)(R_L + r_c) + r_b[R_S + r_e + R_L + r_c(1 - \alpha)]}$$

In this case, series base resistor $R_S = 0$, so

$$A_V = \frac{R_L(r_b + \alpha r_c)}{r_e(R_L + r_c) + r_b[r_e + R_L + r_c(1 - \alpha)]}$$

$$= \frac{-10^5[10^3 + (0.98)(2 \times 10^6)]}{20(10^5 + 2 \times 10^6) + 10^3[20 + 10^5 + 2 \times 10^6(1 - 0.98)]}$$

$$= \frac{-1.961 \times 10^{11}}{4.2 \times 10^7 + 1.4002 \times 10^8}$$

$$= -1077 \text{ ANS}$$

where the minus sign indicates a phase reversal.

PROBLEM 10-3

For a common-emitter configuration, what parameter can be obtained:

(A) From the vertical spacing of the collector characteristics?

(B) From the slope of these characteristics?

Answer

(A) The vertical spacing (see Fig. 10-4A) corresponds to the current gain:

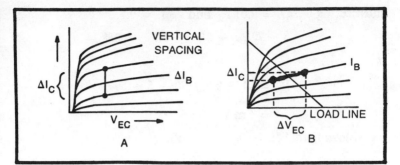

Fig. 10-4.

$$\beta = \frac{\Delta I_C}{\Delta I_B}\bigg|$$
$$V_{EC} = \text{constant}$$

(B) The slope (see Fig. 10-4B) shows the dynamic collector impedance:

$$r_c = \frac{\Delta \dot{V}_{EC}}{\Delta I_C}\bigg|$$
$$I_B = \text{constant}$$

PROBLEM 10-4

If in the circuit shown, it is known that $V_{EB} = 0.7$ volt and $I_B = 0.01 I_E$, find V_0.

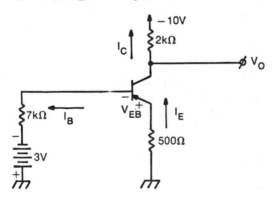

Problem 10-4.

Answer

For the base loop:

$$3 - 7000\,I_B - V_{EB} - 500I_B - 500I_E = 0$$

195

substituting $I_B = 0.01 I_E$ and $V_{EB} = 0.7$,

$$3 - 70I_E - 5I_E - 0.7 - 500I_E = 0$$

or

$$\begin{aligned} I_E &= (3 - 0.7)/(70 + 5 + 500) \\ &= 2.3/575 \\ &= 4\,\text{mA} \end{aligned}$$

Knowing that $I_E = I_C + I_B$, then

$$\begin{aligned} I_C &= I_E - I_B \\ &= I_E - 0.01\,I_F \\ &= 0.99\,I_E \\ &= 0.99(4) \\ &= 3.96\,\text{mA} \end{aligned}$$

The output voltage is then

$$\begin{aligned} V_0 &= -10 + 2000 I_C \\ &= -10 + 2000(0.00396) \\ &= -2.08\text{V ANS} \end{aligned}$$

PROBLEM 10-5

What would the noise voltage be of an amplifier receiving pulsed code from a 70-ohm shielded transmission line operating with a bandwidth of 70 MHz? Assume that the amplifier input impedance matches the line.

Answer

Using the classical equation for a low-noise amplifier,

$$E_N = \sqrt{4kTBR_{IN}}$$

we can assign values of:

$$\begin{aligned} k &= 1.38 \times 10^{-23}\ \text{J/°K} \\ T &= \text{room temperature (300°K)} \\ B &= \text{bandwidth (70 MHz)} \\ R_{IN} &= 70\ \text{ohms} \end{aligned}$$

This yields

$$\begin{aligned} E_N &= [(1.656 \times 10^{-20})\,(7 \times 10^7)\,(70)]^{1/2} \\ &= 9.0 \times 10^{-6} \\ &= 9\,\mu\text{V ANS} \end{aligned}$$

Alternate Answer

Use the nomogram in Fig. 10-5. Enter the bottom of the graph at 70 ohms. Move vertically to 70 MHz (interpolate). Move left and read a noise voltage of slightly less than 10 μV. The answer is approximately 9 μV.

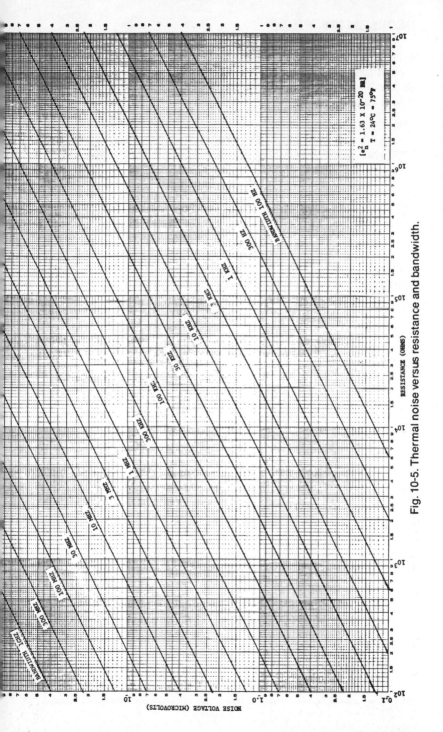

Fig. 10-5. Thermal noise versus resistance and bandwidth.

197

PROBLEM 10-6

A low-noise digital translator amplifier is to be used to receive coded, sub-audible, square-wave pulses at a repetition rate of 100 pps. If the noise voltage is not to exceed 50 nanovolts and a high-quality square wave is desired, what bandwidth would you recommend? What would the input resistance be?

Answer

(A) For a high-quality square wave, the bandwidth should be at least ten times the repetition rate, or 1000 Hz in this case.

(B) The noise voltage is

$$E_N{}^2 = 4kTBR_{IN}$$

where $k = 1.38047 \times 10^{-23}$ J/°K

$t = 300°$K at room temperature

$B =$ bandwidth

$R_{IN} =$ input resistance

Substituting values into the noise equation:

$$(50 \times 10^{-9})^2 = 1.6566 \times 10^{-17} R_{IN}$$

Therefore:

$$R_{IN} = 151\Omega \text{ ANS}$$

Alternate Answer

Use the nomogram in Fig. 10-5. Enter at the left at 0.05 μV (50 nV) and move horizontally to the 1 kHz line. Drop vertically to the resistance line. Read 150Ω.

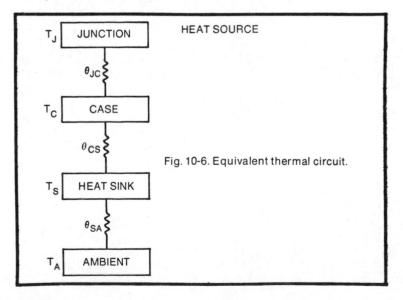

Fig. 10-6. Equivalent thermal circuit.

PROBLEM 10-7

A transistor amplifier requires a heat sink to dissipate 30 watts. The thermal resistance from junction to sink is 1.8°C/W and the thermal resistance from case to sink is 0.6°C/W. This silicon transistor operates in an ambient of 50°C with the heat sink vertically mounted.

(A) Draw the thermal equivalent circuit.

(B) What is the junction-to-case thermal resistance?

(C) What is the thermal heat sink rating (Θ_{SA}) recommended?

Answer

(A) the equivalent thermal circuit is shown in Fig. 10-6.

(B) Using the figure.

$$\Theta_{JS} = \Theta_{JC} + \Theta_{CS}$$

so

$$\begin{aligned}
\Theta_{JC} &= \Theta_{JS} - \Theta_{CS} \\
&= 1.8 - 0.6 \\
&= 1.2°C/W \text{ ANS}
\end{aligned}$$

(B) The thermal rating from the sink to the ambient is:

$$\begin{aligned}
\Theta_{SA} &= (T_{JMAX} - T_A / \text{wattage}) - \Theta_{CS} - \Theta_{JC} \\
&= (200\text{-}50)/30 - 0.6 - 1.2 \\
&= 5 - 1.8 \\
&= 3.2°C/W \text{ ANS}
\end{aligned}$$

PROBLEM 10-8

In the accompanying circuit, a tunnel diode amplifier is to operate between two 70-ohm transmission lines. The negative conductance ($-g_d$) of the diode is 7000 μmhos, the internal resistance R_S is 2Ω, and diode capacitance C_d = 7 pF. What will the values of R and L be when operated at 88 MHz?

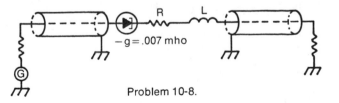

Problem 10-8.

Answer

The equivalent circuit of the tunnel diode amplifier, with its parameters is shown in Fig. 10-7.

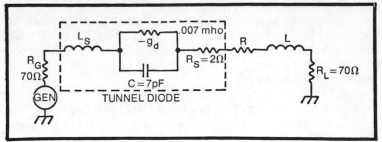

Fig. 10-7. Equivalent circuit.

(A) The total resistance required is

$$R_T = \frac{1}{g(1 + \omega^2 C_d{}^2 / g^2)}$$

where g = negative transconductance
$\omega = 2\pi f$
C_d = diode capacitance

Solving for R_T :

$$1/R_T = (7 \times 10^{-3})\,[1 + (2\pi \times 88 \times 10^6)^2$$
$$(7 \times 10^{-12})^2 / (7 \times 10^{-3})^2$$
$$1/R_T = 9.14 \times 10^{-3}$$
$$R_T = 109\Omega$$

The present circuit resistance is

$$R_P = R_L + R_S$$
$$= 70 + 2$$
$$= 72\Omega$$

The value of R is then

$$R = R_T - R_P$$
$$= 109 - 72$$
$$= 37\Omega \text{ ANS}$$

(B) The value of inductance is:

$$L = R_T\,C/g$$
$$= (109)\,(7 \times 10^{-12})/(7 \times 10^{-3})$$
$$= 1.09 \times 10^{-7}$$
$$= 0.109\,\mu\text{H ANS}$$

PROBLEM 10-9

Construct a truth table that summarizes the following conditions, then draw a circuit that will perform the same function.

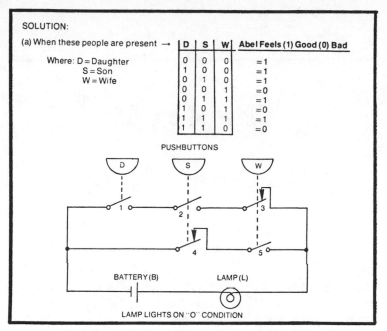

SOLUTION:

(a) When these people are present →

D	S	W	Abel Feels (1) Good (0) Bad
0	0	0	= 1
1	0	0	= 1
0	1	0	= 1
0	0	1	= 0
0	1	1	= 1
1	0	1	= 0
1	1	1	= 1
1	1	0	= 0

Where: D = Daughter
S = Son
W = Wife

PUSHBUTTONS

BATTERY (B) LAMP (L)

LAMP LIGHTS ON "O" CONDITION

Fig. 10-8. Construction of warning circuit.

Abel Baker is a man with a heart condition. His doctor would like to know of the things that bother Abel. For instance, Abel gets along with his daughter alone, and with his son Sonny alone, but when they are together they fight, causing Abel to become upset. He never gets along with his wife except when Sonny is around; Sonny can calm his mother.

When wife and daughter are together they argue, causing tension. But when wife, son, and daughter are all together, there is peace and quiet because the wife prevents fighting between daughter and son, and she will not aggravate Abel in the presence of her children.

The doctor would like to have a circuit that will alert Abel when an undesirable situation exists. The doctor called you (a P.E.) and asked for you to become his consultant. He is willing to pay you in Booleans for the expert services rendered.

Answer

(A) Using the symbols *A* for Abel, *D* for daughter, *S* for Sonny, and *W* for wife, the truth table would be as in Fig. 10-8A. The condition here is that *A* = 1 if Abel feels good, and *A* = 0 if Abel feels bad.

(B) The circuit in Fig. 10-8B will light the lamp when *A* = 0 to warn Abel of undesirable conditions.

201

Chapter 11

Vacuum Tubes

The most common types of vacuum tube amplifiers are the triode and pentode. It is also common to find combinations of diodes, triodes, and pentodes in one envelope. This chapter will review mostly triode and push-pull amplifiers, although there are areas that apply equally to pentodes.

A comparison can be made between certain aspects of the vacuum tube, the gas tube, and semiconductors. The major difference between vacuum and gas tubes is that vacuum tube plate current can be proportionately controlled by the grid, whereas the gas tube is either on or off. However, the gas tube handles higher currents. Both the vacuum tube and the gas tube are considered to be voltage amplifiers, by comparison to semiconductors, which are current amplifiers. Yet, there are solid-state FETs that compare favorably with vacuum tube characteristics, particularly their input impedance. The SCR (thyristor) acts similarly to gas type tubes, but as in all semicondcutors, there are no filaments. (See Chapter 10 on semiconductors.)

The three basic tube factors associated with vacuum tubes are the plate resistance, transconductance, and amplification factor. To review these, it is helpful to draw the familiar plate characteristic curves and associated load line as in Fig. 10-1.

Plate reistance r_p is defined as

$$r_\mathrm{p} = \left.\frac{\Delta e_\mathrm{p}}{\Delta i_\mathrm{p}}\right| \text{ with } e_\mathrm{G} \text{ constant}$$

Transconductance g_m is defined as

$$g_m = \frac{\Delta i_p}{\Delta e_G} \bigg|\ \text{with } e_p \text{ constant}$$

Finally, amplification factor μ is defined as

$$\mu = \frac{\Delta e_p}{\Delta e_G} \bigg|\ \text{with } i_p \text{ constant}$$

These three factors are shown in Fig. 11-1 by the three heavy lines. The factors vary for each tube type, and the values of these factors are found in tube manuals, along with the characteristic curves and typical operating values.

Amplifiers are rated in accordance with their operating classification, ranging from class A to class C. Drawings of the three basic types are illustrated in Fig. 11-2 to show the biasing and comparative waveforms. The general properties of each classification, such as distortion, efficiency, current flow, and bias, are given in Table 11-1. Other classifications are included that display the incremental differences in the properties that exist between the ranges of class A and class C.

BASIC VACUUM TUBE EQUATIONS

The following are the most frequently used vacuum tube equations. Refer to Fig. 11-3 to establish the reference parameters used in the formulas.

Single-Tube Amplifiers

Plate current:

$$i_p = \mu e_G / r_p + R_L)$$

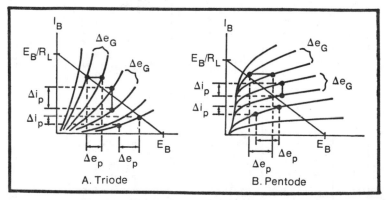

Fig. 11-1. Origin of vacuum tube parameters.

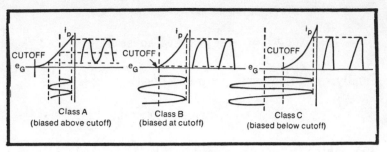

Fig. 11-2. Classes of operation.

voltage gain:

$$A_V = e_O / e_G = \mu R_L / (r_p + R_L)$$

Power output:

$$P_O = i_p{}^2 R_L = \mu^2 e_G{}^2 R_L / (r_p + R_L)$$

Midfrequency voltage gain:

$$A_{VM} \approx -g_m R_E$$

where R_E is paralled equivalent resistance of r_p, R_L, and the R_G of the next stage.

High-frequency voltage gain:

$$A_{VH} = \frac{-\mu r_p}{1/Z_E + 1/r_p} = \frac{A_{VM}}{1 + j\omega C_C R_E}$$

Table 11-1. Amplifier Classifications.

CLASS	APPROXIMATE BIAS	PLATE CURRENT FLOW	EFFICIENCY	DISTORTION
A_1	Linear area above cutoff	Full 360° of cycle	Poorest Less than 20%	Least
A_2	Above cutoff but less than A_1	Full 360° of cycle	Very poor Less than 30%.	Very Low
AB_1	Above cutoff but below A_2	More than 270° but less than 360°	Poor. 25 to 30%	Very Low Better than A_2
AB_2	Slightly above cutoff put less than AB_1	Between 180° and 270° of cycle	Poor. 25 to 35%	Low. Less than AB_1
B_1	Above, but near cutoff	Slightly above 180° (Used in push-pull)	Fair Greater than 40%	Can be very low in push-pull. High in single ended
B_2	At cutoff	About 180° of cycle.	Fair Greater than 40%	Same as above
C	1.5 to 2.5 times below cutoff	Less than 180° (120° to 160°) of full cycle	Highest Greater than 60%	Highest (Used mostly at high frequency)

204

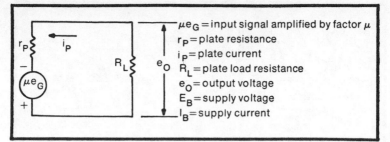

Fig. 11-3. Equivalent circuit.

where R_E is the same as in the preceding formula.

$$Z_E = R_E + j\omega C_C$$
$$C_C = \text{output coupling capacitance}$$

Percentage second harmonic distortion:

$$H_2 = \frac{(100)(I_{MAX} + I_{MIN} - I_p)}{(2)(I_{MAX} - I_{MIN})}$$

where I_{MAX} and I_{MIN} are the maximum and minimum plate currents, and I_p is the quiescent plate current with no signal applied.

Power input:

$$P_{IN} = E_B I_B$$

Plate circuit efficiency:

$$\eta = 100 P_{OUT} / P_{IN}$$

Cathode-follower voltage gain:

$$A_K = e_0 / e_{IN} = \mu R_K / (\mu + 1) R_K + r_p$$

where $R_K = $ AC resistance from cathode to ground.

Push-Pull Triodes

Power input:

$$P_{IN} = 2E_B I_{B(AVE)} \approx (2/\pi) E_B I_{B(MAX)}$$

Power output for two tubes:

$$P_{OUT} = (I_{B(MAX)})^2 R_L / 2 = (I_{B(MAX)} E_{B(MAX)})^{1/2}$$

Plate circuit efficiency:

$$\eta = P_{OUT} / P_{IN} = \pi E_{B(MAX)} / 4E_B = 0.785 E_{B(MAX)} / E_{BB}$$

Plate-to-plate resistance:

$$R_{p-p} = 4(N_1 / N_2)^2 R_L$$

205

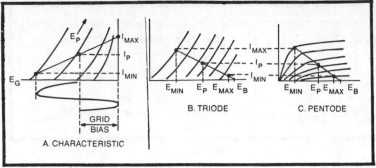

Fig. 11-4. Class A amplifiers.

where N_1 and N_2 are primary and secondary turns, respectively.

LOAD LINE OF CLASS A AMPLIFIERS *(ref. 2, p. 379)*

The response of a class A amplifier is illustrated by the characteristic curves shown in Fig. 11-4. Curves are given for both triode and pentode tubes for comparison.

The basic load line equation is

$$i_p = I_p + (e_p - E_p)/R_L$$

where E_P = plate voltage with no signal applied
I_P = plate current with no signal applied
e_p = instantaneous plate voltage at some point on curve
i_B = instantaneous plate current at some point on curve
R_L = plate load resistance

The output power is given by the formulas

$$P_{OUT} = (E_{MAX} - E_{MIN})(I_{MAX} - I_{MIN})/8$$

and

$$P_{OUT} = \frac{E_p I_p}{2} \left(1 - \frac{E_{MIN}}{E_p}\right) \left(1 - \frac{I_{MIN}}{I_p}\right)$$

The plate efficiency is given by the formulas

$$\eta_p = \frac{(E_{MAX} - E_{MIN})(I_{MAX} - I_{MIN})}{8 E_p I_p}$$

and

$$\eta_p = \frac{1}{2}(1 - E_{MIN}/I_p)(1 - I_{MIN}/I_p)$$

206

The load line in Fig. 11-4 is described by

$$R_L = \frac{E_{MAX} - E_{MIN}}{I_{MAX} - I_{MIN}} = \frac{E_p}{I_p} \left(1 - E_{MIN}/E_p\right)$$

The DC component of the output signal is

$$A_O = (I_{MAX} + I_{MIN} + 2I_p)/4$$

The second harmonic is

$$A_2 = \frac{I_{MAX} + I_{MIN} - 2I_p}{2(I_{MAX} - I_{MIN})}$$

The output power of the fundamental frequency component is

$$P_O = A_1{}^2 R_L /2$$

DESIGN PROCEDURE FOR CLASS A AMPLIFIERS

1—Assume a plate dissipation as the limiting factor.
2—Choose an E_G to give maximum permissible I_p at noted E_p.
3—select an I_{MIN} (usually 0.1 to 0.2 of I_p at operating point).
4—Approximate $I_{MAX} = 2I_p - I_{MIN}$
5—Apprxoimate $E_{MAX} = 2E_p - E_{MIN}$.
6—Calculate R_L and P_O.
7—Draw a load line and check I_{MIN} and I_{MAX}.
8—If modification is needed, increase E_G until R_L is equal to $2r_p$ or more, which often happens when E_P is 250 or less.

PROBLEM 11-1

A vacuum tube draws a plate current of 6 mA when operating at 200V (plate) and −6V (grid). If the grid voltage is reduced to −4V with the same plate voltage, the plate current is doubled. In order to bring the plate current back to its original value, the plate voltage has to be reduced to 150. Find:

(A) The amplification factor.
(B) The mutual conductance.
(C) The plate resistance.

Answer   *(ref. 1, p. 7-12; ref. 2, p. 296)*

(A) We are given the variations in plate current and grid voltage for conditions of constant plate voltage. Therefore:

$$g_m = \Delta I_p / \Delta E_G$$
$$= 0.006/(2)$$
$$= 3000 \ \mu\text{mhos ANS}$$

(B) For constant plate current, the mutual conductance is

$$\mu = \Delta E_p / \Delta E_G$$
$$= 50/(2)$$
$$= 25 \ \text{ANS}$$

(C) Similarly, for constant grid voltage

$$r_p = \Delta E_p / \Delta I_p$$
$$= 50/0.006$$
$$= 8333\Omega \ \text{ANS}$$

As a check,

$$r_p = \mu/g_m$$
$$= 25/0.003$$
$$= 8333\Omega$$

PROBLEM 11-2

The characteristics of a three-element (triode) vacuum tube are as follows: plate voltage, 150V; plate current, 5 mA; amplificaton factor, 10.

(A) Draw a circuit showing how the grid may be biased with a cathode resistor. What resistance would be needed to bias the grid to −3V?

(B) What is the plate resistance of the tube?

(C) If the load resistance is two-thirds of the plate resistance, determined in (B), what would be the voltage amplification of the tube?

Answer *(ref. 2, p. 7-12)*

(A) A suitable biasing arrangement is shown in Fig. 11-5. The cathode bias resistance would be

$$R_K = E_{BIAS} / I_p$$
$$= 3/0.005$$
$$= 600\Omega \ \text{ANS}$$

(B) The plate resistance would be

$$r_p = E_P / I_P$$
$$= 150/0.005$$
$$= 30\text{k}\Omega \ \text{ANS}$$

(C) The load resistance for the calculated value of plate resistance is then

$$R_L = (2/3) \ (30,000)$$
$$= 20\text{k}\Omega$$

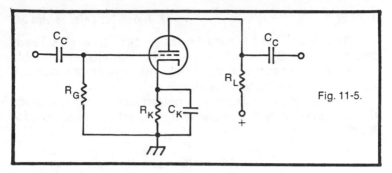

Fig. 11-5.

Therefore, the amplification of the triode circuit would be

$$A = -\mu R_L / (r_p + R_L)$$
$$= (10)(20K)/(30K + 20K)$$
$$= 4 \text{ ANS}$$

PROBLEM 11-3

Determine the plate resistance, amplification factor, and transconductance of the triode whose average plate characteristics are given in the accompanying figure. Assume an operating plate voltage of 200V and a grid bias of −2V.

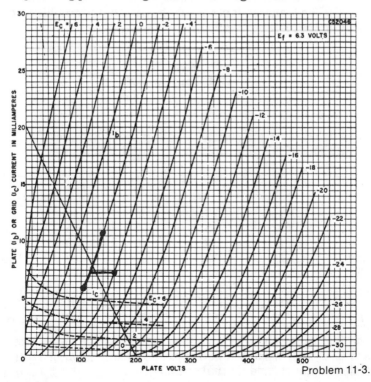

Problem 11-3.

209

Answer *(ref. 1, p. 7-12; refs. 53 and 54)*

First, draw a load line from $E_p = 200V$ through a linear portion of the $E_G = -2V$ curve. This establishes a usable operating point representing typical values of plate resistance and amplification factor (see graph).

(A) The plate resistance is determined by moving along the E_G curve, which represents a constant E_G condition, for which

$$
\begin{aligned}
r_p &= \Delta E_p / \Delta I_p \\
&= (140 - 110)/(11 - 6) \\
&= 30V/(5\text{ mA}) \\
&= 6000\Omega \text{ ANS}
\end{aligned}
$$

(B) The amplification factor is found by moving horizontally from the reference point on the tube's characteristic curves, which represents a constant I_p. Thus,

$$
\begin{aligned}
\mu &= \Delta E_p / \Delta E_G \\
&= (160 - 120)/[(-4) - (-2)] \\
&= 40V/2V \\
&= 20 \text{ ANS}
\end{aligned}
$$

(C) The transconductance is the ratio of the amplification factor to the plate resistance, so

$$
\begin{aligned}
g_m &= \mu/r_p \\
&= 20/6000 \\
&= 3333 \ \mu\text{mhos ANS}
\end{aligned}
$$

PROBLEM 11-4

A triode has the following static characteristics for the range above 1000V plate voltage (I_p in milliamperes):

$E_G = 0$	$E_G = -15$	$E_G = -30$	$E_G = -45$	$E_G = -60$
E_p	I_p	I_p	I_p	I_p	I_p
1000	121	79	45	15	0
1100	138	95	55	25	2
1200	155	113	70	38	8
1300	173	129	91	52	18
1400	189	145	108	69	30
1500	205	162	124	85	44
1600	218	178	141	101	58
1700	230	193	159	118	83

The operating point in a particular circuit is to be $E_P = 1500V$ and E_G −20V. What will be the expected amplification factor, plate resistance, and mutual conductance?

Answer

(A) Amplification factor:

$$\mu = \Delta E_p \ \Delta E_G \text{ with } I_p \text{ constant}$$
$$= (1600 - 1400)/(32 - 14)$$
$$= 11.1 \text{ ANS}$$

(B) Plate resistance:

$$r_p = \Delta E_p / \Delta I_p \text{ with } E_G \text{ constant}$$
$$= (1600 - 1400)(0.165 - 0.135)$$
$$= 6667\Omega \text{ ANS}$$

(C) Mutual conductance:

$$g_m = \Delta I_p / \Delta E_G \text{ with } E_p \text{ constant}$$
$$= (0.160 - 0.140)/(30 - 15)$$
$$= 1333 \ \mu\text{mhos ANS}$$

PROBLEM 11-5

A power triode, with fixed values of excitation and plate voltage, is working into a resistance load as a power converter. If the maximum power output occurs with a load resistance of 2000 ohms, what is the approximate plate resistance of the tube?

Answer *(ref. 11, p. 14)*

This is merely an application of the maximum power transfer theorem: Maximum power transfer takes place when the generator and load resistances are equal. Hence, r_p must be approximately 2000 ohms.

PROBLEM 11-6

Find the expression for the voltage gain of the circuit shown. Use an equivalent circuit.

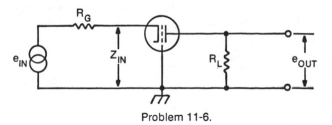

Problem 11-6.

Answer *(ref. 1, p. 3—5 and 3—17)*

The equivalent circuit is given in Fig. 11-6, which is for the grounded-grid amplifier. Cathode resistor R_G will cause de-

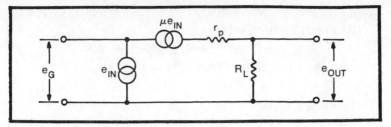

Fig. 11-6.

generation in the amplifier. Without such degeneration, the amplifier gain would be approximately

$$A = -g_m R_L$$

where R_L is assumed to be the total load resistance, being equal to the parallel combination of the plate load and any following input resistance to following stages.

The effect of degeneration is to modify the gain equation to:

$$A = \frac{-\mu R_L}{r_p + R_L + R_G (\mu + 1)} \quad \text{ANS}$$

A simplification results if r_p is much greater than the sum of R_G and R_L, in which case the gain is approximately

$$A = \frac{-\mu R_L}{1 + g_m R_G}$$

where $g_m = \mu/r_p$.

PROBLEM 11-7

Given the plate characteristics and circuit shown in Fig. 11-7, find the gain of the amplifier by the graphical technique.

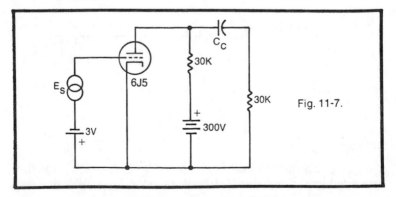

Fig. 11-7.

Assume coupling capacitor C_C offers negligible impedance at signal frequencies.

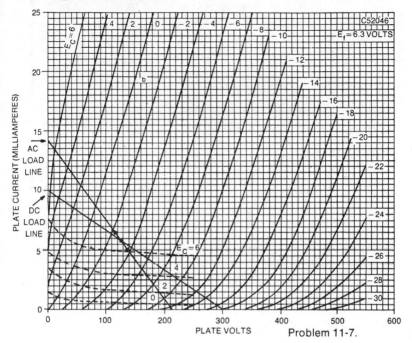

PLATE CURRENT (MILLIAMPERES)

PLATE VOLTS

Problem 11-7.

Answer

Under static conditions the load line would be drawn, as in Fig. 11-7, between the plate voltage of 300V and the maximum plate current determined by the single plate resistor, which would be

$$I_{PMAX} = 300/30K$$
$$= 10 \text{ mA}$$

The quiescent operating point would fall on this load line at the point of grid bias, which from the circuit is −3V.

The dynamic load line would have an increased slope, due to the paralled 30 kΩ load resistance seen through the coupling capacitor, and would pass through the same quiescent operating point.

Assuming a 2V peak-to-peak grid excitation signal, move along the dynamic load line to grid voltage values of −2V and −4V. At these points the corresponding plate voltages are approximately 212V and 241V. Therefore, the voltage gain of the amplifier is approximately

$$A_V = (241 - 212)/[-4 - (-2)]$$
$$= -14.5 \text{ ANS}$$

213

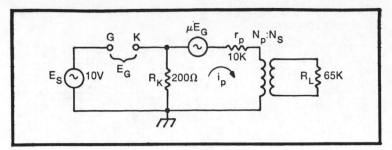

Fig. 11-8.

PROBLEM 11-8

A triode has an amplification factor of 10 and a plate resistance of 10 kΩ. Between ground and cathode is connected a cathode resistor of 200Ω. The plate of the tube and ground are connected to the two primary terminals of an output transformer, the secondary of which is connected to a resistance of 65 kΩ. The input voltage applied between grid and ground is 10V RMS.

(A) What secondary-to-primary turns ratio results in the delivery of maximum power to the 65 kΩ load resistor?

(B) What is the maximum power delivered to the 65 kΩ load resistor?

Answer

(A) The equivalent circuit would appear as in Fig. 11-8 and would form two loop equations:

$$\mu E_G - 200 i_p - R_L i_p (N_p / N_S)^2 - r_p i_p = 0$$
$$E_S - 200 i_p = E_G$$

Substituting for E_G in the first loop equation yields

$$\mu E_S - 200 \mu i_p - 200 i_p - R_L i_p (N_p / N_S)^2 - r_p i_p = 0$$

Substituting values:

$$100 - 2000 i_p - 200 i_p - 65,000 i_p (N_p / N_S)^2 - 10,000 i_p = 0$$

or

$$100 - 12,200 i_p - 65,000 i_p (N_p / N_S)^2 = 0$$

Maximum power transfer occurs when source and load impedances are equal; that is, when

$$12,200 i_p = 65,000 i_p (N_p / N_S)^2$$

The turns ratio must then be

$$N_p / N_S = \sqrt{12,200/65,000}$$
$$= 0.433$$
$$= 1:2.3 \text{ ANS}$$

(B) The maximum power delivered to the load is determined by the plate current, which is found by solving for i_p :

$$100 - 12,200 i_p - 65,000 i_p \, (0.433)^2 \, 0$$
$$i_p = 100/24,400$$
$$= 4.098 \text{ mA}$$

The power delivered to the load is then

$$P_L = i_p{}^2 \, (65,000) \, (0.433)^2$$
$$= 0.2049 \text{W ANS}$$

PROBLEM 11-9

For the circuit shown, find the impedance presented to the source, e_S , in terms of R_1 , R_2 , R_3 , r_p , and μ.

Problem 11-9.

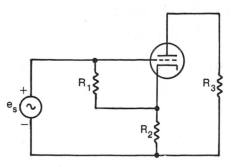

Answer

The equivalent circuit appears in Fig. 11-9, for which two loops equations are defined:

$$e_S - i_S R_1 - (i_p + i_S) R_2 = 0$$

$$\mu e_G + i_p R_2 + I_p R_3 + i_p r_p + i_S R_2 = 0$$

The relationship for e_G is then defined as

$$e_G = i_S R_1$$

so the second loop equation becomes

$$i_S \, (\mu R_1 + R_2) + i_p \, (R_2 + R_3 + r_p) = 0$$

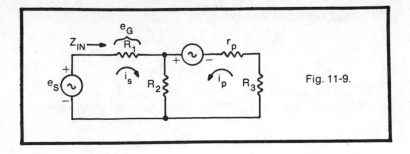

Fig. 11-9.

Solving now for i_S :

$$i_S = -i_p \, (R_2 + R_3 + r_p)/(\mu R_1 + R_2)$$

And for i_p :

$$i_p = -i_S \, (\mu R_1 + R_2)/(R_2 + R_3 + r_p)$$

Returning to the first loop equation and solving for e_S :

$$e_S = i_S R_1 + i_p R_2 + i_S R_2$$
$$= i_p R_2 + i_S (R_1 + R_2)$$

And substituting now for i_p :

$$1e_S = -i_S R_2 (\mu R_1 + R_2)/(R_2 + R_3 + R_p) + i_S (R_1 + R_2)$$

The input impedance is defined as

$$Z_{IN} = e_S / i_S$$

Therefore:

$$Z_{IN} = R_1 + R_2 - R_2 (\mu R_1 + R_2)/(R_2 + R_3 + r_p)$$

$$= \frac{R_1 R_2 (\mu + 1) + R_1 (R_3 + r_p) + R_2 (R_3 + r_p)}{(R_2 + R_3 + r_p)}$$

ANS

Chapter 12
Amplifiers

This chapter on amplifiers differs from Chapters 10 and 11 in that it concentrates on the general category of amplifiers. The review covers operational amplifiers, integrators, differentiators, and the like, while the problem section delves into various amplifier types. For other amplifier types, consult references 11, 29, and 30.

OPERATIONAL AMPLIFIERS

An "ideal" operational amplifier has five noteworthy characteristics. Using the subscripts P for the positive input and N for the negative, these five characteristics can be summarized as follows:

1—No offset voltage. That is, if E_{IN} and E_{ERROR} are zero, then output E_{OUT} must be zero.
2—Open-loop gain approaches infinity. Stated mathematically,

$$A_{OL} = E_{OUT} / (E_p - E_N) \to \infty$$

3—Currents at the input terminals are zero: $I_p = I_N = 0$.
4—Infinite frequency response. That is, the gain must remain constant and the phase shift must remain zero for all frequencies from DC on up.
5—Output voltage is independent of *common mode* input voltage. Common mode refers to input voltages that are applied simultaneously to both input terminals of the amplifier. For this condition to be satisfied, the common mode rejection ratio (CMRR) must be zero.

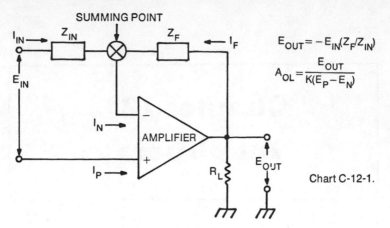

SUMMING POINT

$$E_{OUT} = -E_{IN}(Z_F/Z_{IN})$$

$$A_{OL} = \frac{E_{OUT}}{K(E_P - E_N)}$$

Chart C-12-1.

Table 12-1. Operational Amplifiers.

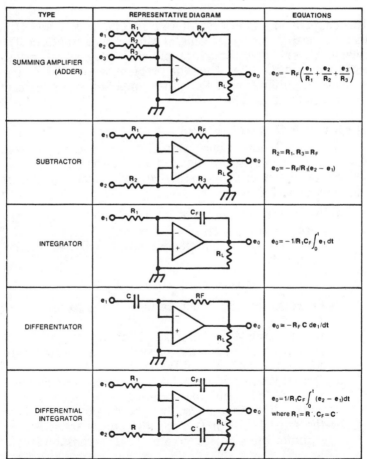

TYPE	REPRESENTATIVE DIAGRAM	EQUATIONS
SUMMING AMPLIFIER (ADDER)		$e_0 = -R_F\left(\dfrac{e_1}{R_1} + \dfrac{e_2}{R_2} + \dfrac{e_3}{R_3}\right)$
SUBTRACTOR		$R_2 = R_1, R_3 = R_F$ $e_0 = -R_F/R_1(e_2 - e_1)$
INTEGRATOR		$e_0 = -1/R_1C_F\displaystyle\int_0^t e_1\,dt$
DIFFERENTIATOR		$e_0 \cong -R_F\,C\,de_1/dt$
DIFFERENTIAL INTEGRATOR		$e_0 = 1/R_1C_F\displaystyle\int_0^t (e_2 - e_1)dt$ where $R_1 = R', C_F = C'$

218

Using the ideal amplifier and its parameters as a guide, Table 12-1 lists a number of op-amp circuits along with pertinent equations.

Various low-frequency response defects are illustrated in Table 12-1 to help pinpoint deficiencies in amplifier frequency response.

PROBLEM 12-1

In the circuit shown, the rectangle represents an amplifier with the following characteristics: gain $= -A = -10^4$; input impedance is infinite; output impedance is negligible. Determine the range of e_0/e_I obtainable by varying the potentiometer.

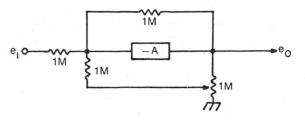

Problem 12-1.

Table 12-2. Low End Distortion of Square Wave.

A. Low Frequency Phase Leading

B. Low Frequency Phase Lagging

C. Low Frequency Amplitude Up

D. Low Frequency Amplitude Down

E. Low End Simple RC Cutoff (A & D)

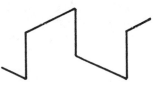

F. Result of Phase Compensating E

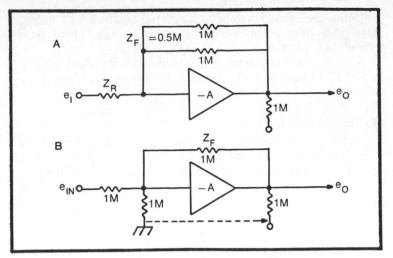

Fig. 12-1.

Answer

For the first case (Fig. 12-1A), the potentiometer is set at maximum, which is closest to e_0. This results in two 1 MΩ resistors in the feedback path, so the gain is

$$G_{MIN} = Z_F / Z_R$$
$$= -0.5/1.0$$
$$= -0.5 \text{ ANS}$$

For the second case (Fig. 12-1B), the potentiometer is set at the other extreme, at ground, which effectively removes the second 1 MΩ resistor from the circuit. The gain then becomes

$$G_{MAX} = -1.0/1.0$$
$$= -1.0 \text{ ANS}$$

PROBLEM 12-2

Given the feedback amplifier circuit illustrated, find the overall gain from the signal input to the load. The amplifier block has a gain of 10 from its input terminals to its output

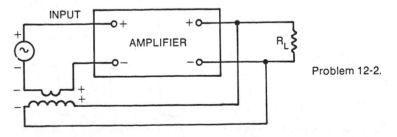

Problem 12-2.

terminals. A feedback with turns ratio of 4:1 is connected as shown to oppose the input signal.

Answer

The basic gain equation is

$$G = A/(1 - BA)$$

where A = overall gain (10)
B = feedback (1/4)

Thus,

$$G = \frac{10}{1 - (-0.25 \times 10)} = \frac{10}{1 + 2.5} = 2.86 \text{ ANS}$$

Alternate Answer

With the feedback voltages as shown in the figure,

$$E_{IN} - E_0/10 - E_0/4 = 0$$

Transposing,

$$E_{IN} = \frac{4E_0 + 10E_0}{40} = 14E_0/40$$

Consequently,

$$E_0/E_{IN} = 40/14 = 2.857 \text{ ANS}$$

PROBLEM 12-3

In the system depicted, the amplifier has a gain of $V_0/V_I = -100$, and it draws negligible current at either input terminal. Determine the system gain, V_0/V_S.

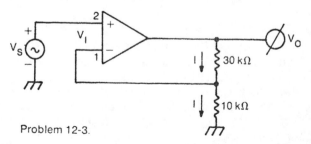

Problem 12-3.

Answer

In a noninverting amplifier, the voltage at input terminal 1 is $V_S + V_I$, and since the current in the 10 kΩ resistor must

221

equal the current in the 30 kΩ resistor, we have the relationship

$$\frac{V_I + V_S}{R_{10K}} = \frac{V_0 - (V_I + V_S)}{R_{30K}}$$

Since we know that $V_0 = -AV_I$, which is to say, $V_I - V_0 / A$, it follows that

$$\frac{-V_0 / A + V_S}{R_{10K}} = \frac{V_0 - (-V_0 / A + V_S)}{R_{30K}}$$

so that

$$-V_0 R_{30K} / A + V_S R_{30K} = V_0 R_{10K} + V_0 R_{10K} / A - V_S R_{10K}$$

$$(R_{10K} + R_{30K}) V_S = (R_{10K} + R_{10K} / A + R_{30K} / A) V_0$$

$$\frac{V_O}{V_S} = \frac{R_{10K} + R_{30K}}{R_{10K} + R_{10K} / A + R_{30K} / A}$$

$$= \frac{4 \times 10^4}{10^4 + 10^4 / (-10^2) + 3 \times 10^4 / (-10^2)}$$

$$= \frac{4 \times 10^4}{10000 - 100 - 300}$$

$$= \frac{4 \times 10^4}{9600}$$

$$= 4.17 \text{ ANS}$$

PROBLEM 12-4

The plate circuit of a triode having a plate resistance of 10 kΩ is coupled to the grid of the next tube by means of a transformer, the primary of which has an inductance of 75 μH. The mutual inductance is also 75 μH. The secondary winding has an effective resistance of 10Ω and an inductance of 300 μH. Across the secondary winding is connected a variable capacitor. Calculate:

(A) The capacitance value of this variable capacitor required to bring about a resonance at 1 MHz.

(B) The voltage across this variable capacitor at resonance, stated at a per volt input to the plate circuit of the preceding tube.

Answer *(ref. 1, p. 7-32)*

The circuit would appear as in Fig. 12-2.

(A) With untuned primary, resonance occurs when the secondary circuit becomes resonant. The tuning capacitance in this situation must then be

$$C = \frac{1}{(2\pi f)^2 \, L_S}$$

$$= \frac{1}{(2\pi \times 10^6)^2 \, (300 \times 10^{-6})}$$

$$= 84.4 \, \text{pF ANS}$$

(B) The gain of this amplifier stage is given by the formula

$$A = \frac{\mu \, \omega_0 \, M Q_0}{r_p + (\omega_0^2 \, M^2 / R_S)}$$

where ω_0 = resonant frequency (rad/sec) = $2\pi f$
  M = mutual inductance
  R_S = secondary series resistance
  Q_0 = Q of the secondary at resonance
  r_p = plate resistance

When r_p is much greater than the reflected impedance of the secondary at resonance,

$$A \cong g_m \, \omega_0 \, M Q_0$$

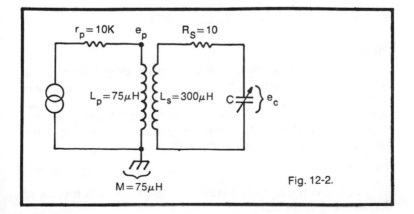

r_p = 10K e_p R_S = 10
L_p = 75μH L_s = 300μH C e_c
M = 75μH

Fig. 12-2.

223

Maximum gain occurs when $M = \sqrt{R_S\, r_p\, /\omega_0}$, at which the Q of the circuit is

$$Q_0 = \frac{(2\pi \times 10^6)\,(300 \times 10^{-6})}{10\Omega} = 188.5$$

Using the exact gain equation,

$$A = \frac{\mu\, 2\pi (75 \times 10^{-6})\,(10^6)\,(188.5)}{10^4 + [(2\pi \times 10^6)^2\,(75 \times 10^{-6})^2\,/10]}$$

$$= \frac{\mu\, 8.88 \times 10^4}{10^4 + 2.22 \times 10^4}$$

$$= 2.76\,\mu$$

$$= E_0\,/E_{IN}\ \text{ANS}$$

Alternate Answer

(B) The gain at resonance is

$$G = \frac{\mu\, M/C}{(r_p + R_1 + j\omega L_{PRi})R_2 + \omega^2 M^2}$$

Now $R_1 + jL_{PRi}$ is much smaller than r_p , so the denominator becomes $r_p\, R_2 + \omega^2\, M^2$. Thus,

$$G = \frac{\mu\, 75 \times 10^{-6}\,/84.4 \times 10^{-12}}{10^4\,(10) + (2\pi \times 10^6)^2\,(75 \times 10^{-6})^2}$$

$$= \frac{\mu\, 8.89 \times 10^5}{10^5 + 2.22 \times 10^5}$$

$$= 2.76\,\mu\ \text{ANS} \quad \text{or} \quad \frac{E_0}{\mu e g} = 2.78$$

PROBLEM 12-5

If you were to design a class C power amplifier to run at maximum efficiency but were forced to use a power supply voltage equal to half of normal voltage for the tubes, what method would you use? Give the method of determining the L/C ratio of the plate tank circuit when the maximum permissible harmonic radiation is 3%.

Answer *(Ref. 1, p. 4-33; ref. 2, p. 461)*

Find: (A) method for maximum efficiency; (B) method of determining L/C ratio.

(A) To operate the amplifier at maximum plate efficiency, it is necessary to optimize the efficiency expression

$$\eta_p = \frac{P_{OUT}}{P_{IN}} = \frac{e_{PMAX}\, I_{H1}}{2E_{BB}\, I_B}$$

where e_{PMAX} = maximum plate voltage
I_{H1} = peak value of fundamental plate current
E_{BB} = DC plate supply voltage
I_B = DC plate current

Rearranging this expression with consideration given to the important variables,

$$I_{BA} = \frac{P_{OUT}/n + kP_{OUT}/n}{E_{BB}}$$

where I_{BA} = average plate current
n = number of tubes used
k = ratio of plate dissipation to P_{OUT} (see Table 12-3).

At half the supply voltage, it is necessary to readjust the bias to give 120 to 150° conduction angle of plate current flow. The output load impedance should be set so that the voltage drop in the load with this plate current pulse gives minimum plate potential moderately greater than the maximum grid potential.

(B) The percentage harmonic distortion is given by the formula

$$D = \frac{67\, I_2\, /I_M}{Q\, I_1\, /I_M}$$

From Fig. 86 (Ref. 1), the plate phase angle of 120° yields

$$I_0\, /I_M = 0.18$$
$$I_1\, /I_M = 0.35$$
$$I_2\, /I_M = 0.27$$
$$I_3\, /I_M = 0.155$$

Solving for the required Q resulting from a 3% distortion figure, we obtain

$$Q = \frac{(67)\,(0.27)}{(0.03)\,(0.35)} = 1723$$

K	Harmonic
0.33	Fund.
1.00	2nd
2.33	3rd
3.44	4th
4.35	5th

Table. 12-3.

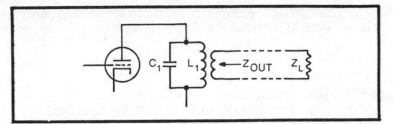

Fig. 12-3. Class C output stage.

To find the L/C ratio, set $Z_{OUT} = Z_L$ (Fig. 12-3). The L value should be such that $\omega L_1 = Z_{OUT} = Z_L$. When properly loaded, tune L_1 and C_1 for minimum plate current. At this point, you will have the correct values for L_1 and C_1 to determine their ratio.

Alternate Answer

Select a peak space current I_M on the basis of filament capability, such as 100 mA/W, with a safety factor of 6.

Select a combination of maximum grid potential and minimum plate potential to draw this total space current. The minimum plate voltage must not be less than the maximum grid potential.

Decide upon a suitable angle of plate current flow θ_p, usually between 120° and 150°, a compromise between high efficiency, small output, large input power for small conduction angles, and the low efficiency, large output, small input power for large conduction angles.

Calculate the grid bias by Fig. 12-4 and the following formulas.

$$E_C = E_B/\mu + \left(E_{MAX} + \frac{E_{MIN}}{\mu}\right)\left[\frac{\cos(\theta_p/2)}{1 - \cos\theta_p/\mu}\right]$$

$$\cos(\theta_p/2) = \frac{E_S - E_{MAX}}{E_S} = E_C/E_S$$

$$\cos(\theta_p/2) = 1\frac{1 + \mu E_{MAX} + EMIN}{\mu E_C - E_B}$$

Calculate the DC plate current, plate dissipation, power output, efficiency, and grid drawing power (ref. 2, p. 446).

Check the solution.

Design the tank circuit (Fig. 12-5) using the formulas (ref. 50, p. 434):

$$Q_{EFF} = (Q_1 - Q_2)/Q_1$$

226

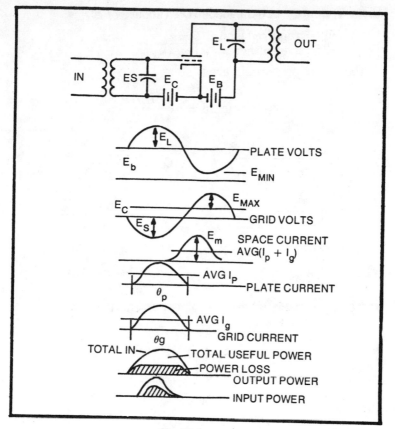

Fig. 12-4.

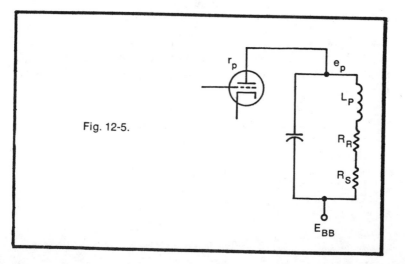

Fig. 12-5.

227

where Q_1 and Q_2 are the loaded Q's (about 8 or 10).

$$\omega L = \text{required load } R/Q_{EFF}$$
$$\text{Effective voltage} = \sqrt{P\omega L Q_{EFF}}$$

where P = power delivered

$$\% \text{ harmonic} = \frac{67\, I_2\, /I_M}{Q\, I_1\, /I_M}$$

$$r_p = \frac{L_p}{(R_R + R_S)C_p} \qquad Q_0 = \frac{2\pi f_0\, L_p}{R_R + R_S}$$

$$r_p = \frac{Q_0}{2\pi f_0\, C_p} \qquad r_p = \frac{(E_{BB} - E_{CC})^2}{2P_0}$$

$$Q_0 \cong 12$$

$$P_{DC} = E_{BB}\, I_{BA} \qquad \eta_p = \frac{e_{PMAX}\, I_{H1}}{2\, E_{BB}\, I_{BA}} \times 100$$

where I_{H1} = peak value of the plate current function

$$e_p{}' = E_{PMAX}$$
$$e_p{}' = 4000$$
$$e_p{}'' = 0.866\, E_{PMAX}$$
$$e_p{}'' = 3664$$
$$e_p{}''' = 0.5\, E_{PMAX}$$
$$e_p{}''' = 2000$$

PROBLEM 12-6

Determine the lower half-power frequency of the circuit shown. Assume a tube is used with $\mu = 20$ and $g_m = 2000$ μmhos.

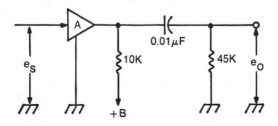

Problem 12-6.

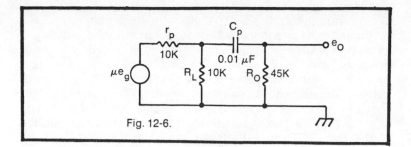

Fig. 12-6.

Answer

Since $\mu = g_m\, r_p$, it follows that $r_p = 10$ kΩ. This then results in the equivalent amplifier circuit shown in Fig. 12-6.

By Thevenin's theorem, the source resistance to the left of the coupling capacitor is the parallel combination of R_L and r_p , which equals 5kΩ. This makes the total circuit resistance $5 + 45 = 50$kΩ.

The half-power point will occur where

$$X_C = 1/2\pi f = 5 \times 10^4$$

This, of course, assumes that the X_C of the capacitor approaches zero at high frequencies, making it a lossless element in the circuit.

Solving for the half-power frequency,

$$f = \frac{1}{(2\pi)\,(5 \times 10^4\,)\,(1 \times 10^{-8}\,)} = 318.3 \text{ Hz ANS}$$

PROBLEM 12-7

A DC voltmeter as in the illustration uses a 100 μA meter having a resistance of 800Ω. The field-effect transistors have a forward transfer admittance of 2000 μmhos. What voltage will have to be applied to terminals A and B in order to obtain full-scale deflection?

Problem 12-7.

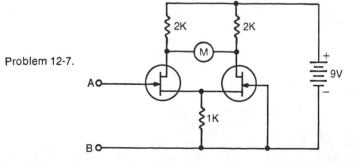

Answer

Assuming other FET parameters negligible, the FET voltmeter operates as a differential amplifier in which the combined source current flowing through the 1kΩ resistor remains essentially constant for small variations in voltage applied between terminals A and B. Such variations in input voltage unbalance the differential amplifier, causing differences in drain current, and consequently differing voltages across the meter terminals. Provided the drain voltage of the two FETs is sufficiently high the FETs will act as constant current sources, thereby eliminating output admittance from consideration and leaving only the two drain resistors and meter resistance to determine meter deflection.

The 100 μA meter requires a voltage difference of

$$\begin{aligned} E_M &= IR \\ &= (10^{-4})(800) \\ &= 0.08V \end{aligned}$$

for full-scale deflection. This deflection voltage is derived from the differential drain currents, but the two drain resistors effectively shunt the meter, appearing as a 4000-ohm shunt across the 800-ohm meter. The Y_{fs} of the FETs is given to be 2000 μmhos, so the required input voltage to produce full deflection is

$$E_{IN} = E_M /(Y_{fs} R_L)$$

$$= \frac{(0.08)(4000 + 800)}{(2 \times 10^{-3})(4000)(800)}$$

$$= 60 \, \text{mV ANS}$$

Chapter 13
Oscillators
and Modulation

This chapter is comprised of typical P.E. problems on solid-state and vacuum-tube oscillators and on methods of amplitude and frequency modulation. It deals with the basics of two areas: First, it reviews the fundamental equations used in determining the frequency of oscillation in some of the popular types of oscillators, along with other pertinent data. Second, it reviews the basic equations and characteristics of AM and FM, plus a mention of PCM.

OSCILLATORS

As a matter of recollection, the two ingredients needed to cause an amplifier to oscillate are a 180° phase shift in the feedback loop and a feedback signal of sufficient magnitude to sustain oscillation. The feedback loop may consist of either tuned circuits (including crystals) or a series of RC networks. In the latter case, a greater amplifier gain is required to overcome amplitude losses.

Figure 13-1 illustrates the typical feedback amplifier and its associated gain equation. The plus-or-minus sign in the equation is controlled by the phase of the feedback signal, which must yield a voltage gain in the circuit that approaches infinity for oscillation to occur.

Examples of tuned oscillators are the Hartley, tuned-plate/tuned-grid (TPTG), Colpitts, and resonant feedback, and Clapp. The crystal also possesses an inherent behavior equivalent to an LC tank circuit, with its frequency controlled principally by the thickness of the piezoelectric material.

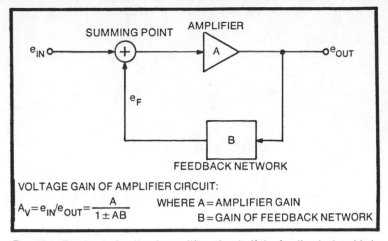

Fig. 13-1. The basic feedback amplifier circuit. If the feedback signal is in phase with the input signal, A_V will be greater than A; this is positive feedback and can lead to oscillation. If the feedback is out of phase, negative feedback results and the circuit gain is reduced.

Untuned oscillators depend on RC phase shift in the feedback loop or on the RC time constant of the frequency-determining components. Some examples of these oscillators are the phase-shift, parallel-T, multivibrator, and unijunction oscillators. The gas discharge type oscillator is a brute-force device with no amplifier, in which the frequency is determined by the RC time constants of the components involved. Modern replacements of such gas discharge devices as the neon lamp include the unijunction transistor, programmable unijunction transistor (PUT), and a wide variety of four-layer solid-state devices and microwave diodes that operate on the negative-resistance principle.

The following is a summary of basic oscillators and their related formulas.

Tuned-Plate/Tuned-Grid *(ref. 1, ch. 6)*

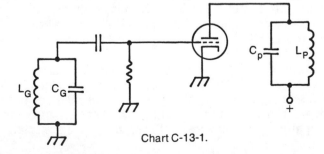

Chart C-13-1.

232

Frequency of oscillation:

$$\omega_O \approx \omega_G = 1/\sqrt{L_G C_G}$$

Plate tank oscillation frequency:

$$\omega_P = 1/\sqrt{L_p C_p}$$

Gain required:

$$\mu \geq \frac{L_G (1 - \omega_O^2/\omega_p^2)}{L_p (1 - \omega_O^2/\omega_p^2)}$$

Hartley

(ref. 1, ch. 6)

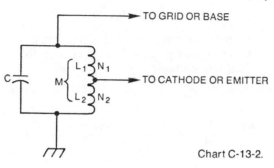

Chart C-13-2.

Frequency of oscillation:

$$\omega_O = 1/\sqrt{LC}$$

Gain required:

$$\mu = (L_1 + M)/(L_2 + M)$$
$$= N_1^2/N_2^2$$

Colpitts

(ref. 1, ch. 6)

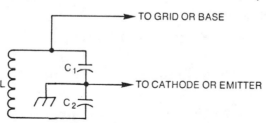

Chart C-13-3.

233

Frequency of oscillation:

$$\omega_0 = 1/\sqrt{LC}$$

when plate or collector resistance is much greater than coil resistance.

Gain required:

$$\mu \geq C_1 / C_2$$

Resonant Feedback (Transistor) *(ref. 8, ch. 8)*

Frequency of oscillation:

$$\omega_0 = 1/\sqrt{LC}$$

Gain required:

$$\mu = h_{fb} = (L_S + M)/(L_p + M)$$

where L_S = transformer secondary inductance at base
L_p = transformer primary inductance at collector
M = mutual coupling

Phase Shift (Transistor) *(ref. 8, ch. 8)*

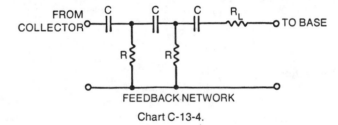

FEEDBACK NETWORK

Chart C-13-4.

Frequency of oscillation:

$$\omega_0 = \frac{1}{\sqrt{6R^2 C^2 + 4R R_L C^2}}$$

Gain required:

$$A = h_{fe} \approx 22 + (30/R_L) + (4R_L/R)$$

Phase Shift (Vacuum Tube) *(ref. 1, ch. 6)*

Frequency of oscillation:

$$\omega_0 = \frac{1}{RC\sqrt{3 + 2/a + 1/a^2} + R_S/R(2 + 2a^{-1})}$$

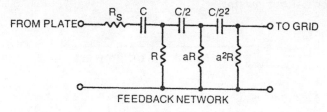

Chart C-13-5.

Gain required:

$$A = 8 + 12/a + 7/a^2 + 2/a^3 + R_S/R(9 \times 11/a + 4/a^2) + (R_S/R)^2 (2 + 2/a)$$

Multivibrator (Transistor)   *(ref. 17)*

Frequency of oscillation:

$$f = \frac{1}{0.69 (R_1 C_1 + R_2 C_2)}$$

where R_1 and C_1 are the base resistor and capacitor of transistor Q_1, and R_2 and C_2 are the base components of Q_2.

Crystal   *(ref. 1, ch. 6; ref. 2, ch. 6)*

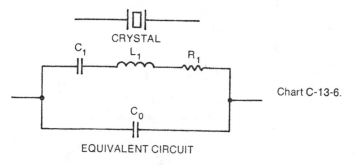

Chart C-13-6.

Series resonant frequency:

$$\omega_S = 1/\sqrt{L_1 C_1}$$

Parallel resonant frequency:

$$\omega_p = 1/\sqrt{L_1 C_1 C_0 /(C_1 + C_0)}$$

For A-T cut crystal:

$$f = \frac{1675}{\text{thickness in millimeters}}$$

For C-T cut crystal:

$$f = \frac{3070}{\text{width (square) in millimeters}}$$

For X cut crystal:

$$f = \frac{2860}{\text{thickness or width in millimeters}}$$

MODULATION

The most commonly used types of broadcast modulators use AM or FM signal processing methods. The AM wave is characterized by variations in amplitude, producing sidebands to either side of the RF carrier, unless one sideband is suppressed as in SSB transmission. The FM wave is produced by the modulation of two frequencies to produce a constant-amplitude, frequency-varying carrier.

Amplitude Modulation

Figure 13-2 illustrates the modulation principles of AM. For upward or positive modulation, the degree of modulation is

$$M_U = (E_{MAX} - E_O)/E_O$$

For downward or negative modulation

$$M_D = (E_O - E_{MIN})/E_O$$

The instantaneous voltage of a sinusoidally modulated wave is

$$e = E_O [1 + M \sin(2\pi f_A t)] \sin(2\pi f_C t)$$

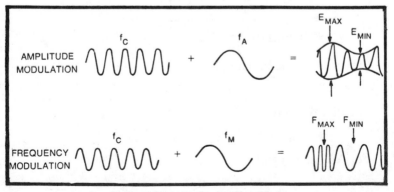

Fig. 13-2. Modulation techniques.

where e = instantaneous amplitude of the wave
$\quad E_O$ = average output amplitude
$\quad M$ = degree of modulation (upward or downward)
$\quad f_A$ = low-frequency modulating signal
$\quad f_C$ = carrier frequency

A modulated carrier signal will contain upper and lower sidebands, which are the sum and difference of the carrier and modulator frequencies. All three signals can be expressed as a sum, producing the instantaneous AM signal

$$e(t = E_O \sin(2\pi f_C\, t) + 0.5ME_O$$
$$\cos[2\pi(f_C - f_A)] - 0.5ME_O \cos[2\pi(f_C + f_A)]$$

Frequency Modulation

FM is characterized by the modulation of two frequencies to produce a constant-amplitude, frequency-varying carrier, as illustrated in Fig. 13-2. When modulated by a sine wave, the modulated output frequency is expressed as

$$f = f_C + \Delta f \cos(2\pi f_M\, t)$$

and the instantaneous phase angle is determined by

$$\theta = 2\pi f_C\, t + (\Delta f/f_M) \sin(2\pi f_M\, t) + \theta_0$$

where f_C = carrier frequency
$\quad f_M$ = modulating frequency
$\quad \Delta f$ = peak frequency deviation
$\quad \Delta f/f_M$ = modulation index (or deviation ratio)
$\quad \theta_0$ = unshifted phase angle

The combination of the preceding expressions provides the instantaneous signal voltage $e(t)$ of a carrier signal modulated by a single sinusoidal wave:

$$e(t) = E_O \sin[(2\pi f_C\, t) + (\Delta f/f_M) \sin(2\pi f_M\, t) + \Theta]$$

where E_O is the average amplitude of the carrier.

Pulse Code Modulation

Although PCM is not a form of sine-wave modulation, it is mentioned here as a matter of reference. Square-wave pulses are transmitted by either pulse division multiplexing (PDM) or by frequency division multiplexing (FDM), and these pulses are usually synchronized by either a prearranged code or by a separate reference signal. These pulses can be only in one of two states, *1* or *0*. Table 13-1 shows some of the more popular PCM methods and waveforms. (See Chapter 10 for some digital circuits and problems.)

Table 13-1. Pulse Modulation.

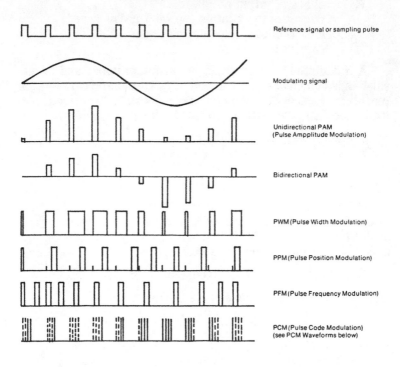

Reference signal or sampling pulse

Modulating signal

Unidirectional PAM
(Pulse Amplitude Modulation)

Bidirectional PAM

PWM (Pulse Width Modulation)

PPM (Pulse Position Modulation)

PFM (Pulse Frequency Modulation)

PCM (Pulse Code Modulation)
(see PCM Waveforms below)

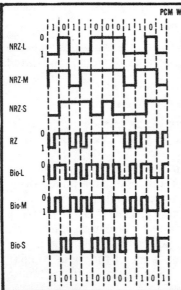

PCM WAVEFORMS

NRZ-L
NRZ-Level (or NRZ Change)
"One" is represented by one level
"Zero" is represented by the other level

NRZ-M
NRZ-Mark
"One" is represented by a change in level
"Zero" is represented by no change in level

NRZ-S
NRZ-Space
"One" is represented by no change in level
"Zero" is represented by a change in level

RZ
RZ
"One" is represented by a half-bit wide pulse
"Zero" is represented by no pulse condition

Bio-L
Bi-Phase-Level (or Split Phase, Manchester 11 +180°)
"One" is represented by a 10
"Zero" is represented by a 01

Bio-M
Bi-Phase-Mark (or Manchester 1)
A transition occurs at the beginning of every bit period
"One" is represented by a second transition ½ bit period later
"Zero" is represented by no second transition

Bio-S
Bi-Phase-Space
A transition occurs at the beginning of every bit period
"One" is represented by no second transition
"Zero" is represented by a second transition ½ bit period later

RADAR

Radar is a specialized form of transmission. The modulation qualities of radar are not nearly as valuable to the P.E. candidate as are the transmitting and receiving characteristics. Sophisticated modulation techniques do exist, but these are employed primarily in military systems, which are not of concern here.

Radar transmission consists of periodic bursts of continuous-wave emissions, and the information contained therein is readily available and not complex. Of primary concern, however, is the transmission, propagation, and reception characteristics that determine major system design requirements.

Consider a radar transmitter/receiver with a parabolic antenna having a power gain of G, transmitting power P, wavelength λ, for distance D, to a target with cross section C.

The power intensity at the target area is

$$S_T = \frac{\text{effective radiated power}}{4\pi D^2}$$

The signal power intensity at the receiver is

$$S_A = \frac{\text{radar cross-sectional area of target}}{4\pi D^2}$$

The signal power delivered by the receiving antenna is

$$S_R = \frac{(\text{power gain of antenna})\lambda^2}{4\pi}$$

Combining these basic equations results in an expression for the total received signal power:

$$S_{PR} = S_T S_A S_R$$
$$= \frac{PG^2 \; \lambda^2 \; C}{(4\pi)^3 \; D^4}$$

The maximum power gain attainable is then

$$G_{MAX} = 4\pi AK/\lambda^2$$

where A is the antenna aperture and K is a constant ranging typically from 0.5 to 0.6.

Maximum power gain under G_{MAX} conditions is

$$S_{PR} = \frac{PA^2\,CK^2}{4\pi D^4\,\lambda^2}$$

The maximum range under weak signal conditions (minimum received signal) is

$$D_{MAX} = \sqrt{\frac{PA^2\,CK^2}{4\pi\lambda^2\,S_{MIN}}}$$

where S_{MIN} is the minimum usable signal power.

PROBLEM 13-1

In the circuit shown, a 2N2417 unijunction transistor is to be used to generate pulses at 15,750 pps. Find values for R_2, R_T, and C_T, assuming that $\eta = 0.6$.

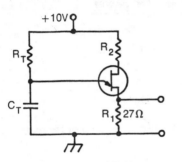

Problem 13-1.

Answer

Find (A) R_2, (B) R_T, and (C) C_T. From data books it is seen that channel resistance R_{BB} is usually from 4.7 to 6.8 kΩ, and this will be used in lieu of the measured value or the manufacturer's recommended value.

(A) The selection of R_2 is made through the use of the equation

$$R_2 \approx 0.4R_{BB}\,/\eta E$$

The selection of R_2 determines the effective temperature compensation of the UJT junction diode. Assuming that R_{BB} is the average of 4.7 and 6.8 kΩ we will use $R_{BB} = 5.75$ kΩ in the approximation equation. Thus,

$$R_2 = 0.4(5.75 \times 10^3\,)/(0.6)\,(10)$$
$$= 383\Omega\ \textbf{ANS}$$

(B) From the general equation

$$f \cong \frac{1}{R_T \, C_T \, \ln[1/(1 - \eta)]}$$

we can solve for the RC time constant of the oscillator:

$$R_T \, C_T \; = \; \frac{1}{15750 \ln[1/(1 - 0.6)]} = \frac{10^{-3}}{15.75 \ln(2.5)} = 69.39 \times 10^{-6}$$

Select a reasonable capacitor value and solve for R_T. Letting $C_T = 0.05 \mu\mathrm{F}$ (ANS),

$$R_T = (69.39 \times 10^{-6})/0.05 \times 10^{-6}$$
$$= 1.39 \, \mathrm{k\Omega} \; \text{ANS}$$

Alternate Answer

A quick solution to this problem is to use the nomogram in Fig. 13-3. In this case, start by setting a straight edge from $\eta = 0.6$ to $f = 15{,}750$. At the crossover point (center line), mark an index point, and then reposition the straight edge to find C_T and R_T. For example, choose any convenient value of C_T, follow the straight edge through the index point you marked, and read off the resulting value of R_T. Following the selection of $C_T = 0.05 \mu\mathrm{F}$ as in the preceding solution, the result will be $R_T = 1.39 \, \mathrm{k\Omega}$.

To find the frequency, or pulse repetition rate, the reverse procedure is used. The values of R_T and C_T establish the first line, and you mark the intersection point with the center line as your index point. Then pivot about this point to find η or f.

PROBLEM 13-2

In the astable symmetrical multivibrator shown, $C_1 = C_2$, $R_{C1} = R_{C2}$, $R_{B1} = R_{B2}$, and the transistors are matched.

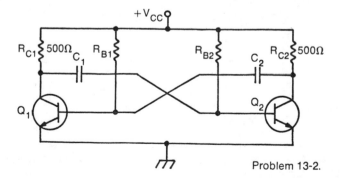

Problem 13-2.

(A) What would be the *on* duration time (assuming equal time constants) if $R_{B1} = 1\,k\Omega$ and $C_2 = 0.01\,\mu F$?

(B) What would be the operating frequency or repetition rate?

(C) What minimum current gain is required by the transistors?

(D) What are the parameter requirements for producing good rectangular waveforms?

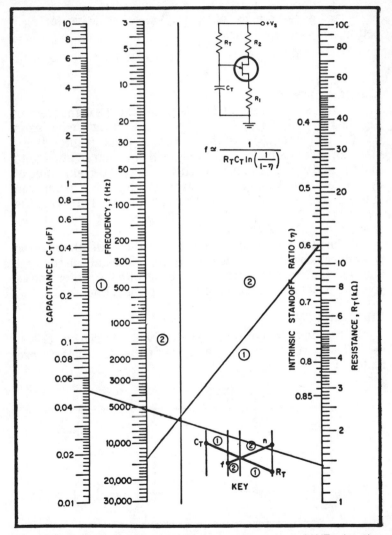

$$f \simeq \frac{1}{R_T C_T \ln\left(\frac{1}{1-\eta}\right)}$$

Fig. 13-3. Nomogram for calculating output frequency of UJT relaxation oscillator. (Courtesy EDN).

Answer

(A) For equal time constants and a supply voltage V_{CC} much greater than base-emitter voltage V_{BE} of the transistors,

$$\tau = \tau_2 \approx R_{B1} C_1 \ln(2)$$
$$= (10^3)(0.01 \times 10^{-6})(0.69)$$
$$= 6.9 \,\mu\text{sec ANS}$$

(B) The operating frequency is approximately

$$f = \frac{1}{(0.69)(R_{B1} C_1 + R_{B2} C_2)}$$

$$= \frac{1}{\tau_1 + \tau_2}$$

$$= \frac{1}{(0.69 + 0.69) \times 10^{-6}}$$

$$= 725 \,\text{kHz ANS}$$

(C) For transistor saturation to occur, $R_{B1} \approx \beta R_{C1}$; so

$$\beta \approx R_{B1}/R_{C1}$$
$$= 10^3/500$$
$$= 2 \,\text{ANS}$$

(D) In order to obtain high-quality square waves at the output:

1—R_{B1} and R_{B2} must be much greater than the *on* resistance of the transistor and much less than the *off* resistance.

2—C_1 and C_2 should be much greater than associated input and output capacitances of the transistors or output circuit.

3—Time constant $R_{C1} C_1$ should be much less than $R_{B1} C_2$, and $R_{C2} C_2$ should be much less than $R_{B2} C_1$

4—R_{C1} and R_{C2} should be kept relatively small, particularly with respect to output loads placed on the multivibrator.

PROBLEM 13-3

The neon lamp in the figure breaks down at 70V and establishes a glow. This glow is extinguished when the voltage across the neon bulb is reduced to 50V.

(A) Determine the frequency of oscillation.

(B) Draw a graph showing the voltage across C.

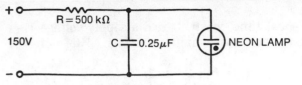

Problem 13-3.

Answer *Ref. 10*

(A) The voltage across the capacitor is

$$V_C = E_{DC} [1 - \exp(-t/RC)]$$

So

$$f = \cfrac{1}{RC \ln(E_{DC} - V_E / E_{DC} - V_I)}$$

$$= \cfrac{1}{(5 \times 10^5)(0.25 \times 10^{-6}) \ln(100/80)}$$

$$= \cfrac{1}{0.125 \ln(1.25)}$$

$$= \cfrac{1}{0.125 (0.223)}$$

$$= 35.9 \text{ Hz ANS}$$

(B) Voltage across C is graphed in Fig. 13-4.

PROBLEM 13-4

For the tuned-plate oscillator shown, prove that the frequency of oscillation is given by

$$\omega^2 = 1/LC (1 + R/r_p)$$

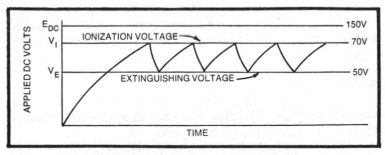

Fig. 13-4. Sawtooth oscillator operation.

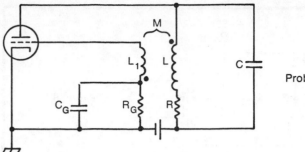

Problem 13-4.

Answer

(ref. 10, p. 388)

Assuming critical coupling between the grid and tank circuits, the equivalent circuit of the oscillator would appear as in Fig. 13-5. The impedance of this common tuned circuit is

$$Z(s) = \frac{1}{[1/(Ls + R)] + Cs + 1/r_p}$$

Multiplying by $Ls + R$ to obtain a polynomial denominator,

$$Z(s) = \frac{Ls + R}{(1 + R/r_p) + s(RC + 1/r_p) + s^2 LC}$$

The frequency of oscillation is then obtained from the real parts of the polynomial, so that

$$\begin{aligned} s^2 LC + (1 + R/r_p) &= 0 \\ s^2 &= -1/LC \, (1 + R/r_p) \\ &= (-j\omega)^2 \\ &= -\omega^2 \end{aligned}$$

Therefore,

$$\omega^2 = +(1 + R/r_p)/LC$$

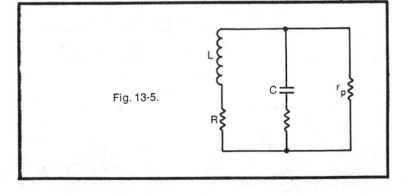

Fig. 13-5.

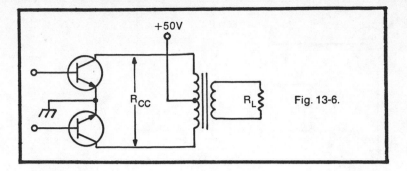

+50V

R_{CC}

R_L

Fig. 13-6.

PROBLEM 13-5

In a transformer-coupled modulator circuit with push-pull primary, the DC collector voltage is 50V and the DC collector current is 8 mA. If the secondary turns are 2500, what must the primary turns be to produce a load of 4000Ω in the collector circuit of the modulator?

Answer *(ref. 10, p. 253, ref. 13, p. 188)*

The transformer circuit would appear as in Fig. 13-6. The collector-to-collector resistance across the transformer primary is

$$R_{CC} = 4R_L (N_p/N_S)^2$$
$$= 4V_{CC}/I_C$$

Solving for the number of primary turns,

$$N_p = N_S \sqrt{V_{CC}/R_L I_C}$$
$$= 2500\sqrt{50/(4000)(0.008)}$$
$$= 3125 \text{ turns ANS}$$

PROBLEM 13-6

A class-A RF amplifier stage consists of a pentode operating into an RF transformer. The transformer secondary is tuned by means of a variable capacitor. The mutual conductance and plate resistance of the pentode are 1000 μmhos and 1MΩ, respectively. The Q factor of the secondary coil is 100 at its operating frequency of 1 MHz, and its inductance is 100 μH. Assume negligible losses in the tuning capacitor and succeeding circuits.

(A) Calculate the value of mutual inductance required in the transformer to yield maximum gain for this stage.

(B) Is this optimum value of mutual inductance readily realizable? Give reasons for your answer.

(A) In this circuit (Fig. 13-7) the Q is 100, so the secondary coil resistance is

$$R_S = \omega L / Q = 6.28 \Omega$$

Knowing this, the mutual inductance for maximum gain is

$$M = \sqrt{\frac{R_S \, r_p}{\omega_0}} = \frac{\sqrt{6.28 \, (10^6)}}{6.28 \times 10^6} = 399 \, \mu\text{H ANS}$$

Therefore

$$A \approx g_m \, \omega_0 \, MQ$$
$$= (10^3 \times 10^{-6})(6.28 \times 10^6)(399 \times 10^{-6})(100)$$
$$= 250.6 \, \text{ANS}$$

(B) This optimum value of mutual inductance is not readily realizable. The plate resistance is too high.

If coupling is increased above optimum, both bandwidth and gain will decrease. If coupling is decreased below optimum, the bandwidth will increase and gain will decrease.

Let $g_m = k$ and $A = 251$:

$$A/k = \omega_0 \sqrt{R_s \, r_p} \, (1/\omega_0) \approx \sqrt{r_p}$$

Thus

$$r_p = (A/k)^2 = 6.28 \times 10^{10}$$

which is an impossible plate resistance requirement to get maximum gain.

PROBLEM 13-7

Write the mathematical forms expressing:

(A) Amplitude modulation.
(B) Frequency modulation.
(C) Phase modulation.

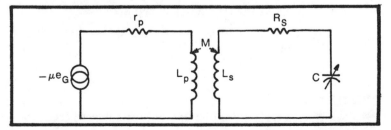

Fig. 13-7.

(A) Amplitude modulation: The instantaneous signal is given by the formula

$$e(t) = E[1 + M \cos(\omega_m t)] \sin(\omega_C t + \phi)$$

where E = peak amplitude of the carrier
ω_M = frequency of modulation voltage (rad/sec)
ω_C = frequency of carrier voltage (rad/sec)
ϕ = phase angle of carrier (rad)
M = modulation factor

For upward or positive modulation,

$$M = (E_{MAX} - E)/E \text{ ANS}$$

and for downward or negative modulation

$$M = (E - E_{MIN})/E \text{ ANS}$$

(B) Frequency modulation: The instantaneous signal is given by the formula

$$e(t) = E_C \sin[(\omega_C t) + (\Delta f/f_M) \sin(2\pi f_M t) (\Delta f/f_M) + \theta_0$$

where f_C = carrier frequency
f_M = modulation frequency
$f = (1/2\pi)(d\Theta/dt)$

The instantaneous output frequency in FM is

$$f = f_C + \Delta f \cos(\omega_M t)$$

and the instantaneous phase angle is

$$\Theta = 2\pi f_C t + (\Delta f/f_M)[\sin(2\pi f_M t)] + \Theta_0$$

(C) Phase modulation: The instantaneous signal is given by the formula

$$e(t) = E_C \sin[(\omega_C t) + \phi + \Delta\phi \cos(\omega_M t)]$$

where $\Delta\phi$ = peak value of phase deviation
ω_M = modulating frequency in rad/sec

Frequency modulation or phase modulation is a form of angle modulation.

PROBLEM 13-8

(A) What is the distinguishing feature between frequency modulation and phase modulation?

(B) Make up a table comparing the essential characteristics of amplitude modulation and frequency modulation.

Answer

(A) In phase modulation, the instantaneous phase angle of the carrier is varied by the amplitude of the modulating signal. In frequency modulation the instantaneous frequency of the carrier (time derivative of the phase angle) is made to vary in accordance with the amplitude of the modulating signal.

(B) Table 13-2 compares the difference between amplitude modulation and frequency modulation for nine important characteristics.

PROBLEM 13-9

What is the basis of the advantage of frequency modulation over amplitude modulation in noise-free transmission of radio signals? Explain fully.

Answer

FM is easier to minimize interference from other nearby stations operating on the same frequency. And because of the arrangement of the circuits in the FM transmitter, it is more

Table 13-2. Comparison of AM and FM.

CHARACTERISTIC	AMPLITUDE MODULATION	FREQUENCY MODULATION
Effective range	Intermediate, determined by noise threshold.	Short, since FM is usually high frequency.
Bandwidth	Twice the highest modulating frequency.	Same as AM plus nominal deviation of FM wave.
Signal-to-noise performance	Poor, since carrier power does not contribute to output signal speech process by preemphasis and clipping.	Good. Improvement is possible by increasing deviation.
Interference rejection	Performed by selective filtering.	Rejects interference by capture effect.
Linearity distortion	Fair.	Can be excellent.
Overmodulation distortion	Limited to four times the carrier power.	Peak limited by channel bandwidth.
Transmitter power	Average is carrier power plus average modulating power. Peak is four times carrier power.	Constant RF power. Efficiency is high.
Transmitter complexity	More complex than FM. High level modulation most complex.	Intermediate complexity.
Receiver complexity	Relatively uncomplicated	Somewhat more complex.

economical to produce a given wattage signal with FM techniques than in a comparable AM transmitter, since with FM very little modulation power is required, while with AM the audio modulation power is generally 50% of the carrier power. Also, with FM the output power does not vary with modulation so no additional provision must be made to handle peak power levels during modulation peaks.

In terms of fidelity, AM is able to deliver just as much as with FM; it is only the crowding of the broadcast band that prevents bandwidth allocations to provide a full 10 or 15 kHz frequency response.

An FM wave is relatively immune to natural and man-made sources of noise. The FM wave is constant in amplitude, and superimposed noise pulses are easily removed in the receiver. The addition of amplitude pulses has been found to have little effect on the frequency content of the wave, provided that the signal-to-noise ratio does not deteriorate too badly.

The basis for the immunity that FM has from noise also comes from the larger bandwidth and more complex nature of the FM signal. The basic equation is

$$e = E_O \sin[\omega t + M_F \sin(pt)]$$

where E_O = crest amplitude of wave
ω = radian frequency of carrier in absence of modulation
p = radian frequency at which carrier frequency is varied

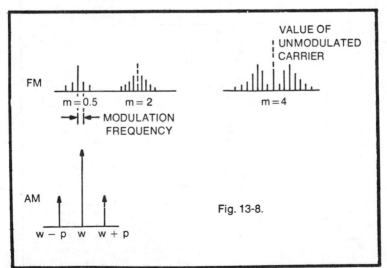

Fig. 13-8.

250

M_F = modulation index, which is the ratio of the change in carrier frequency to the modulating frequency

This equation illustrates that the intelligent portion of the FM signal resides in the instantaneous frequency and phase of the carrier, rather than in the amplitude (E_O), which is a constant.

The FM equation can also be written in terms of its spectral components, several of which are illustrated in Fig. 13-8 for different values of modulation index. The general form of the equation involves Bessel functions of the first kind and nth order (J_n), having the modulation index as the argument:

$$e/E_O = J_O (M_F) \sin(\omega t)$$

$$+J_1 (M_F [\sin(\omega + p)t - \sin(\omega - p)t]$$

$$+J_2 (M_F)[\sin(\omega + 2p)t + \sin(\omega - 2p)t]$$

$$+J_3 (M_F)[\sin(\omega + 3p)t - \sin(\omega - 3p)t]$$

$+ \bullet \bullet \bullet$

PROBLEM 13-10

A 40 MHz signal of 1V peak amplitude is frequency modulated by $\pm$ 75 kHz at a 60 Hz rate. Both the frequency and instantaneous magnitude are maximum at time $t = 0$. Write an expression for the instantaneous voltage.

Answer *(ref. 1, p. 5-38; ref. 2, p. 579; ref. 6, p. 21-7)*

A maximum at $t = 0$ represents a cosine function. Let:

$e(t)$ = instantaneous voltage
$\omega_I (t)$ = instantaneous frequency of modulated signal
ω_M = modulated frequency
$\Delta\theta$ = peak ϕ deviation
$\Delta\omega$ = peak frequency deviation
ω_C = angular frequency of unmodulated carrier
ϕ_C = carrier ϕ angle
ϕt = instantaneous phase angle
$\Delta\omega/\omega_M$ = peak ϕ deviation = β
β = modulation index

In this problem $A = 1$ and the modulation index is

$$\beta = \frac{\Delta\omega}{\omega_M} = \frac{2\pi(75 \times 10^3)}{2\pi(60)} = 1250$$

The equation for frequency modulation is

$$e(t) = A \cos[(\omega_C \, t) + \beta \sin(\omega t)]$$

where $\phi t = \omega_C \, t + \beta \sin(\omega t)$.

Frequency is a maximum at $t = 0$, so

$$f = 40 \times 10^6 + 75 \times 10^3 = 40 \text{ Hz}$$

Therefore,

$$\begin{aligned} \phi t &= \int \omega \, dt \\ &= 2\pi \int [f_C + \Delta\omega_C \, \cos(\omega_M \, t)] \, dt \\ &= 2\pi \{(40 \times 10^6) + \frac{75 \times 10^3}{2\pi(60)} \, \sin[2\pi(60)t]\} + C \end{aligned}$$

If $e(t)$ is to be a maximum at $t = 0$,

$$\phi t = 2\pi[0 + 0] + C$$

The constant C must be 0 (or π). Therefore,

$$e(t) = A \cos \{2\pi(40 \times 10^6)t + \frac{75 \times 10^3}{60} \, \sin[2\pi(60) \, t]\}$$

$$= \cos[(251 \times 10^6)t + 1250 \sin(377t)]$$

PROBLEM 13-11

A supressed-carrier, amplitude-modulated signal has the expression

$$e(t) = 100 \cos(50t) \cos(1000t)$$

What is the carrier level injection required to produce a 50% modulated amplitude signal?

Answer (ref. 1, p. 5-2)

The problem expression is in the general form

$$e(t) = E[1 + M \cos(\omega_M \, t) \sin(\omega_C \, t) + \phi]$$

where $e(t)$ = instantaneous AM voltage signal
E = peak amplitude of unmodulated carrier
$M = (E_{MAX} - E)/E$ for upward modulation
$M = (E - E_{MIN})/E$ for downward modulation
ω_M = frequency of modulating voltage (rad/sec)
ω_C = carrier frequency (rad/sec)
ϕ = carrier phase angle (rad)

Comparing the two equations, we get

cos(50t) = modulating frequency term
cos(1000t) = carrier frequency term

So the 1000 rad/sec carrier frequency is modulated by a 50 rad/sec wave. The peak amplitude is indicated by the multiplying factor 100, therefore, $EM = 100$.

At 50% modulation, $E = 100$ and $E_{MAX} = 150$. To find E, we use the relationship for the upward modulation factor. Thus, for 50% modulation.

$$M = \frac{E_{MAX} - E}{E} = \frac{150 - 100}{100} = 0.5$$

Since $EM = 100$,

$E = 100/M$
  $= 100/0.5$
  $= 200$ ANS

PROBLEM 13-12

The phase of a 60 Hz source shifts sinusoidally by $\pm 30°$ once a minute. By how much does the frequency vary?

Answer

It is assumed that the frequency source is changing to produce a frequency difference above and below 60 Hz. Since

$$\omega = \theta/t = 2\pi f \text{ (rad/sec)}$$

at 60 Hz,

$$\theta/t = 120\pi$$

The phase angle $\theta = 30° = (\pi/6)$ rad/sec. So at $+30°$ the frequency is

$\omega(+30°) = 120\pi + \pi/6 = 377.515$ rad/sec
$f(+30°) = 60.0834$ Hz ANS

And at $-30°$ the frequency is

$\omega(-30°) = 120\pi - \pi/6 = 376.468$ rad/sec
$f(-30°) = 59.9167$ Hz ANS

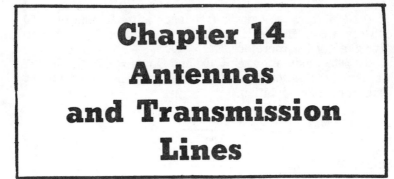

Chapter 14
Antennas
and Transmission
Lines

The impedance parameters of a transmission line of reasonable length can be determined by classical open-circuit and short-circuit methods. Using the parameters defined in Fig. 14-1, it can be stated that

$$Z_{SC} = Z_0 \tanh(\gamma l)$$

and

$$Z_{OC} = Z_0 / \tanh(\gamma l)$$

where Z_0 = characteristic impedance
Z_{SC} = input impedance with output short circuited
Z_{OC} = input impedance with output open circuited
γ = propagation constant (hyperbolic angle in feet)
l = length of line in feet

From the preceding two equations it can also be shown that

$$Z_0 = \sqrt{Z_{SC} Z_{OC}}$$

and

$$\tanh(\gamma l) = \sqrt{Z_{SC} / Z_{OC}}$$

When a line appears to possess little or no resistive component, which is to say that Z_{OC} is highly capacitive or Z_{SC} is highly inductive, then

$$Z_0 = \sqrt{L_{SC} / C_{OC}}$$

where L_{SC} = short-circuit inductance (measured at input)
C_{OC} = open-circuit capacitance (measured at input)

254

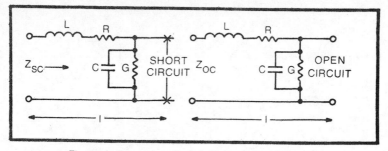

Fig. 14-1. Open circuited and short-circuited lines.

In terms of admittance ($Y_{OC} = 1/Z_{OC}$) and conductance (G_0), this hyperbolic relationship is expressed as

$$\tanh(\gamma l) = \sqrt{Z_{SC} \, Y_{OC}} = \sqrt{R_S + j\omega L_{SC}} \sqrt{G_0 + j\omega C_{OC}}$$

where R_S = series resistance

$\omega = 2\pi f$

$\gamma = \alpha + j\beta$

α = attenuation constant (nepers per unit length)

β = phase constant (radians per unit length)

Admittance can also be expressed as

$$Y_{OC} = G_0 + j\omega C_{OC}$$

and short circuit impedance as

$$Z_{SC} = R_S + j\omega L_{SC}$$

For relatively short transmission lines, attenuation α becomes (in decibels per 100 feet)

$$\alpha = 434/l[1/(1 + \omega^2 \, L_{SC} \, C_{OC})] \, [R_S/Z_0 + G_0 \, Z_0]$$

and phase constant β becomes (in degrees per foot)

$$\beta = \frac{\tan^{-1} (\omega\sqrt{L_{SC} \, C_{OC}})}{l}$$

As an approximation,

$$\tan(\beta l) \approx \omega\sqrt{L_{SC} \, C_{OC}}$$

TRANSMISSION LINE PARAMETERS

Characteristic impedance: see Figs. 14-2 and 14-3.
Sending end impedance: see Fig. 14-4.

255

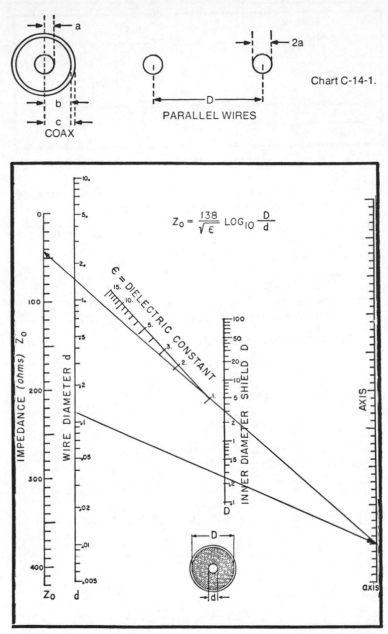

Chart C-14-1.

PARALLEL WIRES

COAX

$$Z_0 = \frac{138}{\sqrt{\epsilon}} \, LOG_{10} \, \frac{D}{d}$$

ϵ = DIELECTRIC CONSTANT

IMPEDANCE (ohms) Z_0

WIRE DIAMETER d

INNER DIAMETER SHIELD D

AXIS

Z_0 d   D   axis

Fig. 14-2. Characteristic impedance of lines. To determine impedance Z_0, draw line from wire diameter d through inner diameter shield D to axis scale. From axis scale, then draw line through dielectric constant to intersect impedance scale Z_0. To find diameters, reverse procedure. (From Radio-Electronics, April 1950).

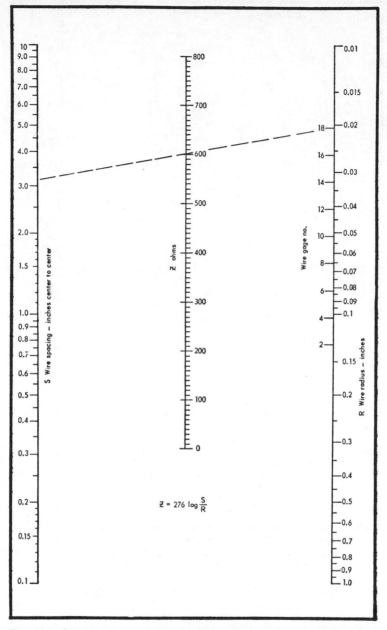

Fig. 14-3. Characteristic impedance of parallel wires. For regular wires, use scales as in above example for 18-gauge wires spaced for 600-ohm transmission line. The S and R scales may be multiplied or divided by the same scale factor for extended calculations. For parallel coax lines, wire radius R is approximately half of coax diameter. (Courtesy EDN).

257

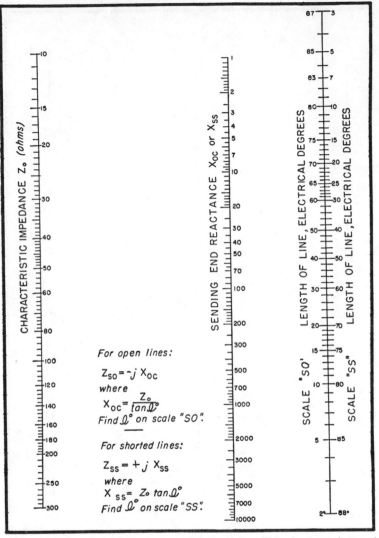

Fig. 14-4. Sending end impedance of uniform line. This chart may be used to obtain the input resistance of lossless shorted and open transmission lines. (From Radio-Electronics, Jan. 1950).

Capacitance of coaxial cable (in microfarads per mile):

$$C = \frac{0.03888 \, k/k_0}{\log(b/a)}$$

Capacitance (approximate) of parallel, round wires (in microfarads per mile):

$$C = \frac{0.01944 \, k/k_0}{\log (D/a)}$$

Capacitance (exact) of parallel, round wires (in farads per meter):

$$C = \frac{k/k_0}{(35.96 \times 10^9) \ln[D/2a + \sqrt{(D/2a)^2 - 1}]}$$

Inductance of coaxial cable at low frequencies (in henries per meter):

$$L = \times 1)^{-7} \{\mu_i /4\mu_0 + \ln(b/a) + \mu_i /\mu_0$$

$$[c^4 /(c^2 - b^2)^2 [\ln(c/b)] - (3c^2 - b^2)/4(c^2 - b^2)]\}$$

Inductance of parallel, round wires at low frequencies (in millihenries per mile):

$$L = 0.0805 \, \mu_i /\mu_0 + 0.7411 \log (D/a)$$

PROPAGATION CONSTANTS

Propagation constant:

$$\gamma = \alpha + j\beta$$

where α = attenuation constant
β = phase constant

Basic relationships:

$$\beta\lambda = 2\pi \text{ radians}$$
$$\lambda = 2\pi/\beta \text{ meters}$$
$$\beta = 2\pi/\lambda \text{ radians per meter}$$
$$V = 2\pi f \text{ meters per second}$$
$$\nu = 2\pi f = \omega/\beta$$

For a lossless line ($R = 0, G = 0$):

$$Z_0 = \sqrt{L/C} \text{ ohms}$$
$$\gamma = j\omega\sqrt{LC}/l \text{ per meter}$$

If $\alpha = 0$,

$$\beta = 2\pi f\sqrt{LC}/l$$
$$\lambda = 2\pi/\beta = l/(f\sqrt{LC})$$
$$\nu = \lambda f - 1/\sqrt{LC} \text{ meters per second}$$

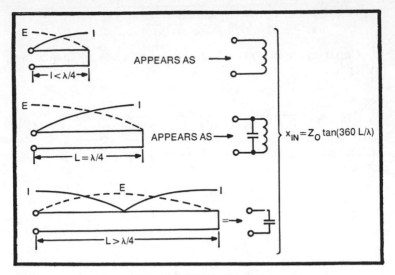

Fig. 14-5. Shorted stubs.

STUBS AND QUARTER-WAVE SECTIONS

The general characteristics of shorted and open stubs are shown in Figs. 14-5 and 14-6.

Quarter-wave stubs and sections have many desirable properties. The equivalent circuit of a quarter-wave open stub is shown in Fig. 14-7B. When tapping along the length of an open stub, it behaves as a voltage transformer as illustrated in Fig. 14-7C and D. Impedance transformations are also easily accomplished by inserting a quarter-wave matching section

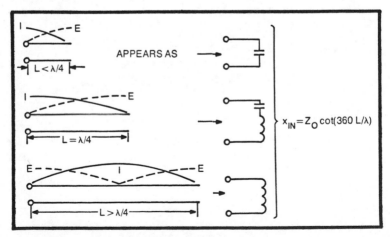

Fig. 14-6. Open stubs.

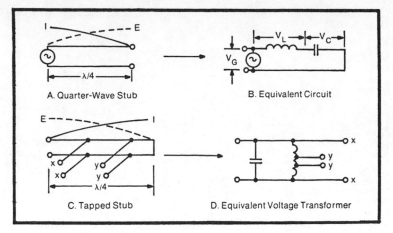

Fig. 14-7. Properties of quarter-wave stubs.

between two lines having different impedances, and the characteristic impedance of the quarter-wave section may be found using Fig. 14-8.

PROBLEM 14-1

In the accompanying diagram, the signal current is found to be 15.5 mA with switch S open and 44 mA with the switch closed. Find the radiation resistance and radiated power.

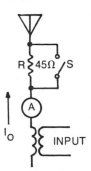

Problem 14-1.

Answer

Signal current I_O is the current flow that would dissipate the same energy as actually radiated from the antenna. Since resistor R is in series with the antenna when the switch is open, the radiated power would be

$$P = I_O{}^2 R =$$
$$= (15.5 \times 10^{-3})^2 (45)$$
$$= 240 \times 10^{-6} \times 45$$
$$= 10.8 \, \text{mW ANS}$$

261

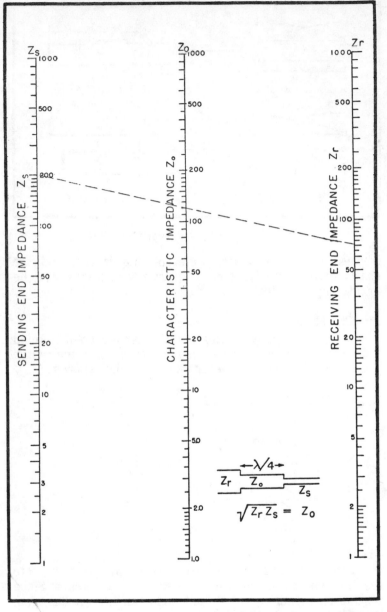

Fig. 14-8. Quarter-wave matching sections. When using a quarter-wave matching section as an impedance transfomer from one real impedance to another, the required characteristic impedance of the quarter-wave section is found by drawing a line between the sending end impedance and the desired receiving end impedance. (From Radio-Electronics, May 1950).

Knowing the radiation power, radiation resistance Z_O is found with the current value given for the switch closed:

$$Z_O = P/I_0{}^2$$
$$= (10.8 \times 10^{-3})/(44 \times 10^{-3})^2$$
$$= 5.58\Omega \text{ ANS}$$

PROBLEM 14-2

Calculate the signal-to-noise ratio at the input to the IF stage for each of the block diagrams shown. In each case, the induced antenna voltage contains 22 μV of signal and 0.8 μV of noise. The mixer provides a gain of 32 and produces an internal noise of 4.8 μV. Each RF stage provides a gain of 14 and produces 1 μV of noise.

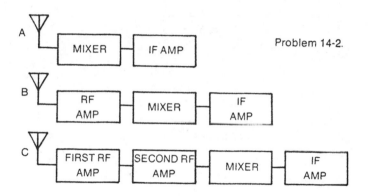

Problem 14-2.

Answer

(A) At the mixer:

$$\text{Signal} = (22\ \mu\text{V})\ (32) = 704\mu\text{V}$$
$$\text{Noise} = (0.8\ \mu\text{V})\ (32) = 25.6\ \mu\text{V}$$
$$\text{Stage noise} = 4.8\ \mu\text{V}$$
$$\text{Total noise} = 25.6 \times 4.8 = 30.4\ \mu\text{V}$$
$$\text{S/N} = 704/30.4 = 23.2\ \text{ANS}$$

(Also see noise nomograph in Fig. 10-5 for similar problems in which noise levels are to be calculated.)

(B) At the RF amplifier:

$$\text{Signal} = (22\ \mu\text{V})\ (14) = 308\ \mu\text{V}$$
$$\text{Noise} = (0.8\ \mu\text{V})\ (14) = 11.2\ \mu\text{V}$$
$$\text{Stage noise} = 1.0\ \mu\text{V}$$
$$\text{Total noise} = 11.2 \times 1.0 = 12.2\ \mu\text{V}$$
$$\text{S/N (RF amplifier)} = 308/12.2 = 25.2$$

At the mixer:

$$\text{Signal} = (308 \, \mu\text{V}) \, (32) = 9856 \, \mu\text{V}$$
$$\text{Noise} = (12.2 \, \mu\text{V}) \, (32) = 390 \, \mu\text{V}$$
$$\text{Mixer noise} = 4.8 \, \mu\text{V}$$
$$\text{Total noise} = 390 + 4.8 = 394.8 \, \mu\text{V}$$
$$\text{S/N (mixer)} = 9856/394.8 = 24.9 \, \text{ANS}$$

(C) At the first RF output:

$$\text{Signal} = 308 \, \mu\text{V}$$
$$\text{Noise} = 12.2 \, \mu\text{V}$$

At the second RF output:

$$\text{Signal} = (22 \, \mu\text{V}) \, (14)^2 = 4310 \, \mu\text{V}$$
$$\text{Noise} = (0.8 \, \mu\text{V}) \, (14)^2 = 163.5 \, \mu\text{V}$$
$$\text{Stage noise} = 1 + 1 = 2 \, \mu\text{V}$$
$$\text{Total noise} = 163.5 + 2 = 165.5 \, \mu\text{V}$$
$$\text{S/N} = 4310/165.5 = 26$$

At the mixer:

$$\text{Signal} = (22 \mu V) \, (14)^2 \, (32) = 138 \text{m}V$$
$$\text{Noise} = (0.8 \, \mu\text{V}) \, (14)^2 \, (32) = 5017.6 \, \mu\text{V}$$
$$\text{Mixer noise} = (1 + 14 + 1) \, (32) + 4.8 = 50.8 \, \mu\text{V}$$
$$\text{Total noise} = 5017.6 + 50.8 = 5.068 \, \text{mV}$$
$$\text{S/N} = 138/5.068 = 27.3 \, \text{ANS}$$

PROBLEM 14-3

A grounded vertical aerial is inductively fed at the base. The velocity of propagation is 95% of the velocity of light. The frequency is 2.8 MHz. What is the height of the aerial in meters if it is to be as tall as possible but not over 300 meters high?

Answer

$$\lambda \text{ (in free space)} = 300/f \text{ MHz}$$
$$= 300/2.8$$
$$= 107.1 \text{ meters}$$
$$\lambda \text{ (antenna)} = 107.1 \, (0.95) = 101.7 \text{ meters}$$

At the top of the antenna, the voltage is maximum and current is minimum. At the base, feed current is maximum. Therefore, the height (h) is an odd number of $\lambda/4$ multiples.

Taking $\lambda/4 = 101.7/4 = 25.4$ meters, then in 300 meters there are $300/25.5 = 11.8$ multiples of $\lambda/4$. Therefore, use 11 quarter wavelengths. The height is then

$$h = 11(\lambda/4) = 11(25.4) = 279.4 \text{ meters}$$

PROBLEM 14-4

The two vertical transmitting antennas shown are each one-half wavelength long and spaced one-half wavelength apart in free space.

(A) Draw a horizontal radiation pattern with respect to the location of the antennas when fed *in phase*

(B) Draw a horizontal radiation pattern as in (A) but when the antennas are fed 180° out of phase.

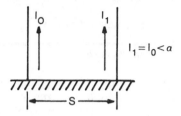

$I_1 = I_0 < \alpha$

Problem 14-4.

Answer   *(ref. 2, p. 804)*

The radiation patterns are shown in Fig. 14-9. (See Problem 14-7 for patterns of two other common situations.)

PROBLEM 14-5

The height of a vertical antenna is 617 ft. The operating frequency is 1590 kHz. Assume the velocity of propagation is 299 m/μsec. The antenna is grounded through a coupling network.

(A) What fraction of a wavelength is the height of the antenna?

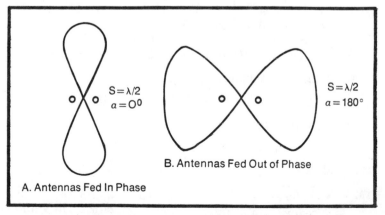

Fig. 14-9. Antennas radiation patterns.

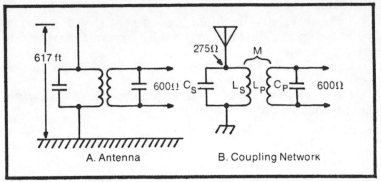

Fig. 14-10. Designing a coupling network.

(B) Sketch a coupling network to connect the transmission line from the broadcast station to the antenna. Assume that the antenna is equivalent to a 275Ω pure resistance. Select approximate values of inductance, capacitance, and mutual inductance if inductive coupling is used. Assume 600Ω transmission line and coupling networks with Q of 20.

Answer *(ref. 2 p. 153)*

(A) Refer to Fig. 14-10A.

$$\lambda = (299 \times 10^6 \text{ m/sec}) \text{..so} (1.59 \times 10^6 \text{ Hz})$$
$$= (984 \times 10^6 \text{ ft/sec})/(1.59 \times 10^6 \text{ Hz})$$
$$= 617 \text{ ft}$$

Therefore, the antenna is one wavelength in height.

(B) Coupling network is shown in Fig. 14-10B.

$$Q_{SEC} = \omega L_S / R_S$$

$$L_S = Q_S R_S / \omega$$
$$= 20(275)/6.28(1.59 \times 10^6)$$
$$= 550 \, \mu H$$
$$X_L = \omega L_S = 6.28(1.59 \times 10^6)(550 \times 10^{-6}) = 5.5 \, k\Omega$$

Knowing that at resonance $X_L = X_C$,

$$C_S = 1/\omega X_C = 10^{-6}/10(5500) = 18.2 \, pF$$

E_C/E is maximum when

$$k_C = 1/\sqrt{Q_p Q_S} = 1/20 = 0.05$$

Also, $\omega M = \sqrt{R_p R_S}$ so

$$M = \sqrt{R_p R_S}/\omega \qquad\qquad L_p = Q_p R_p/\omega$$
$$= \sqrt{600(275)}/10 \times 10^6 \qquad = 20(600)/10 \times 10^6$$
$$= 40.5 \, \mu H \qquad\qquad = 1200 \, \mu H$$

The coefficient of coupling

$$k = M/\sqrt{L_p \, L_S}$$
$$= 40.5 \times 10^{-6} /\sqrt{(1200 \times 10^{-6})(550 \times 10^{-6})}$$
$$= 0.05$$

Coupling is maximum when $k = k_C$, which is the case here.

PROBLEM 14-6

If two vertical half-wave antennas are spaced 3/4 wavelength apart and fed 90° out of phase, where will the nulls and peaks appear in the horizontal field pattern? Give directions measured in degrees from the axis of the array.

Answer *(ref. 1, p. 21-9; ref. 2, p. 789)*

$$d_R = (3\lambda/4)(2\pi/\lambda) = 3\pi/2$$

$$G_F(\phi) = k \cos[d_R \cos(\phi) + \delta]/2$$
$$= k \cos[(3\pi/2)\cos(\phi) + 90/2]$$
$$= k \cos[(3\pi/4)\cos(\phi) + 45]$$

Maximum value occurs when

$$(3\pi/4)\cos(\phi) + 45° = 2\pi$$

or

$$\cos(\phi) = (7\pi/4)(4/3\pi) = 2.33$$

so that

$$\phi = 134°$$

Thus, one maximum is at +134°, and since the pattern is symmetrical, the other maximum is at −134°.

One null appears at 180°. The other nulls appear where

$$(3\pi/4)\cos(\phi) + 45° = \pi/2$$
$$\cos(\phi) = (\pi/4)(4/3\pi) = 0.33$$
$$\phi = \pm19°$$

The field pattern is illustrated in Fig. 14-11.

PROBLEM 14-7

Two vertical radiators are over a perfect ground and carry equal currents differing only in phase shift. Sketch their radiation pattern in the horizontal plane for each of the following cases:

(A) The spacing is $\lambda/2$ and $\alpha = 0°$.
(B) The spacing is $\lambda/2$ and $\alpha = 180°$.
(C) The spacing is $\lambda/4$ and $\alpha = -90°$.
(D) The spacing is λ and $\alpha = 0°$.

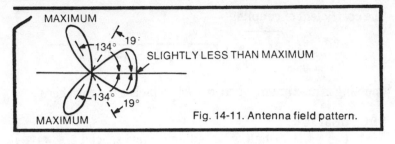

Fig. 14-11. Antenna field pattern.

Answer *(ref. 1, p. 21-13; ref. 2, p. 804)*

The radiation patterns are shown in Fig. 14-12.

PROBLEM 14-8

A broadcast station desires to increase its coverage in diametrically opposite directions without appreciably affecting the signal strength at right angles to this line. The present antenna is a vertical quarter-wavelength radiator. Design an antenna array to supplement the existing radiator, and draw a diagram indicating the antenna placements relative to the radiation pattern.

Answer *(ref. 2, p. 792; ref. 6, p. 25-23; ref. 32, p. 656—663)*

A diagram of the antenna placements is in Fig. 14-13.

$$A = (D/2) \sin(\theta)$$
$$\phi = \omega A/c \text{ (phase shift)}$$

where c is the speed of light. The phase shift is

$$\phi = (\omega D/2c) \sin(\theta)$$
$$= (\pi D/\lambda) \sin(\theta)$$

where $\lambda f = c$

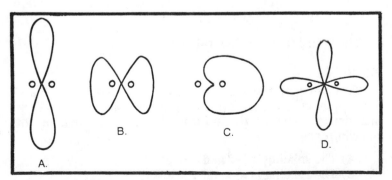

Fig. 14-12. Antennas radiation patterns in horizontal plane.

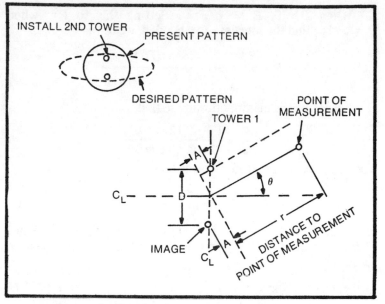

Fig. 14-13. Antenna tower placement.

Now, $E_1 = (k/r) \exp[j(\omega t + \phi)]$

$E_2 = (k/r) \exp[j(\omega t - \phi)]$

$E = E_1 + E_2$

$E = (2k/r)\epsilon^{j\omega t} [(\epsilon^{jd} + \epsilon^{-jd})/2]$

$\quad = (2k/r)\epsilon^{j\omega t} \cos(\phi)$

$E^2 = (4k^2/r^2) \cos^2(\phi)$

$\quad = (2k^2/r^2)[1 + \cos(2\phi)]$

$p = \int_0^{2\pi} E^2 (r^2/2) d\theta$

$\quad = k^2 \int_0^{2\pi} \{1 + \cos[(2\pi D/\lambda) \sin(\theta)]\} d\theta$

$\quad = 2\pi k^2 [1 + J_0 (2\pi D/\lambda)]$

Solving for k and substituting into the equation above for E^2,

$$E^2 = \frac{2P \cos^2 [(\pi D/\lambda) \sin(\theta)]}{\pi r^2 [1 + J_0 (2\pi D/\lambda)]}$$

269

When $D = 0$, $E^2 = p/\pi r^2$. Setting this to E^2 when $\theta = \pi/2$ to find the increase in p needed, we have

$$E^2 = \frac{2P \cos^2 (\pi D/\lambda)}{\pi r^2 [1 + J_0 (2\pi D/\lambda)]}$$

Let P_0 be the original power, so that

$$\frac{P}{P_0} = \frac{1 + J_0 (2\pi D/\lambda)}{2 \cos^2 (\pi D/\lambda)}$$

The increase in radiated power is then

$$\Delta P = \frac{1}{\cos^2 (\pi D/\lambda)}$$

so that $\pi D/\lambda = \pi/4$, or $D = \lambda/4$ wavelength. Thus, $P/P_0 \approx 1.52$ or 52% greater radiation in the desired direction.

Alternate Answer *(ref. 1, p. 21-17)*

$$E_\beta = 60\, I/r \left[\frac{\cos[(2\pi l/\lambda) \sin(\beta)] - \cos(2\pi l/\lambda)}{\cos(\beta)} \right]$$

where I = RMS value of current at maximum input
β = elevation angle from horizontal
l/λ = antenna height in wavelengths above the ground plane;
r = distance to the measuring point

Refer to Fig. 14-14. By changing the length of the antenna, the pattern can be altered to suit the desired coverage. When the antenna is grounded, $\lambda/8$ or less, radiation is approximately equal to the cosine of the angle of elevation.

PROBLEM 14-9
Find the impedance and admittance of a slotted line, using a Smith Chart. For the slotted line, $V_{MIN}/V_{MAX} = 0.6$ and $x/\lambda = 0.4$, where x is the distance from the load to the first V_{MIN}, and λ is the line wavelength.

Answer
Refer to Fig. 14-15.
(1) Using a straightedge, pivot about the resistive component 1.0 at the center of the Smith Chart. Rotate the straightedge toward the load from the zero mark at the top of the chart, continuing counterclockwise until 0.4 is reached, which is x/λ. Draw a line across the chart through the center.

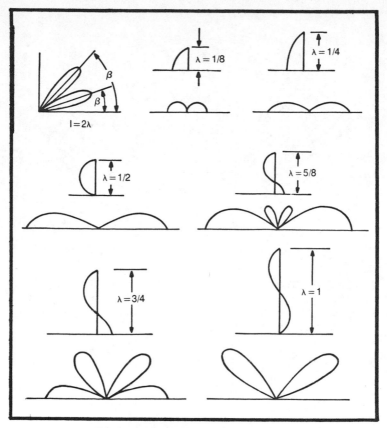

Fig. 14-14. Tailoring antenna patterns.

(2) Since $V_{MIN}/V_{MAX} = 0.6$, to determine the impedance, move up the chart, along the resistive component line, and locate 0.6. Again, using the 1.0 as the pivotal point, use a compass to draw a circle that passes through the point 0.6, counterclockwise till it intersects the line for 0.4.

(3) The impedance is determined by noting the point at which the circle intersects the line drawn toward the 0.4 mark. Read the resistive component, 0.77, by moving along the arc to the point where it intersects the resistive component axis. Read the imaginary component, $+j0.39$, by moving outward along the arc to the point where it intersects the positive reactance component axis. The impedance is then $Z = Z_0 (0.77 + j0.39)$ ANS.

(4) Since $V_{MIN}/V_{MAX} = 0.6$, then $V_{MAX}/V_{MIN} = 1.67$. The procedure for finding admittance is identical to that described in step (2), except that the circle is drawn through

271

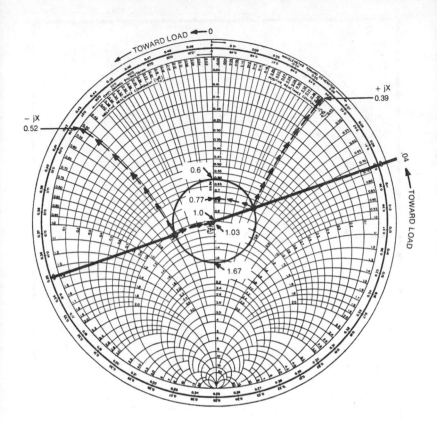

Fig. 14-15. Smith Chart.

1.67, rather than 0.6, counterclockwise till the circle intersects the line opposite the point found in step (2). (The circle need not actually be drawn in this case because the circles have the same radius.)

(5) The admittance is determined by noting the intersection point described in step (4). Following the procedure outlines in step (3), the resistive component is found to be 1.03 and the reactive component, $-j0.52$. The admittance is then $Y = (1/Z_0) = (1.03 - j0.52)$ **ANS.**

PROBLEM 14-10

Calculate the electric field intensity and potential at point P, which is equally spaced between the three charges shown in the diagram.

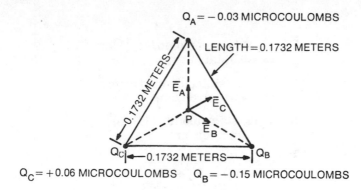

Q_A = −0.03 MICROCOULOMBS

$Q_A = -0.03$ MICROCOULOMBS

LENGTH = 0.1732 METERS

0.1732 METERS

$\overline{E}_A$

E_C

P

$\overline{E}_B$

Q_C ← 0.1732 METERS → Q_B

$Q_C = +0.06$ MICROCOULOMBS $Q_B = -0.15$ MICROCOULOMBS

Problem 14-10.

Answer

Assume that the three field intensities are $\overline{E}_A$, $\overline{E}_B$, and E_C. Noting the signs of the three charges, these field intensities would point in the directions shown.

Knowing that a stat coulomb is $e\ (3 \times 10^9)$,

$$Q_A = (-0.03 \times 10^{-6})(3 \times 10^9) = -90 \text{ statcoulombs}$$
$$Q_B = (-0.15 \times 10^{-6})(3 \times 10^9) = -450 \text{ statcoulombs}$$
$$Q_C = (0.06 \times 10^{-6})(3 \times 10^9) = 180 \text{ statcoulombs}$$
$$\overline{E}_A = Q_A / (kr_A{}^2)$$

where Q_A = charge in statcoulombs, as above
   $\overline{E}_A$ = field intensity in statvolts/cm
   k = 1 in air
   r_A = distance from Q_A to p.

Being an equilateral triangle. (See Fig. 14-16),

$$h = \frac{17.32 \text{ cm}}{2 \cos(\theta)} = 10 \text{cm}$$

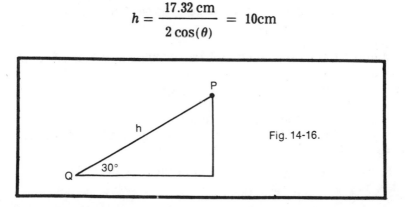

Fig. 14-16.

Now,

$$\overline{E}_A = -90 \text{ statcoulomb}/(1)(10\text{cm})^2$$
$$= -0.9 \text{ statvolt/cm}$$

$$\overline{E}_B = -450 \text{ statcoulomb}/(10\text{cm})^2$$
$$= -4.5 \text{ statvolt/cm}$$

$$\overline{E}_C = 180 \text{ statcoulomb}/(10 \text{ cm})^2$$
$$= 1.8 \text{ statvolt/cm}$$

$$\overline{E}_p = Q_p /k r$$
$$= (-450 - 90 + 180)/(1)(10)$$
$$= 36 \text{ statvolt/cm ANS}$$

PROBLEM 14-11

A uniform electric field of 10^4 volts/meter is parallel to a magnetic field of 5 milliwebers/meter^2. A charge of $+e$ is placed in the field with the following initial conditions at $t = 0$: coordinates $x = y = z = 0$; $V_Y = V_Z = 0$; $V_X = 5.93 \times 10^6$ meters/sec. Find the position of the particle at $t = 21.42 \times 10^{-9}$ sec if $e = 1.602 \times 10^{-19}$ coulombs and $m = 9.1085 \times 10^{-31}$ kilogram.

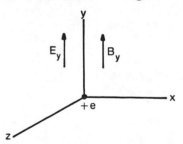

Problem 14-11.

Answer

The force exerted on a charge by an electric field is

$$\mathbf{F} = e\mathbf{E}$$

And the force exerted on a charge moving at velocity $\mathbf{V}$ in a magnetic field is

$$\mathbf{F} = e\mathbf{V} \times \mathbf{B}$$

where $\mathbf{F}$, $\mathbf{E}$, $\mathbf{V}$, and $\mathbf{B}$ are vector quantities.

(A) During time t, the charge will travel in the x direction a distance

$$X = V_X t = (5.93 \times 10^6)(21.42 \times 10^{-9})$$
$$= 0.127\text{m ANS}$$

(B) The electric field will at the same time have accelerated the charge in the y direction a distance

$$y = \frac{eE_y\, t^2}{2m} = \frac{(1.602 \times 10^{-19})\,(10^4)\,(21.42 \times 10^{-9})^2}{(2)\,(9.1085 \times 10^{-31})}$$

$$= 0.403\text{m ANS}$$

(C) Since the magnetic field is in the y direction while the charge is moving in the x direction, the vector product will be in the z direction. The position of the charge will then be

$$Z = \frac{eV_X\, B_y\, t^2}{2m}$$

$$= \frac{(1.602 \times 10^{-19})\,(5.93 \times 10^6)\,(5 \times 10^{-3}) = (21.42 \times 10^{-9})^2}{(2)\,(9.1085 \times 10^{-31})}$$

$$= 1.196\text{m ANS}$$

PROBLEM 14-12

A rectangular wave guide is 12 cm by 18 cm. The electric intensity is in the direction of the smaller dimension.

(A) Assuming the simplest mode of propagation, determine the lowest frequency that can be transmitted through the guide.

(B) What is the phase constant in radians per meter?

(C) Sketch a method of inserting or extracting energy from the wave guide having this mode of excitation.

Answer *(ref. 1, p. 20-36; ref. 2, p. 256; ref. 32, p. 596)*

(A) Refer to Fig. 14-17A.

$$f = c/\lambda = 3 \times 10^4/\lambda$$
$$\lambda_C = 2a = 2(18) = 36\text{cm}$$
$$f_{CO} = 3 \times 10^4/36 = 833\text{ MHz ANS}$$

However, waveguides are rarely used at their cutoff frequency. The lowest recommended operating frequency should be

$$2 f_{CO} = 2(833) = 1.66\text{ GHz}$$

But $4f_{CO}$, or 3.3 GHz, is recommended.

(B) Phase constant:

$$\phi_C = \beta m \pi^2$$

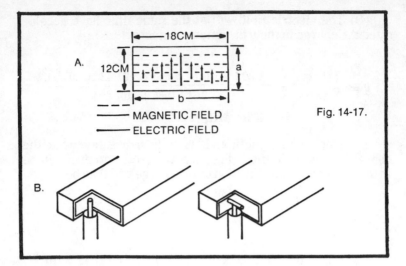

Fig. 14-17.

MAGNETIC FIELD
ELECTRIC FIELD

Since

$$\lambda_C = \frac{2}{\sqrt{(m/a)^2 + (n/b)^2}}$$

$$(2/\lambda_C)^2 = m^2/a^2 + n^2/b^2$$

The phase constant is

$$\begin{aligned}
\phi_C &= \beta mn^2 \\
&= \omega^2/c^2 - \pi^2(2/\lambda_C)^2 \\
&= \frac{(2\pi)^2(1.67 \times 10^9)^2}{(3 \times 10^{10})^2} - \pi^2(2/36)^2 \\
&= 0.1223 - 0.03 \\
&= .0923 \, \text{rad/cm} \\
&= 9.23 \, \text{rad/m} \textbf{ ANS}
\end{aligned}$$

(C) A method of inserting or extracting energy is sketched in Fig. 14-17B.

PROBLEM 14-13

A lossless transmission line having a characteristic impedance of 50Ω feeds power to an antenna. The continuous-wave power absorbed by the antenna is 1000W and the VSWR is 2.5. What will be:

(A) The minimum instantaneous voltage between lines?
(B) The maximum instantaneous voltage between lines?
(C) The minimum RMS voltage between lines?
(D) The maximum RMS voltage between lines?

Answer   *(ref. 1, p. 20; ref. 6, p. 22-4)*

Refer to Fig. 14-18.

$$P_O = E_{MAX} \, E_{MIN} / Z_O$$

$$P_O \, Z_O = 1000(50) = 50$$

$$\frac{E_{MAX}}{E_{MIN}} = 2.5$$

$$E_{MIN} = E_{MAX} / 2.5 \text{ ANS}$$

(D) The maximum RMS voltage is

$$P_O \, Z_O = E_{MAX} \, (E_{MAX} / 2.5)$$

$$\begin{aligned}
E_{MAX} &= \sqrt{2.5 P_O \, Z_O} \\
&= \sqrt{2.5(50,000)} \\
&= 354\text{V ANS}
\end{aligned}$$

(C) The minimum RMS voltage is

$$E_{MIN} = 354/2.5 = 141.5\text{V ANS}$$

(A) Assuming that the phases of the line voltages coincide, the minimum instantaneous voltage between lines is

$$e_{MIN} = \sqrt{2} \, (r E_{MIN})$$

Being a lossless line, $r = 1$. Therefore,

$$e_{MIN} = \sqrt{2}(141.5) = 200\text{V ANS}$$

(B) The maximum instantaneous voltage is

$$\begin{aligned}
e_{MAX} &= \sqrt{2} \, (r E_{MAX}) \\
&= \sqrt{2}(354) \\
&= 501\text{V ANS}
\end{aligned}$$

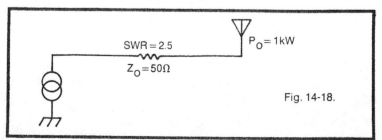

Fig. 14-18.

PROBLEM 14-14

The propagation constant of a telephone line operating at 1000 Hz is $0.005 + j0.349$, where the real part is in nepers/mile and the complex part is in radians/mile. The line is 400 miles long. The input voltage is 3.32V at 1000 Hz.

(A) What is the wavelength in miles?

(B) If the output is terminated in the characteristic impedance of the line, what will be the value of the output voltage?

Answer

(A) The propagation constant is of the form

$$\gamma = \alpha + j\beta$$
$$= 0.005 + j0.349$$

The wavelength of the line is

$$\lambda = 2\pi/\beta$$
$$= 2\pi/0.0349$$
$$= 180 \text{ miles ANS}$$

(B) The attenuation for a 400-mile line is described by the equation

$$E_{IN}/E_{OUT} = \exp(400\alpha + 400\beta j)$$
$$= \exp(400\alpha)\exp(400\beta j)$$

The imaginary term merely contributes a phase shift, so it can be ignored in determining the attenuation. For an input of 3.32 volts, the output is

$$E_{OUT} = E_{IN}\exp(-400 \times 0.005)$$
$$= 3.32\exp(-2)$$
$$= 0.449\text{V ANS}$$

PROBLEM 14-15

A DC ground-return telephone line is 80 miles long and has a resistance of 10Ω per mile and a shunt conductance of 1×10^{-5} mho per mile. A relay having a 500Ω coil is to be located at the receiving end, and it requires 100V for operation. What input voltage is required at the sending end to operate the relay?

Answer

Refer to Fig. 14-19A. The current required to operate the relay at the receiving end is

$$I_R = 100\text{V}/500\Omega = 0.2\text{A}$$

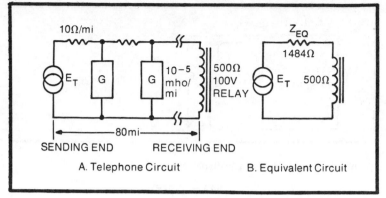

Fig. 14-19.

The reactance per mile is $1/G = 1/10^{-5} = 100k\Omega$. So the total reactance per 80 miles is then

$$X_T = 10^5/80 = 1250\Omega$$

And the total resistance is

$$R_T = (10\Omega/\text{mi})\ (80\ \text{mi}) = 800\Omega$$

Now Z_{EQ} can be calculated and placed in series with the load:

$$Z_{EQ} = \sqrt{R_T{}^2 + X_T{}^2}$$
$$= \sqrt{(800)^2 + (1250)^2}$$
$$= 1484\Omega$$

Now Z_{EQ} is drawn (see Fig. 14-19B).

The total series circuit is then $1485 + 500 = 1985\Omega$, and the total voltage required at the sending end is

$$V_T = (0.2A)\ (1984\Omega) = 397V\ \text{ANS}$$

PROBLEM 14-16

The input impedance of a 95-mile telephone line is substantially 680Ω pure resistance. If 1.5V at 1000 Hz is impressed across this line, calculate the current through the distant termination of 680Ω if the line loss is 0.04 decibels per mile.

Answer

Refer to Fig. 14-20.

$$S = (0.04\ \text{dB/mi})\ (95\ \text{mi}) = 3.8\ \text{dB}$$

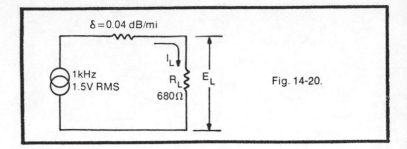

δ = 0.04 dB/mi

1kHz
1.5V RMS

I_L

R_L
680Ω

E_L

Fig. 14-20.

which corresponds to a voltage ratio of 1.549 to 1.

$$E_L = 1.5/1.549 = 0.968V$$
$$I_L = 0.968/680\Omega = 1.42 \text{ mA ANS}$$

PROBLEM 14-17

The maximum input power to a 600Ω line is +30 dB referred to a 0.006W level. It is desired to indicate the input power to this line on a thermal ammeter calibrated in decibels.

(A) What should the maximum current rating be for this instrument?

(B) If the deflection of the meter needle is proportional to the square of the current, what would the decibel indication be at midscale?

Answer

(A) See Fig. 14-21. A +30 dB level corresponds to a power ratio of 1000 to 1, so the maximum power input is 6 watts. This power is equal to

$$P = I^2 Z \cos(\theta)$$

where $Z = 600\Omega$, the impedance of the line. Assuming a resistive circuit, $\cos(\theta) = 1$. Then at full scale, the maximum ammeter current should be

$$I_A = \sqrt{P/Z}$$
$$= \sqrt{6/600}$$
$$= 0.1 \text{ A ANS}$$

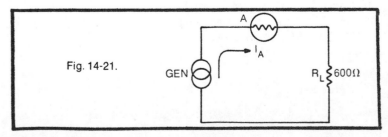

Fig. 14-21.

A

I_A

GEN

R_L 600Ω

(It would, however, be desirable to allow a 25 to 50 percent safety factor for burnout protection.)

(B) Since the meter deflection is proportional to the square of the meter current, let the full-scale deflection reading be $kI_F{}^2 = 1$, where k is a proportionality constant. Then at full scale,

$$k = 1/I_F{}^2$$
$$= 1/(0.1)^2$$
$$= 100$$

At midscale, the reading should be $kI_M{}^2 = 0.5$, so

$$I_M = \sqrt{0.5/k}$$
$$= \sqrt{0.5/100}$$
$$= 0.0707A$$

A current ratio of 0.1 to 0.0707 corresponds to a power change of 3.01 dB, so the midscale power reading of the ammeter would be

$$P_M = (30\ \text{dB}) - (3\ \text{dB})$$
$$= 27\ \text{dB ANS}$$

PROBLEM 14-18

A certain open-wire line has the following constants per mile of line: $R = 11.9\Omega$, $C = 0.0084\ \mu F$, $L = 0.00376H$, and $g = 1 \times 10^{-6}\ \Omega$. What should the value of the terminating impedance be of such a line in order to minimize reflection losses? The angular velocity of the impressed voltage is 5000 rad/sec.

Answer

To minimize reflections, the line must be terminated in its characteristic impedance, which, ignoring losses, is approximately

$$Z_O = \sqrt{L/C}$$
$$= \sqrt{(3.76 \times 10^{-3})/8.4 \times 10^{-9})}$$
$$= 669\Omega$$

For more accuracy, use

$$Z_O{}^2 = (R + j\omega L)/(g + j\omega C)$$
$$= (11.9 + j18.8)/(1 \times 10^{-6} + j4.2 \times 10^{-5})$$
$$= (22.25\ \angle\ 57.67)/(4.201 \times 10^5\ \angle 88.64)$$
$$= 5.296 \times 10^5\ \angle -30.97$$

Therefore, for a resistive termination,

$$\boxed{Z_O} = \sqrt{5.296 \times 10^5}$$
$$= 727.7\Omega\ \text{ANS}$$

and for an ideal termination,

$$Z_O = 727.7 \angle -(-30.97/2)$$
$$= 727.7 \angle 15.49$$
$$= 710 + j194 \text{ ANS}$$

where the complex phase angle has been reversed in sign as required to neutralize phase shift.

PROBLEM 14-19

A telephone line consists of two hard-drawn copper wires having a diameter of 165 mils and resistivity of 10.7Ω per circular-mil foot at 20°C. The center-to-center separation of the wires is 12 inches, and the length of the line is 23.3 miles. The operating temperature is 20°C. The line is terminated in a telephone receiver having a nearly pure resistance of 620Ω.

(A) What is the length of the line in wavelengths for a frequency of 1000 Hz?

(B) If a sine-wave voltage of 1000 Hz and one-volt RMS value is applied to the input end, will there be an appreciable reflection—and therefore an appreciable standing wave-along the line?

Answer

(A) One wavelength along the line is

$$\lambda = \frac{984 \times 10^3 \text{ ft}}{f(\text{kHz})} = \frac{984 \times 10^3 \text{ ft}}{(1)\ (5280 \text{ ft/mi})} = 186 \text{ mi}$$

Therefore

$$l/\lambda = (23.3 \text{ mi})/(186 \text{ mi}) = 0.125 \text{ ANS}$$

and the line is then 1/8 wavelength long.

(B) From resistance vs wire diameter tables, the area of wire is 27×10^3 circular mils (CM). So the resistance of wire is

$$\rho l/A = \frac{(10.7)\ (23.3 \text{ mi})\ (5280 \text{ ft/mi})}{27 \times 10^3 \text{ CM}} = 49\Omega$$

which should cause few problems in the circuit (see Fig. 14-22A).

The impedance of the open-wire lines (Fig. 14-22B) is

$$Z_O = 276 \log(2D/d)$$
$$= 276 \log (24/0.165)$$
$$= 276(2.161)$$
$$= 597\Omega$$

282

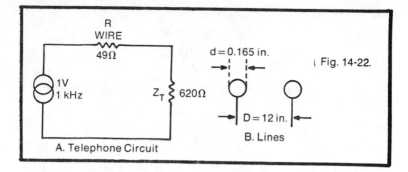

Fig. 14-22.

A. Telephone Circuit

B. Lines

Knowing that $Z_T = 620\Omega$, the VSWR is

$$V_{MAX}/V_{MIN} = Z_T/Z_O$$
$$= 620/597$$
$$= 1.039$$

Therefore the VSWR is nearly unity, and there will be no appreciable reflections along the line.

PROBLEM 14-20

A lossless transmission line has a characteristic impedance of 300Ω and is a quarter wavelength long. What will the voltage be at the open-circuited receiving end when the sending end is connected to a generator with 50Ω internal impedance and output of 10V?

Answer   *(ref. 1, p. 20-10)*

Refer to Fig. 14-23. For open-circuit terminations, the input impedance is capacitive for $0< X <\lambda 4$, where X is the length. For a quarter-wave line

$$Z_{IN} = Z_O^{\,2}/Z_R$$

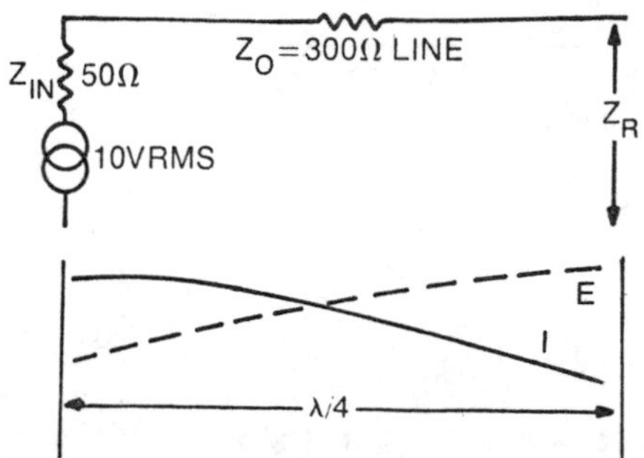

$Z_O = 300\Omega$ LINE

Fig. 14-23.

so

$$Z_R = ZO^2/Z_{IN}$$
$$= (300)^2/50$$
$$= 1800\Omega$$

The maximum power available at the generator is

$$P_{IN} = E^2/R$$
$$= (10)^2/50$$
$$= 2W$$

At the receiving end of the line, the voltage obeys the formula

$$P_{IN} = E_{MAX} E_{MIN}/Z_O$$

so

$$E_{MAX} = P_{IN} Z_O/E_{MIN}$$

$$= (2)(300)/10$$

$$= 60V \text{ RMS } \textbf{ANS}$$

PROBLEM 14-21

In a lossless transmission line, the following measurements were noted, using a 10-foot section of RG58G/U at 100 MHz. Short-circuit measurements: $L_{SC} = 0.1 \ \mu H$ and $R_S = 2\Omega$. Open-circuit measurements: $C_{OC} = 50$ pF and $G_O = 400 \ \mu$mhos.

Find Z_O, α, and β.

Answer

(A) For a lossless line,

$$Z_O = \sqrt{L_{SC}/C_{OC}}$$
$$= \sqrt{(0.1 \times 10^{-6})/(50 \times 10^{-12})}$$
$$= 45\Omega \text{ ANS}$$

(B) The attenuation coefficient in decibels per 100 ft is

$$\alpha = 434/10\left[\frac{1}{1 + (2\pi f)^2 \ L_{SC} \ C_{OC}}\right]2/45 + 45 \ G_O$$

$$= (43.4)(0.044 + 0.018)/(1 + 1.974)$$
$$= 0.905 \text{ dB}/100 \text{ ft}$$

(C) $\tan(\beta l) = \omega\sqrt{L_{SC}\, C_{OC}}$

$\qquad = 2\pi f\sqrt{(0.1 \times 10^{-6})\,(50 \times 10^{-12})}$

$\qquad = 1.405$

$\beta l = 54.6°$

$\beta = 54.6/l$

$\qquad = 54.6/10$

$\qquad = 5.46°/\text{ft}$ **ANS**

PROBLEM 14-22

Derive the expression for the Q of an open-end, half-wave transmission line.

Answer   *(ref. 1, p. 20-26; ref. 6, p. 22-13)*

Refer to Fig. 14-24.

$$1/Q = 2\alpha/\beta = R/\omega L + G/\omega C$$

where $\beta = \omega\sqrt{LC}$ in radians per unit length

$\alpha = R/2Z_0 + GZ_0/2$ in nepers per unit length

If $R \approx \omega L$ or $G \approx \omega C$, then $\alpha = 4.34\,R/Z_0$, where R is the resistance per unit length.

PROBLEM 14-23

Compute the line constants per unit length (R, L, C, G, α, β, and Z_0) for each of the following lines:

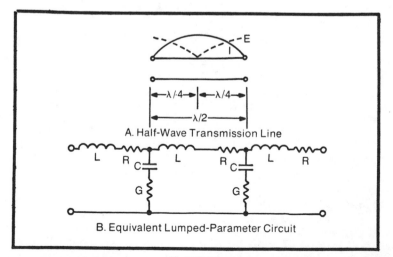

A. Half-Wave Transmission Line

B. Equivalent Lumped-Parameter Circuit

Fig. 14-24.

285

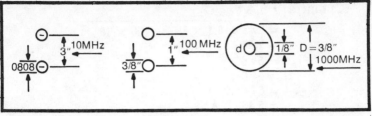

Fig. 14-25.

(A) Number 12 wires (0.0808 in. dia.) spaced 3 in. apart at 10 MHz.

(B) Two rods, 3/8 inch in diameter, spaced 1 inch apart at 100 MHz.

(C) A coaxial line having a 1/8 inch diameter inner conductor and a 3/8 inch diameter outer conductor, operated at 1000 MHz.

Answer

Refer to Fig. 14-25.

(A) Spaced wires:

$$R = \rho l / A$$
$$= (10.7)\ (1000\text{ft}) / (80.8\ \text{mils})^2$$
$$= 1.67 \Omega \text{ per 1000 ft}$$
$$L = 0.281 \log (2D/d) \textbf{ ANS}$$

where D is the distance between wires and d is the diameter of the wire.

$$L = 0.281 \log [2(3)/0.0808]$$
$$= 0.281\ (1.87)$$
$$= 0.523 \mu\text{H/ft ANS}$$

$$C = 3.677 / \log(2D/d)$$
$$= 3.677 / 1.87$$
$$= 1.97\ \text{pF/ft ANS}$$
$$G = 2\pi f C \tau$$

where τ is the power factor. If $\tau = 0$, then $G = 0$, and this is true when the insulation is air. Otherwise, $Y = G + j\beta$, and if purely resistive,

$$R/\omega L = G/\omega C$$

Therefore

$$G = RC/L$$
$$= (1.67 \times 10^{-3})\ (1.96 \times 10^{-12}) / 0.51 \times 10^{-6}$$
$$= 0.0064\ \mu\text{mhos/ft ANS}$$

$$Z_O = 276 \log (2D/d)$$
$$= 276 (1.87)$$
$$= 516\Omega \text{ ANS}$$

(Or, use the nomograph as shown in the review section to find Z_O .)

$$\alpha = R/2Z_O + GZ_O /2$$
$$= 1.67 \times 10^{-3} /2(515) + 0(515)/2$$
$$= 1.62 \times 10^{-6} \text{ neper/ft}$$
$$= (1.62 \times 10^{-6}) (8.686)$$
$$= 14.1 \times 10^{-6} \text{ dB/ft ANS}$$

For small attenuation,

$$\alpha = 0.5R\sqrt{C/L} + 0.5G\sqrt{L/C}$$
$$= 0.5(1.67 \times 10^{-3})\sqrt{(1.96 \times 10^{-12})/(0.51 \times 10^{-6})}$$
$$+ 0.5(0)\sqrt{(0.51 \times 10^{-6})/(1.96 \times 10^{-12})}$$
$$= 1.64 \times 10^{-6} \text{ neper/ft}$$
$$= 14.2 \times 10^{-6} \text{ dB/ft ANS}$$

At radio frequencies when the skin effect is great, and neglecting the proximity effect,

$$\alpha = \frac{3.62 \times 10^{-6} \sqrt{f}}{d \log (2D/d)}$$

$$= \frac{3.62 \times 10^{-6} \sqrt{10}}{0.0808 (1.87)}$$

$$= 75.7 \times 10^{-6} \text{ neper/ft}$$
$$= 664 \times 10^{-6} \text{ dB/ft ANS}$$

This does not take into account spacing.

Finally,

$$\beta = 2\pi f\sqrt{CL}$$
$$= 2\pi(10 \times 10^6)\sqrt{(0.51 \times 10^{-6}) (1.96 \times 10^{-12})}$$
$$= 62.8 \times 10^{-3} \text{ rad/ft.}$$

(B) Spaced rods:

$$L = 0.281 \log (2D/d)$$
$$= 0.281 \log (2/0.375)$$
$$= 0.281 (0.728)$$
$$= 0.2046 \mu\text{H/ft ANS}$$

$$C = 3.677/\log(2D/d)$$
$$= 3.677/0.728$$
$$= 5.05 \text{ pF/ft ANS}$$

$$R = \rho l/d^2$$
$$= 10.72 \,(1000 \text{ ft})/(375 \text{ mils})^2$$
$$= 0.0762 \,\Omega/1000\text{ft ANS}$$

$$G = 0 \text{ ANS}$$

$$Z_0 = 276 \log(2D/d)$$
$$= 276 \,(0.728)$$
$$= 201\Omega \text{ ANS}$$

Since $G = 0$,
$$\alpha = R/2Z_0$$
$$= (0.0762 \times 10^{-3})/2(201)$$
$$= 0.1896 \times 10^{-6} \text{ neper/ft}$$
$$= 1.646 \times 10^{-6} \text{ dB/ft ANS}$$

For very small α,

$$\alpha = 0.5R\sqrt{C/L}$$
$$= 0.5(0.0762 \times 10^{-3})\sqrt{(5.05 \times 10^{-12})/(0.2046 \times 10^{-6})}$$
$$= 0.1893 \times 104\text{su}-6 \text{ neper/ft}$$
$$= 1.644 \times 10^{-6} \text{ dB/ft ANS}$$

$$\beta = 2\pi f\sqrt{LC}$$
$$= 2\pi(100 \times 10^6)\sqrt{(0.2046 \times 10^{-6})\,(5.05 \times 10^{-12})}$$
$$= 0.639 \text{ rad/ft ANS}$$

$$Z_0 = 138 \log(D/d)$$
$$= 138 \log(0.375/0.125)$$
$$= 138(0.477)$$
$$= 65.8\Omega \text{ ANS}$$

(C) Coaxial line:

$$L = 0.140 \log(D/d)$$
$$= 0.140(0.477)$$
$$= 0.0668 \,\mu\text{H/ft ANS}$$

$$C = 7.35/\log(D/d)$$
$$= 7.35/0.477$$
$$= 15.4 \text{ pF/ft ANS}$$

The resistance R of the inner conductor, from wire tables, is $0.641\Omega/1000$ ft. Or, calculating as in parts (A) and (B),
$$R = \rho l/A$$
$$= 10.72(1000\text{ft})/(125 \text{ mils})^2$$
$$= 0.686\Omega/1000\text{ft ANS}$$
$$G = 0 \text{ ANS}$$

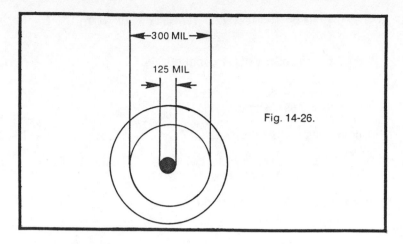

Fig. 14-26.

$$\alpha = R/2Z_O$$
$$= 0.686 \times 10^{-3}/2(65.8)$$
$$= 5.21 \times 10^{-6} \text{ neper/ft}$$
$$= 45.3 \times 10^{-6} \text{ dB/ft } \textbf{ANS}$$

$$\beta = 2\pi f\sqrt{LC}$$
$$= 2\pi(1000 \times 10^6)\sqrt{(0.0668 \times 10^{-6})(15.4 \times 10^{-12})}$$
$$= 6.37 \text{ rad/ft } \textbf{ANS}$$

PROBLEM 14-24

A coaxial cable has an internal copper conductor with 125 mil radius, and an external copper conductor with 350 mil inside radius. The copper has 97% conductivity. The operating temperature is 20°C.

(A) What is the DC resistance in ohms of one mile of the inner conductor?

(B) What is the DC resistance in ohms of one mile of the external conductor?

(C) What is the inductance of the coaxial cable in microhenries per mile?

(D) What is the capacitance of the coaxial cable in picofarads per mile?

(E) What is the characteristic impedance in ohms at very high frequency?

Answer   *(ref. 6, sec. 22)*

Refer to Fig. 14-26.

(A) The inner conductor is 0.125 inch in diameter. This is No. 8 wire, which has a resistance of 0.6282 Ω/1000 ft., from the wire tables. Resistance per mile is

$$R = (0.6282\Omega/10^3 \text{ ft}) (5.28 \times 10^3 \text{ ft/mi})$$
$$= 3.32\Omega/\text{mi ANS}$$

(B) The area of the outer conductor is

$$A = \pi(0.35 - 0.25) (0.3)$$
$$= 0.09425 \text{ sq in.}$$

Converting to circular mils (CM),

$$A = (0.09425 \text{ sq in.}) (1.27 \times 10^6 \text{ CM/sq in.})$$
$$= 119.7 \times 10^3 \text{ CM}$$

Assuming a 75% effective mesh,

$$A = 0.75(119.7 \times 10^3 \text{ CM})$$
$$= 89.78 \times 10^3 \text{ CM}$$

Resistance per mile is then

$$R = \rho l/A$$
$$= (10.37 \times 10^{-3}) (5.28 \times 10^3)/89.78$$
$$= 0.6097\Omega/\text{mi ANS}$$

(E) Because Z_0 is needed to solve parts (C) and (D), solve next for Z_0, which can be found either by using the nomograph as shown in the review section or by calculation:

$$Z_0 = 138 \log (D/d)$$
$$= 138 \log(0.25/0.125)$$
$$= 42.4\Omega \text{ ANS}$$

(C) Inductance:

$$L = 1.016 R_0 \sqrt{\epsilon}(10^{-3})$$

where R_0 is the resistive component of Z_0 and ϵ in air is equal to 1.

$$L = 1.016 (42.4) (10^{-3})$$
$$= 0.0431\mu\text{H/ft}$$
$$= (0.0431\mu\text{H/ft}) (5.28 \times 10^3 \text{ ft/mi})$$
$$= .225\mu\text{H/mi. ANS.}$$

(D) Capacitance:

$$C = 1.016(\sqrt{\epsilon}/R_0) (10^{-3})$$
$$= 1.016(1/42.4) (10^{-3})$$
$$= 2.396 \times 10^{-5} \mu\text{F/ft}$$
$$= (2.396 \times 10^{-5} \mu\text{F/ft}) (5.28 \times 10^3 \text{ ft/mi})$$
$$= 0.1265 \mu\text{F/mi ANS}$$

Chapter 15

Transient Analysis

Transient analysis involves the calculation of time-varying parameters in electrical circuits, as opposed to the more simple calculation of steady-state parameters. The most common problems involve the determination of the voltages and currents arising in response to a step input function.

The following review covers the basic circuits used for transient analysis, and in particular the response of fundamental filter circuits. For further review, see references 4, 5, 7, 12, 48, 55, and 56.

THE CHARGING OF A CAPACITOR

Figure 15-1 illustrates the basic capacitor charging circuit. When switch SW is closed, at time $t = 0$, the capacitor begins to charge. The charge, current, and voltage relationships are also shown in the figure.

THE DISCHARGE OF A CAPACITOR

Figure 15-2 shows the basic capacitor circuit during the discharge phase. When switch SW is closed, at time $t = 0$, the capacitor discharges. The charge, current, and voltage relationships are shown in the figure.

THE CHARGING OF AN INDUCTOR

Figure 15-3 shows the charging response of an inductor circuit. When switch SW is closed, at $t = 0$, the inductor begins to charge. The current and voltage relationships are given.

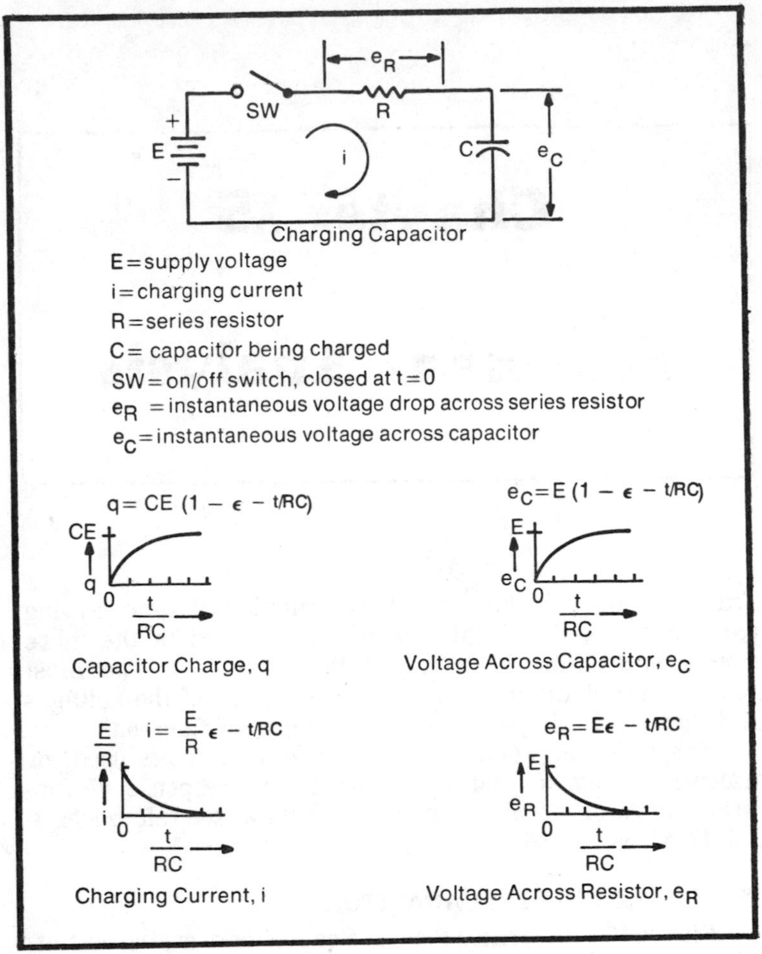

eR
SW
R
+
E
–
i
C
eC
Charging Capacitor
E=supply voltage
i=charging current
R=series resistor
C= capacitor being charged
SW=on/off switch, closed at t=0
eR =instantaneous voltage drop across series resistor
eC=instantaneous voltage across capacitor
q= CE (1 – ϵ – t/RC)
CE
q
0
t
RC
Capacitor Charge, q
eC=E (1 – ϵ – t/RC)
E
eC
0
t
RC
Voltage Across Capacitor, eC
E
R
i= E/R ϵ – t/RC
i
0
t
RC
Charging Current, i
eR= Eϵ – t/RC
E
eR
0
t
RC
Voltage Across Resistor, eR

Fig. 15-1. Charging capacitor characteristics.

THE DISCHARGING OF AN INDUCTOR

Figure 15-4 shows the discharge characteristic of an inductor circuit. When switch SW is closed, at $t = 0$, the inductive field collapses. The discharging current and voltage relationships are shown in the figure.

TYPICAL STEP FUNCTION RESPONSES

The text and Table 15-1 that follow are provided by Hewlett-Packard and serve to summarize the typical step function responses together with the amplitude or frequency characteristics of the systems that have these responses. In these response equations, the symbol p is used to represent the time derivative operator d/dt. The 18 responses given in Table 15-1 require a few explanatory remarks and will now be covered on a case-by-case basis.

Case 1.

This is the typical simple low-frequency cutoff such as might be produced by a series-condenser/shunt-resistor combination. The step response shows an abrupt rise to unity followed by an exponential decay. Usually encountered in amplifier interstages and so-called "differentiating networks." In interstages, f_0 is typically a few cycles; in differentiating networks, f_0 may be as high as several megahertz, in which case the step response is very nearly an impulse.

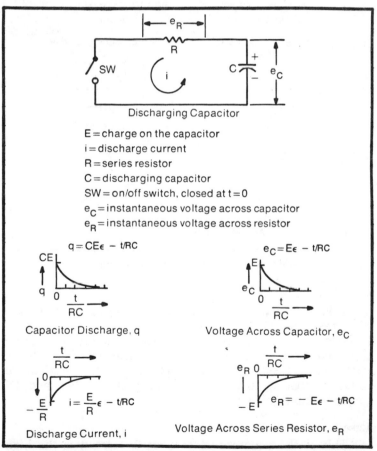

Discharging Capacitor

E = charge on the capacitor
i = discharge current
R = series resistor
C = discharging capacitor
SW = on/off switch, closed at $t = 0$
e_C = instantaneous voltage across capacitor
e_R = instantaneous voltage across resistor

$q = CE\epsilon - t/RC$

Capacitor Discharge, q

$e_C = E\epsilon - t/RC$

Voltage Across Capacitor, e_C

$i = \dfrac{E}{R}\epsilon - t/RC$

Discharge Current, i

$e_R = -E\epsilon - t/RC$

Voltage Across Series Resistor, e_R

Fig. 15-2. Discharging capacitor characteristics.

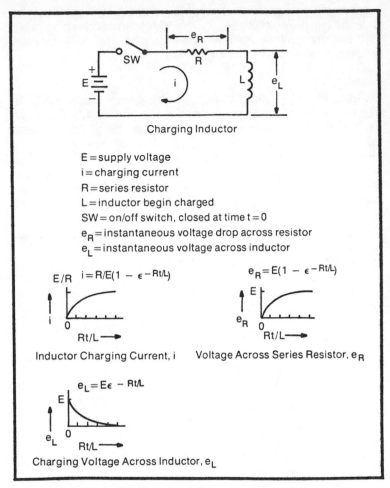

E = supply voltage
i = charging current
R = series resistor
L = inductor begin charged
SW = on/off switch, closed at time t = 0
e_R = instantaneous voltage drop across resistor
e_L = instantaneous voltage across inductor

$i = R/E(1 - \epsilon^{-Rt/L})$

Inductor Charging Current, i

$e_R = E(1 - \epsilon^{-Rt/L})$

Voltage Across Series Resistor, e_R

$e_L = E\epsilon^{-Rt/L}$

Charging Voltage Across Inductor, e_L

Fig. 15-3. Charging inductor characteristics.

Case 2.

This is a rising simple step in the *frequency* characteristic. Step response rises initially to amplitude determined by high-frequency transmission, falls exponentially to level determined by low-frequency (or DC) transmission. This is commonly encountered in improperly compensated resistance-capacity dividers, such as scope probes and DC amplifier interstages.

Case 3.

This is the counterpart of Case 2. Here it is the high-frequency transmisison which is deficient.

294

Case 4.

A typical simple high-frequency cutoff such as this is produced by a parallel RC combination. The step response rises exponentially to the final value determined by the low-frequency (or DC) transmission. It is commonly encountered in simple (not "peaked") interstages, and wherever shunt capacity (as from connecting cables) loads down a resistive source.

Case 5.

Two simple RC high-frequency cutoffs in tandem provide a typical rise characteristic of two-stage resistance-coupled

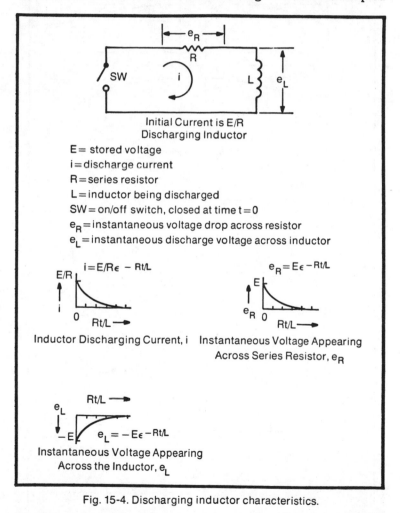

Initial Current is E/R
Discharging Inductor

E = stored voltage
i = discharge current
R = series resistor
L = inductor being discharged
SW = on/off switch, closed at time t = 0
e_R = instantaneous voltage drop across resistor
e_L = instantaneous discharge voltage across inductor

$i = E/R\epsilon^{-Rt/L}$

Inductor Discharging Current, i

$e_R = E\epsilon^{-Rt/L}$

Instantaneous Voltage Appearing
Across Series Resistor, e_R

$e_L = -E\epsilon^{-Rt/L}$

Instantaneous Voltage Appearing
Across the Inductor, e_L

Fig. 15-4. Discharging inductor characteristics.

Table 15-1. Table of Step Function Responses.

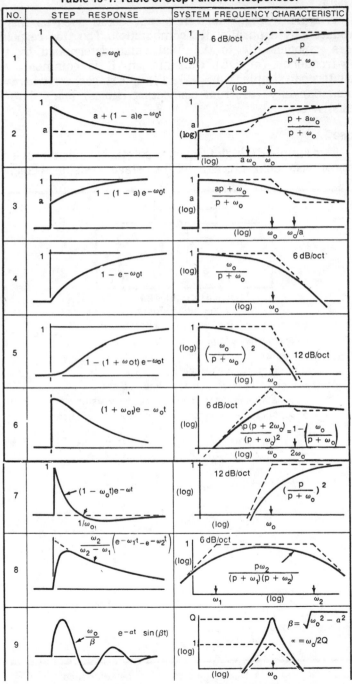

Table of Step Function Responses (Cont'd)

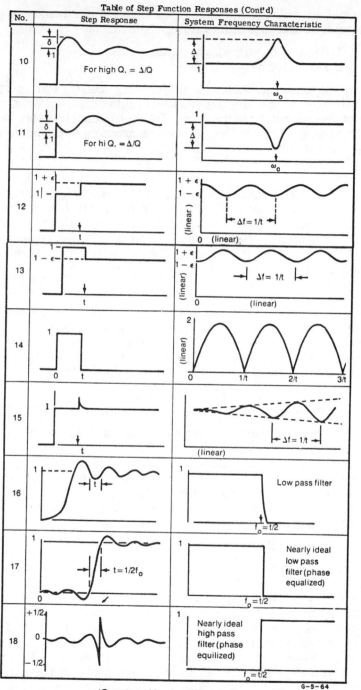

(Courtesy Hewlett-Packard)

G-S-64

amplifier without "peaking." Principal differences are compared with case 4, showing:

1. Greater rise time for same ω_0 .
2. Zero slope at $t = 0$.

For each additional high-frequency cutoff, one more derivative of the step response vanishes at $t = 0$. Thus, if high-frequency transmission falls (ultimately) at $6n$ dB/octave, all derivatives of step response up to the nth are zero at $t = 0$.

Case 6.

Phase-compensated low-end cutoff: The step function response falls to zero eventually, but the initial slope is zero. As a result, the square-wave response shows little or no tilt. This may be produced in a single network, or by two networks (Cases 1 and 3) in tandem, and is often found in video amplifiers.

Case 7.

Two simple low-frequency cutoffs (Case 1) in tandem: This is the typical low-frequency transient response of a single-stage resistance-coupled amplifier with input blocking capacitor or two-stage amplifier with no input blocking capacitor. Principal differences compared with Case 1:

1. Faster initial rate of fall for same ω_0 .
2. Response goes negative crossing axis at $t = 1/\omega_0$.

With each additional low-end cutoff, one additional axis crossing is produced. Thus, if the low-end response falls off (ultimately) at $6n$ dB/octave, there will be $n - 1$ axis crossings. They do not occur at regular intervals—each successive half-cycle takes longer. All step function responses produced by n similar simple low-frequency cutoffs are members of the family of LaGuerre functions.

Case 8.

Simple high- and low-frequency cutoff: The step response rises exponentially at a rate determined by high-frequency cutoff, then falls exponentially at a rate determined by low-frequency cutoff. Typical of complete resistance-coupled interstage response. If $\omega_2 / \omega_1 >> 1$, then on a slow time scale, the response looks like Case 1, and on a fast time scale, the response looks like Case 4. If $\omega_2 = \omega_1 = \omega_0$, we have the case of a critically damped RLC circuit; the response then becomes

$$ V = \omega_0 \, t \epsilon^{-\omega_0 t} $$

298

Case 9.

Typical damped oscillation: Exact response shown is current in series RLC circuit in response to series voltage step, or voltage across parallel RLC circuit in response to applied current step. The dotted lines in the frequency characteristic are the asymptotes which the actual characteristic approaches for $\omega/\omega_0 >> 1$ and $\omega/\omega_0 >> 1$. The peak of the resonance curve is Q times as high as the intersection of these asymptotes. For reasonable Qs, such that $\beta \approx \omega_0$, the Q of circuit may be readily found from the fact that the envelope of the oscillation decays to $1/\epsilon$ in Q/π cycles. Thus $Q = \pi n$, where n is the number of cycles to the $1/\epsilon$ point.

Case 10.

Small resonance in an otherwise flat characteristic: The response consists of unit step due to flat transmisson plus damped oscillation due to resonance—a simple superposition. Initial amplitude of oscillation is related to amplitude of hump in frequency characteristic as indicated in the figure. For the same amplitude of hump, increasing the Q decreases the amplitude of oscillation but the oscillation persists longer. If the hump is near top of band, the time scale will be such that the initial rise of response will not appear so abrupt, but will blend with oscillation to give a response like that of over-peaked interstage. Midband resonances as shown in Case 10 often occur as a result of stray feedback paths, such as heater leads, or from attempting to bypass electrolytics with small mica capacitors. (Electrolytics become inductive at high frequencies.)

Case 11.

Similar to Case 10, but here we have a resonant dip: Note that the effect of a complete null ($\Delta = 1$) is no worse than that of a 6 dB hump. The pilot separation filters used in the coaxial television system produce this type of dip—a complete null—because the Q is so high (several thousand)), and the disturbance they produce, while it persists for a long time, is of such low amplitude as to be invisible in the picture.

Case 12.

Positive echo: The associated frequency characteristic has nearly sinusoidal ripple in amplitude and phase. The frequency interval between successive maxima or minima is the reciprocal of the echo delay. The longer the delay, the closer the ripples. This is commonly encountered in systems having faulty or misterminated delay lines, and also in measurements where multipath transmissions can exist, such as acoustic measurements. The reason most speaker characteristics look so ragged (fine structure) is that multipath reflections with long delay are present.

Case 13.

Negative echo: This has the same frequency ripples as in Case 11, but reversed 180°. DC transmission is now $1 - \epsilon$ rather than $1 + \epsilon$.

Case 14.

Rectangular pulse response: This can be considered a 100% negative echo. The minima of the frequency ripples have now become nulls. The shape of the amplitude characteristic is that of rectified sine wave, and the phase characteristic is a sawtooth decreasing from $\pi/2$ linearly to $-\pi/2$ and jumping back to $\pi/2$ at each null. Such a characteristic can be obtained by using a delay line as an interstage with the near end terminated and the far end shorted.

Case 15.

Differentiated echo: This is the sort of disturbance produced when a delay line is terminated is such a way that the reflection coefficient increases with frequency. Typical causes are:

1. Series inductance or shunt capacitance in the termination of a smooth line.
2. Termination of a constant-k filter in simple resistances.

With both ends matched at low frequencies, the transmitted echo involves two reflections, both of which increase with frequency, and so tend to be "double differentiated" and smaller.

Case 16.

Rise characteristic (qualitative only) of a low pass filter *without* phase correction: The initial part of the rise $(t = 0 + \epsilon)$ is a high power of time, the exponent depending on the number of sections. Following the rise there is a ripple whose period is not constant but approaches $2\tau = 1/f_0$. That is, the apparent frequency approaches the cutoff frequency after several cycles. With an increasing number of sections this ripple increases in amplitude and duration. The "ringing sound" so often attributed to sharp cutoff filters is not due to exaggeration of frequencies near cutoff, nor to the sharp cutoff per se, but rather to the delay distortion which exists near

300

cutoff, causing those upper frequencies which are passed to arrive too late and thus be separately audible. The effect is noticeable only in extreme cases as in long telephone circuits with many channel filters in tandem. With proper delay equalization the effect disappears.

Case 17.

The "ideal" low-pass filter passes all frequencies below f_0 with the same amplitude and delay while attenuating completely those above f_0. Its step response is the sine integral, $\int_0^t (\sin x)/X\,dx$. This function differs from zero (except at regular points) for all $t > -\infty$. Hence the ideal filter cannot be realized without infinite delay. A practical approximation will have a finite delay, and its step response therefore will execute only a finite number of wiggles before the main rise. The approximation can be quite good, however. Here again, the ripples in the step response do not indicate high-frequency enhancement, but are the "Gibb's effect" encountered in Fourier series, and are properly called band elimination ripples. The rise time from the last zero crossing to the first crossing of the final amplitude level is $1/2f_0$; that is, one-half cycle of the cutoff frequency.

Case 18.

The ideal high-pass filter: By superposition, the response of this filter is obtained by subtracting the response of the ideal low-pass filter from an equally delayed unit step.

PROBLEM 15-1

A 100Ω resistor and an $8\ \mu$F capacitor are connected in series across a 100V DC supply. What is the charging current 800 μsec after energizing the circuit? Express the answer as a percentage of the initial value.

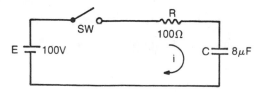

Problem 15-1.

Answer

The loop equation is

$$iR + (1/C)\int i\,dt = 100$$
$$iR + i/pC = 100$$

The differentiation of this equation will bring it into homogeneous form:

$$p(iR + i/pC) = 0$$

This suits the characteristic equation

$$m(iR + i/mC) = 0$$

Therefore, the extra root of the characteristic equation is $m = 0$ and the steady-state current i_S is constant. The derivative of this will be 0. Substituting i_S in the first equation:

$$i_S R + i_S /pC = 100$$
$$i_S /pC = 100 - i_S R$$
$$i_S = 100pC - pRCi_S = 0$$

For i transient, i_T, the first-order characteristic equation,

$$iR + i/mC = 0$$
$$m = -1/RC$$
$$i_T = k \exp(-t/RC)$$

So,

$$i = i_S + i_T$$
$$= 0 + k \exp (-t/RC)$$
$$= k \exp(-t/RC)$$

At $t_0{}^t$ (just after switch closure), C acts as a short circuit. At $t = 0$, $i_T = 1$. So,

$$1 = k\epsilon^0$$
$$k = 1$$

Therefore,

$$i = \exp (-t/RC)$$
$$i/i_0 = \exp(-t/RC)/1$$
$$= \exp(-800 \times 10^{-6} /100(8 \times 10^{-6})$$
$$= \epsilon^{-1}$$
$$= 1/\epsilon$$
$$= 0.368$$
$$= 36.8\% \text{ ANS}$$

Alternate Answer

For the simple charging circuit,

$$i_C (t) = (E_{DC} /R)\epsilon^{-t/RC}$$

Now, $RC = (100) \ (8 \times 10^{-6}) = 8 \times 10^{-4}$ sec, which is equal to the 800 μsec time called for. Therefore,

$$
\begin{aligned}
i_c(t) &= (100/100)\epsilon^{-1/1}) \\
&= \epsilon^{-1} \\
&= 0.368 \\
&= 36.8\% \text{ ANS}
\end{aligned}
$$

PROBLEM 15-2

In the circuit shown, what is the current in each resistive element 0.005 sec after the switch is opened?

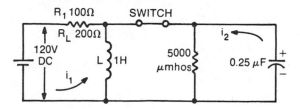

Problem 15-2.

Answer

When the switch is opened, the left side will appear as Fig. 15-5A, where

$$
\begin{aligned}
R_T &= R_1 + R_L \\
&= 100 + 200 \\
&= 300\Omega
\end{aligned}
$$

The inductor will charge to the voltage applied, which is 120V. When the switch is opened, the collapsing field will add to the supply voltage to give $E_1 + E_2 = 240$V. Using the classic equation:

$$
\begin{aligned}
i_{R1} &= (E/R_T)\epsilon^{-R_T \ t/L} \\
&= (240/300) \exp[-300(5 \times 10^{-3})/1] \\
&= 0.8\epsilon^{-1.5} \\
&= 0.8/4.49 \\
&= 178 \text{ mA ANS}
\end{aligned}
$$

In a series circuit, $i_{R1} = i_{RL} = 178$ mA so when the switch opens, the right side will appear as in Fig. 15-5B, where

$$
\begin{aligned}
R_2 &= 1/G \\
&= 10^6/(5 \times 10^3) \\
&= 0.2 \times 10^3 \\
&= 200\Omega
\end{aligned}
$$

303

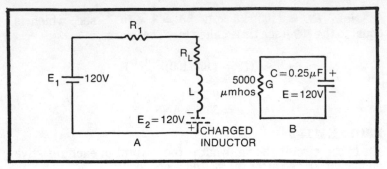

Fig. 15-5.

Thus

$$i_{R2} = (-E/R_2)\epsilon^{-t/R_2 C}$$
$$= (-120/200) \exp[-5 \times 10^{-3}/200(0.25 \times 10^{-6})]$$
$$= -0.6\,\epsilon^{-100}$$
$$\approx 0 \text{ ANS}$$

PROBLEM 15-3

Switch SW_2 is initially closed. If SW_1 is closed at $t = 0$ and SW_2 is opened at $t = t_1$, determine the voltage across C, assuming that its initial charge is zero.

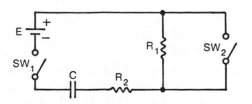

Problem 15-3.

Answer

Find the charge equation for values of $t > t_1$, $t = 0$, then $t = t_1$.

$$E - R_1\,dq/dt - R_2\,dq/dt - q/C = 0$$

Let $p = d/dt$ and factor out q.

$$(R_1 p + R_2 p + 1/C)q = E$$
$$p(R_1 p + R_2 p + 1/C)q = 0$$

This is in the form of

$$m(R_1 m + R_2 m + 1/C) = 0$$

Now, $m = 0$ and the steady-state solution of q is a constant; that is $dq_S/dt = 0$. Substituting the steady state solution into the original equation,

$$E - 0 - 0 - q_S/C = 0$$

So $q_S = CE$.

Using the characteristic equation to find the transient q, or q_T,

$$R_1 m + R_2 m + 1/C = 0$$
$$m(R_1 + R_2) = -1/C$$
$$m = -1/[C(R_1 + R_2)] = -1/\tau_1$$
$$\text{Therefore } q_T = A\epsilon^{-t/\tau_1}$$

Now, $q = q_S + q_T$, so

$$q = CE + A\epsilon^{-t\tau_1}$$

To find the value of q at $t = t_1$, first find the value of A by solving for $t = 0$. The equivalent circuit is now illustrated in Fig. 15-6.

$$E - R_2\ dq/dt - q/C = 0$$
$$E - R_2\ pq - q/C = 0$$
$$(R_2\ p + 1/C)q = E$$
$$p(R_2\ p + 1/C)q = 0$$
$$m(R_2\ m + 1/C = 0$$

Thus $m = 0$ and the steady-state solution of q is a constant; that is, $dq_S/dt = 0$.

Substitute values of the steady state solution into the original equation to obtain

$$E - 0 - q_S/C = 0$$
$$q_S = CE$$

Finding q_T:

$$R_2\ m + 1/C = 0$$
$$m = -1/R_2\ C = -1/\tau_2$$
$$q_T = B\epsilon^{-t/\tau_2}$$

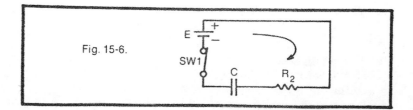

Fig. 15-6.

Thus,
$$q = CE + B\epsilon^{-t/\tau_2}$$

Now solve for B at $t = 0$.
$$q_0 = 0 = CE + B\epsilon^0$$
$$B = -CE$$
$$q = CE - CE\epsilon^{-t/\tau_2}$$
$$= CE(1 - \epsilon^{-t/\tau_2})$$

At $t = t_1$,

$$q = CE(1 - \epsilon^{-t_1/\tau_2})$$
$$q = CE + A\epsilon^{-t_1/\tau_1}$$
$$CE - CE\epsilon^{-t_1/\tau_2} = CE + A\epsilon^{-t_1/\tau_1}$$

Now, solve for coefficient A:

$$A = -CE_\epsilon \exp\left[\frac{(-t_1) - (-t_1)}{R_2 C - C(R_1 + R_2)}\right]$$

$$= -CE_\epsilon \exp\left[(-t_1)\frac{R_1 + R_2 - R_2}{R_2 C(R_1 + R_2)}\right]$$

$$= -CE_\epsilon \exp\left[\frac{-t_1 R_1}{R_2(R_1 + R_2)C}\right]$$

Since $E_C = q/C$, the capacitor voltage at time $t = t_1$ is

$$E_C = E - E_\epsilon \exp\left[\frac{R_2 t - t_1 R_1}{R_2(R_1 + R_2)C}\right] \quad \textbf{ANS}$$

PROBLEM 15-4

Given a series RLC circuit in which C is initially charged and then discharged through R and L by closing switch S, and for which the equation of the discharge voltage across C is

$$e_C = 100E - 200t\cos(300t)$$

Find:

(A) the values of R and L
(B) the natural frequency
(C) the circuit current i at $t = 0$
(D) e_L and e_R at $t = 0$

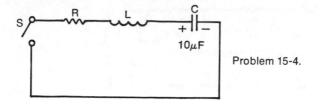

Problem 15-4.

Answer

The equation given in the problem statement is obviously a first-order approximation for the circuit shown, but it is satisfactory to determine the circuit parameters required. The general expression for the voltage across the capacitor in a damped LC circuit of this type is

$$e_C = E\epsilon^{-t/RC} \cos(\omega t)$$

Comparing this equation with that given in the problem statement, it can be seen that the initial capacitor voltage is $E = 100$, that the initial time constant of decay is $RC = 1/200$ and that the natural frequency is $\omega = 300$.

(A) Knowing that $C = 10\ \mu F$, it then follows that

$$R = 1/200C = (200 \times 10 \times 10^{-6})^{-1} = 500\Omega\ \text{ANS}$$
$$L = 1/\omega^2 C = (300^2 \times 10 \times 10^{-6})^{-1} = 1.11H\ \text{ANS}$$

(B) The natural frequency of the circuit is found from ω, since

$$f = \omega/2\pi = 47.75\ \text{Hz ANS}$$

(C) The equation given in the problem for the discharge voltage might lead you to believe that at $t = 0$ the discharge current would be $i = E/R = 0.2A$. But this cannot be because of the inductor in the circuit. The initial discharge current would be zero in a real circuit, so $i = 0$ at time $t = 0$ ANS

(D) Since the initial discharge current is assumed to be zero due to the inductive reactance in the circuit, the voltage drop across L is $e_L = 100$ and the voltage drop across R is $e_R = iR = 0$ ANS.

PROBLEM 15-5

This circuit is in steady state at $t = 0$ when one resistor is short circuited. Find the transient expression for current i when $t > 0$.

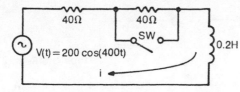

Problem 15-5.

Answer

The loop equation for the circuit is

$$200 \cos(400t) - 40i - 0.2 \, di/dt = 0$$

Let $p = d/dt$, so

$$-200 \cos(400t) + 40i + 0.2p = 0$$
$$(40 + 0.2p)i = 200 \cos(400t)$$
$$p(40 + 0.2p)i = -200 \sin(400t) \, (400)$$
$$= -8 \times 10^4 \sin(400t)$$
$$p^2 (40 + 0.2p)i = -8 \times 10^4 \cos(400t) \, (400)$$
$$= (-16 \times 10^4) \, 200 \cos(400t)$$
$$[p^2 - (-16 \times 10^4)] \, [40 + 0.2p]i = 0$$

The first factor in the preceding equation provides the roots of the steady-state solution; the second provides the roots of the transient solution.

Solving first for the steady-state solution,

$$m^2 + 16 \times 10^4 = 0$$
$$m = \pm\sqrt{-16 \times 10^4} = \pm 400j$$
$$i_{SS} = A \cos(400t) + B \sin(400t)$$
$$p \, i_{SS} = A[-\sin(400t)](400) + B[\cos(400t)](400)$$
$$= -400 \, A \sin(400t) + 400 \, B \cos(400t)$$

Substituting i_{SS} and $p i_{SS}$ back into the original equation:

$$-200 \cos(400t) + 40[A \cos(400t) + B \sin(400t)]$$
$$+ 0.2[-400 \, A \sin(400t) + 400 \, B \cos(400t)] = 0$$
$$(40A + 80B) \cos(400t) + (40B - 80A) \sin(400t)$$
$$= 200 \cos(400t)$$

Thus,

$$40B - 80A = 0$$
$$B = 80A/40 = 2A$$
$$40A + 80B = 200$$
$$40A + 80(2A) = 200$$
$$A = 200/200 = 1$$
$$B = 2(1) = 2$$

Therefore,

$$i_{SS} = \cos(400t) + 2 \sin(400t)$$

Solving for the transient solution

$$40 + 0.2m = 0$$
$$m = -40/0.2 = -200$$

From the characteristic equation,

$$i_{TR} = k\epsilon^{-200t}$$

so the full equation is

$$i = i_{SS} + i_{TR}$$
$$= \cos(400t) + 2\sin(400t) + k\epsilon^{-200t}$$

When the switch is closed, i cannot change instantly because of L. Therefore, when the switch is closed ($t = 0$),

$$i_0 = \frac{200\cos[400(0)]}{40 + 40 + 0.2(400)j}$$

$$= \frac{200}{80(1 + j)}\left(\frac{1 - j}{1 - j}\right)$$

$$= 2.5(1 - j)/(1 + 1)$$
$$= 1.25 - 1.25j$$

Substituting back into the full equation,

$$1.25 - 1.25j = \cos[400(0)] + 2\sin[400(0)] + k\epsilon^{200(0)}$$
$$1.25 - 1.25j = 1 + 0 + k$$
$$k = -.875 - j.125$$

So, at $t > 0$, i becomes

$$i = \cos(400t) + 2\sin(400t) + (-.875 - 1.25j)\epsilon^{-200t}$$

PROBLEM 15-6

Solve for $V_1(t)$ in the circuit shown for the case in which $V(t)$ is a unit step function.

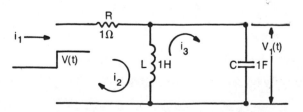

Problem 15-6.

The loop currents are:

$$i_1 = i_2 + i_3 = (V - V_1)/R$$
$$V_1 = L\, di/dt$$
$$i_3 = dq/dt = C\, dV_1/dt$$
$$i_2 = (1/L) \int V_1\, dt$$
$$i_1 = (1/L) \int V_1\, dt + C\, dV_1/dt = (V - V_1)/R$$

Solving for V,

$$V = (R/L) \int V_1\, dt + RC\, dV_1/dt + V_1$$

Taking the derivative,

$$dV/dt = (R/L)V_1 + RC(d^2 V_1/dt^2) + dV_1/dt$$

When $t = 0$, V is constant, so $dV/dt = 0$. Therefore,

$$0 = RC\,\frac{d^2 V_1}{dt} + \frac{dV_1}{dt} + \frac{R}{L}V_1$$

Knowing that $V_1 = A\epsilon^{\alpha t}$ where

$$\alpha = \frac{-1 \pm \sqrt{1 - 4R^2 C/L}}{2RC}$$

$$= \frac{-1 \pm \sqrt{1 - 4(1)(1)/1}}{2(1)(1)}$$

$$= -1/2 \pm j\sqrt{3}/2$$

Therefore,

$$V_1 = A\exp[(-1/2 + j\sqrt{3}/2)t] + B\exp[(-1/2 - j\sqrt{3}/2)t]$$

When $t = 0$, $V = 0$, $dV/dt = 0$, $i = V/R = 1$, $i_2 = 0$, $i_3 = -C\, dV_1/dt$

$$V_1 = (j/\sqrt{3})\epsilon^{-t/2}(\epsilon^{\sqrt{3}jt/2} - \epsilon^{-\sqrt{3}jt2})$$

Knowing that

$$\sin(x) = (\epsilon^{+jx} - \epsilon^{-jx})/2j$$

Multiply both numerator and denominator by $2j$ to get

$$V_1 = (2/\sqrt{3})\epsilon^{-t/2} \sin(\sqrt{3}t/2)$$
$$= 1.555\,\epsilon^{-t/2} \sin(0.865t) \text{ **ANS**}$$

PROBLEM 15-7

In the given circuit, the vacuum tube parameters are $\mu = 20$ and $g_m = 200$ μmhos. Determine the time expression for e_0 when $e_S = 10\ u(t)$, a step function. Neglect tube and wiring capacitance.

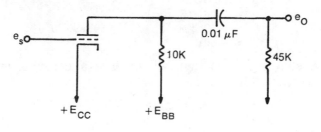

Problem 15-7.

Answer

Since bias values are not shown in the figure, we will assume that the tube operates linearly, that e_S is a positive step voltage, and that the resulting e_0 will be a negative spike as shown in Fig. 15-7. For the equivalent circuit in Fig. 15-7, the equivalent plate resistance is

$$r_p = \mu/g_m$$
$$= (20 \times 10^6)/(2 \times 10^3)$$
$$= 10\ k\Omega$$

The equivalent load resistance is then

$$R_{EQ} = \frac{1}{(1/r_p) + (1/R_L) + (1/R_G)} = 4.5\ k\Omega$$

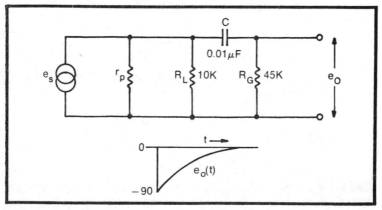

Fig. 15-7.

Since

$$i_p = e_S \, g_m$$

The resulting step in the output voltage is then

$$
\begin{aligned}
e_0 &= i_p \, R_{EQ} \\
&= -e_S \, g_m \, R_{EQ} \\
&= -10(2 \times 10^3)(10^6)(4.5 \times 10^3) \\
&= -90V
\end{aligned}
$$

A step input function will produce a decaying output voltage spike of the form

$$e_0 = e_{MAX} \, \epsilon^{-t/\tau}$$

where

$$
\begin{aligned}
e_{MAX} &= -90V \\
\tau = R_G \, C_C &= (45 \times 10^3)(0.01 \times 10^{-6}) = 0.045
\end{aligned}
$$

Therefore,

$$
\begin{aligned}
e_0 &= 90 \, \epsilon^{-t/0.045} \\
&= 90 \, \epsilon^{-2200t} \quad \textbf{ANS}
\end{aligned}
$$

PROBLEM 15-8

In the circuit shown, capacitor C_1 is charged to voltage V_0. At time $t = 0$ the switch is closed. Solve for the charge as a function of time.

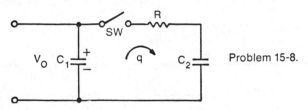

Problem 15-8.

Answer

Assuming that C_2 is not initially charged, the charge in the circuit has the loop equation

$$-q/C_1 - R \, dq/dt - q/C_2 = 0$$

Or, using the p operator,

$$-q/C_1 - Rpq - q/C_2 = 0$$

By factoring out the q and removing it from the equation, we are left with

$$-1/C_1 - Rp - 1/C_2 = 0$$

or

$$p = -(C_1 + C_2)/C_1 C_2 R$$

The generic equation for charge is

$$q = k\epsilon^{-t/RC}$$

But the equation for p can be written as

$$p = -1/RC_T$$

where

$$C_T = (C_1 + C_2)/C_1 C_2$$

Substituting for RC in the generic equation yields

$$q = k\epsilon^{-t/p}$$

or

$$q = k\exp[-(C_1 + C_2)t/C_1 C_2 R]$$

Differentiating to find dq/dt results in

$$dq/dt = -[k(C_1 + C_2)/C_1 C_2 R]$$
$$\exp[-(C_1 + C_2)t/C_1 C_2 R]$$

Now when $t = 0$ we know that $q/C_1 = V_0$ and $q/C_2 = 0$, so

$$-V_0 - R[-k(C_1 + C_2)/C_1 C_2 R] - (0) = 0$$

and we find that

$$k = -V_0 C_1 C_2/(C_1 + C_2)$$

Therefore, the equation for charge as a function of time is

$$q = \frac{-V_0 C_1 C_2}{C_1 + C_2}$$

$$\exp\left[\frac{-(C_1 + C_2)t}{RC_1 C_2}\right] \text{ ANS}$$

Chapter 16

Laplace Transforms

The Laplace transform is an operational method that can be used for solving linear differential equations. To the P.E. candidate, Laplace transforms become an advantageous time-saving tool, one that will convert many functions, such as damped sinuoids and exponential functions, into simple algebraic functions of a complex variable. Such converted expressions can then be treated by straightforward algebraic operations in the complex plane, rather than by time-consuming differentiation or integration. The review provided here is a summary of Laplace transforms, but the P.E. candidate can find more detailed information in references 1, 3, 4, 5, 7, 18, 48, 55, and 56.

TABLES OF LAPLACE TRANSFORMS

In working with logarithms you learned to convert from one mathematical form to another for the purpose of easier mathematical manipulation. That is, you solved multiplication and division problems containing exponents by means of simple addition and subtraction. Tables were used to find the log and the antilog of the numbers, and by such means the conversion process was reduced to a look-up table.

Laplace transform tables are used in much the same way as log tables, and they greatly simplify the conversion and inversion processes. For convenience, a list of Laplace transform pairs is provided in Table 16-1, while Table 16-2 lists a number of basic time and frequency functions as well as graphical illustrations of their time and frequency domains.

TIME AND FREQUENCY VARIABLES

The complex variable used in Laplace transforms is s. This variable has a real component σ and an imaginary (frequency) component $j\omega$. Thus, $s = \sigma + j\omega$. Basically, the Laplace transform associates a function $F(s)$ having a complex variable with a function $f(t)$ having a time variable.

The complex function is derived by multiplying the time function $f(t)$ by the factor ϵ^{-st}. The resulting time function $f(t)\,\epsilon^{-st}$ is then integrated from zero to infinity; that is

$$F(s) = \int_0^\infty f(t)\,\epsilon^{-st}\,dt$$

Since the limits of this integration are fixed, the integral does not depend on time t but only on parameter s. Thus, the Laplace transform of a given expression, as a function of time, is defined as

$$\mathcal{L}[f(t)] = F(s)$$

where $\mathcal{L}$ is the Laplace transform symbol, which indicates that the time function or expression it precedes is transformed by the Laplace integral as defined, and for which the result will be an expression in which s is the complex variable. A restriction is placed on $f(t)$, the function of time, such that $f(t) = 0$ for all values of t less than zero; that is, $f(t)$ is defined to be zero for all negative values of time.

Consider, for example, the Laplace transform of the unit step function $u(t)$. The conditions are that $u(t) = 0$ when $t < 0$ and that $u(t) = 1$ when $t > 0$. Thus,

$$
\begin{aligned}
F(s) &= \int_0^\infty u(t\,\epsilon^{-st}\,dt \\
&= \int_0^\infty \epsilon^{-st}\,dt \\
&= 1/s
\end{aligned}
$$

Another function that the P.E. candidate will often encounter is $\epsilon^{-\alpha t}$, a function expressing an exponential decay with time if α is positive. This function is often combined with sinusoidal terms to express the characteristic exponential decay of a damped sine wave. For this function the Laplace tranform is

$$
\begin{aligned}
F(s) &= \int_0^\infty \epsilon^{-\alpha t}\,\epsilon^{-st}\,dt \\
&= \int_0^\infty \epsilon^{-(s+\alpha)t}\,dt \\
&= 1/(s \times \alpha)
\end{aligned}
$$

This result and many others can be found in Table 16-1. The conversion to and from the time function $f(t)$ and the complex function $F(s)$ is easily achieved with the assistance of the transform pairs found in the table. A pictorial view of these

Table 16-1. Laplace Transform Pairs.

$f(t)$	$F(s)$
Unit impulse	1
Unit step $u(t) = 0$ for < 0 — $u(t) = 1$ for $t > 0$	$1/s$
t	$1/s^2$
ϵ^{-at}	$1/(s + a)$
$1/\omega \sin(\omega t)$	$1/(s^2 + \omega^2)$
$1/\omega\, \epsilon^{-at} \sin(\omega t)$	$1/(s + a)^2 + \omega^2$
$\cos(\omega t)$	$s/s^2 + \omega^2$
$\epsilon^{-at} \cos(\omega t)$	$s + a/(s + a)^2 + \omega^2$
$t\epsilon^{-at}$	$1/(s + a)^2$
$t \cos(at)$	$s2 - a^2/(s^2 \quad a^2)^2$
$t \sin(at)/2a$	$s/(s^2 + a^2)^2$
$\epsilon^{-at} f(t)$	$F(s + a)$
$(1 - at)\,\epsilon^{-at}$	$s/(s + a)^2$
$df(t)/dt$	$s\mathcal{L}[f(t)] - f(0)$
$\int f(t)\, dt$	$\mathcal{L}[f(t)]/s + f^{-1}(0)/s$
$\int_0^t f(t)\, dt$	$\mathcal{L}[f(t)]/s$

time and frequency functions is given in Table 16-2 to help visualize the resulting functions when the transforms are expressed in terms of frequency ω or of the differential operator p.

PROBLEM 16-1

The circuit shown has reached a steady-state condition when switch SW_1 is opened at $t = 0$. Using Laplace tranforms, find the current through coil L_2 as a function of time.

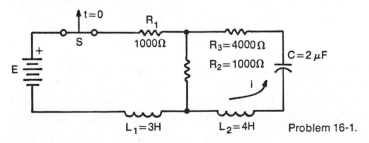

Problem 16-1.

Answer

At $t = 0^-$, capacitor C is charged to a steady-state condition of $E_0/2$, owing to the divider consisting of resistors R_1 and R_2, and the current through L_2 is zero.

At $t = 0^+$, the current through L_2 is

$$(R_2 + R_3)i + 1/C \int i\, dt + E_0/2 + L_2\, di/dt = 0$$

The Laplace transform equation is

$$I(s)(R_2 + R_3) + I(s)/sC + E_0/2s + L_2\, s\, I(s) = 0$$

$$I(s)(R_2 + R_3 + 1/sC + L_2\, s) = -E_0/2s$$

$$I(s) = \frac{-E_0 \times sC}{2s \times (R_2 + R_3)\, Cs + L_2\, Cs^2 + 1}$$

$$= \frac{-E_0\, C}{2(L_2\, C)[s^2 + (R_2 + R_3)s/L_2 + 1/L_2\, C]}$$

$$= \frac{-E_0}{2(4)\,[s^2 + (10^3 + 4 \times 10^3)s/4 + 1/(4)(2 \times 10^{-6})]}$$

$$= \frac{-E_0}{8(s^2 + 1250s + 12.5 \times 10^4)}$$

$$= \frac{-E_0}{8(s + 1140)(s + 110)}$$

The inverse of this Laplace transform is

$$i(t) = -E_0/8(1/110 - 1140)\,[\epsilon^{-1140t} - \epsilon^{-110t}]$$

$$= E_0/8240\,[\epsilon^{-1140t} - \epsilon^{-110t}] \text{ ANS}$$

PROBLEM 16-2

In the accompanying circuit, find $v(t)$ when SW_1 becomes open. Solve the problem by using Laplace transforms.

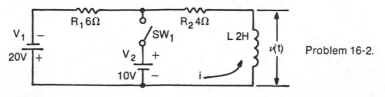

Problem 16-2.

Table 16-2: Table of Important Transforms

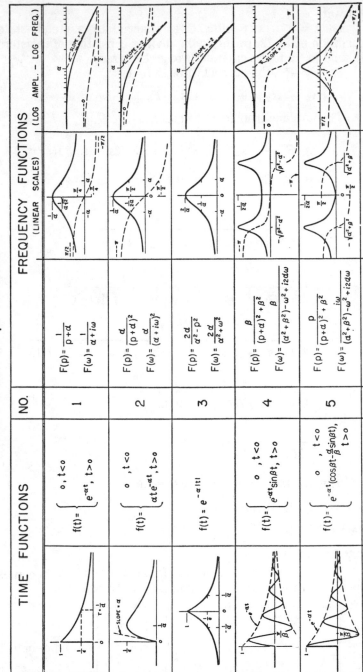

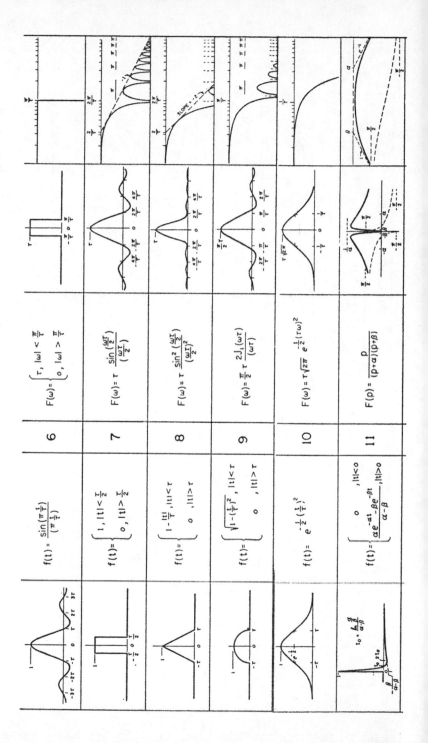

	$f(t)$		$F(\omega)$		
	$f(t)=\dfrac{\sin(\pi\frac{t}{\tau})}{(\pi\frac{t}{\tau})}$	6	$F(\omega)=\begin{cases}\tau, & |\omega|<\frac{\pi}{\tau}\\ 0, & |\omega|>\frac{\pi}{\tau}\end{cases}$		
	$f(t)=\begin{cases}1, & |t|<\frac{\tau}{2}\\ 0, & |t|>\frac{\tau}{2}\end{cases}$	7	$F(\omega)=\tau\,\dfrac{\sin\left(\frac{\omega\tau}{2}\right)}{\left(\frac{\omega\tau}{2}\right)}$		
	$f(t)=\begin{cases}1-\frac{|t|}{\tau}, & |t|<\tau\\ 0, & |t|>\tau\end{cases}$	8	$F(\omega)=\tau\,\dfrac{\sin^2\left(\frac{\omega\tau}{2}\right)}{\left(\frac{\omega\tau}{2}\right)^2}$		
	$f(t)=\begin{cases}\sqrt{1-\left(\frac{t}{\tau}\right)^2}, & |t|<\tau\\ 0, & |t|>\tau\end{cases}$	9	$F(\omega)=\dfrac{\pi}{2}\,\tau\,\dfrac{2J_1(\omega\tau)}{(\omega\tau)}$		
	$f(t)=e^{-\frac{1}{2}\left(\frac{t}{\tau}\right)^2}$	10	$F(\omega)=\tau\sqrt{2\pi}\;e^{-\frac{1}{2}(\tau\omega)^2}$		
	$f(t)=\begin{cases}0, & |t|<0\\ \dfrac{\alpha e^{-\alpha t}-\beta e^{-\beta t}}{\alpha-\beta}, & |t|>0\end{cases}$	11	$F(p)=\dfrac{p}{(p+\alpha)(p+\beta)}$		

Table 16-2 continued.

TIME FUNCTIONS		NO.	FREQUENCY FUNCTIONS		
				(LINEAR SCALES)	(LOG AMPL. – LOG FREQ.)
	$f(t) = \begin{cases} \cos \omega_0 t, & \lvert t \rvert < \frac{\tau}{2} \\ 0, & \lvert t \rvert > \frac{\tau}{2} \end{cases}$	12	$F(\omega) = \frac{\tau}{2} \left[\frac{\sin\left(\frac{\omega-\omega_0}{2}\right)\frac{\tau}{2}}{\left(\frac{\omega-\omega_0}{2}\right)\frac{\tau}{2}} + \frac{\sin\left(\frac{\omega+\omega_0}{2}\right)\tau}{\left(\frac{\omega+\omega_0}{2}\right)\tau} \right]$		
	$f(t) = \lim_{\tau \to 0} \begin{cases} \frac{1}{\tau}, & \lvert t \rvert < \frac{\tau}{2} \\ 0, & \lvert t \rvert > \frac{\tau}{2} \end{cases}$ $= \delta(t) \; \text{(DELTA FUNCTION)}$	1S	$F(p) = F(\omega) = 1$		
	$f(t) = \int_{-\infty}^{t} \delta(\lambda)d\lambda = \begin{cases} 0, & t<0 \\ 1, & t>0 \end{cases}$ $= u(t) \; \text{(UNIT STEP)}$	2S	$F(p) = \frac{1}{p}$		
	$f(t) = \int_{-\infty}^{t} u(\lambda)d\lambda = \begin{cases} 0, & t<0 \\ t, & t>0 \end{cases}$ $= S(t) \; \text{(UNIT SLOPE)}$	3S	$F(p) = \frac{1}{p^2}$		
	$f(t) = \begin{cases} 0, & t<0 \\ 1 - e^{-\alpha t}, & t>0 \end{cases}$	4S	$F(p) = \frac{\alpha}{p(p+\alpha)}$		
	$f(t) = \cos \omega_0 t$	5S	$F(\omega) = \pi[\delta(\omega + \omega_0) + \delta(\omega - \omega_0)]$		
	$f(t) = \sum_{-\infty}^{\infty} \delta(t - n\tau)$	6S	$F(\omega) = \frac{2\pi}{\tau} \sum_{-\infty}^{+\infty} \delta\left(\omega - \frac{n 2\pi}{\tau}\right)$		

(Courtesy Hewlett-Packard)

320

Answer

When SW_1 is opened, the equivalent circuit becomes a single loop, the equation of which is

$$V_1 - L\,di/dt - i(R_1 + R_2) = 0$$

Taking the Laplace transform we get

$$V_1/s - [sLI(s) - Li(0)] - (R_1 + R_2)I(s) = 0$$

where $Li(0)$ represents the energy stored in L before SW_1, opens. This current is $V_2/R_2 = -10/4$, or -2.5 amps, so

$$V_1/s - sLI(s) + L(-2.5) - (R_1 + R_2)I(s) = 0$$

Combining $I(s)$ terms,

$$I(s)[(-sL) - (R_1 + R_2)] = -V_1/s + 2.5L$$
$$= (-V_1 + 2.5sL)/s$$

$$I(s) = \frac{-V_1 + 2.5Ls}{s[-sL - (R_1 + R_2)]}$$

$$= \frac{V_1/L - 2.5s}{s[s + R_1 + R_2/L]}$$

Assigning values

$$I(s) = \frac{20/10 - 2.5s}{s(s + 10/2)}$$

$$= \frac{10 - 2.5s}{s(s + 5)}$$

This can also be expressed as

$$A/s + B/s + 5 = \frac{A(s + 5) + Bs}{s(s + 5)}$$

Further

$$I(s) = \frac{As + 5A + Bs}{s(s + 5)} = \frac{5A + (A + B)s}{s(s + 5)}$$

By comparison,

$$I(s) = \frac{10 + (-2.5)s}{s(s + 5)}$$

and from this, $5A = 10$ or $A = 2$. Also,

$$A + B = -2.5$$
$$B = -2.5 - A$$
$$= -2.5 - 2$$
$$= -4.5$$

Substituting these values for A and B,

$$I(s) = 2/s - 4.5/s + 5$$

From the transform tables,

$$i(t) = 2 - 4.5\,\epsilon^{-5t}$$

Substituting into the original equation for i,

$$V_1 - L\,di/dt - (R_1 + R_2)(2 - 4.5\,\epsilon^{-5t}) = 0$$
$$v(t) = L\,di/dt$$
$$= V_1 - (R_1 + R_2)(2 - 4.5\,\epsilon^{-5t})$$

Assigning values,

$$v(t) = 20 - (10)(2 - 4.5\,\epsilon^{-5t})$$
$$= 20 - 20 + 45\,\epsilon^{-5t}$$
$$= 45\,\epsilon^{-5t} \quad \text{ANS}$$

PROBLEM 16-3

Solve the circuit for $v_0(t)$ by the node method when $v_{\text{IN}}(t)$ is a unit step function.

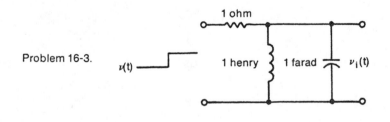

Problem 16-3.

Answer *(ref. 1, p. 23-3)*

If v_N is the node voltage, then

$$[v_N - v(t)] + v_N/j\omega L + j\omega C v_N = 0$$

Substituting $s = \tau + j\omega$,

$$V_N(s) - 1/s + V_N(s)/sL + sCV_N(s) = 0$$

Multiplying through by sL,

$$sLV_N(s) - L + V_N(s) + s^2 LCV_N(s) = 0$$

$$V_N(s)(sL + 1 + s^2 LC) = L$$

$$V_N(s) = \frac{L}{s^2 LC + sL + 1}$$

Dividing numerator and denominator by LC,

$$V_N(s) = 1/C\left[\frac{1}{s^2 + s/C + 1/LC}\right]$$

Substituting for $L = 1$ henry and $C = 1$ farad,

$$V_N(s) = \frac{1}{s^2 + s + 1}$$

Using the quadratic formula,

$$s^2 + s + 1 = (s + 0.5 + 0.867j)(s + 0.5 - 0.867j)$$

Therefore

$$V_N(s) = \frac{1}{(s + 0.5 + 0.867j)(s + 0.5 - 0.867j)}$$

$$= \frac{A}{5 + 0.5 + 0.867j} + \frac{A^*}{s + 0.5 - 0.867j}$$

Multiplying both equations by $s + 0.5 + 0.867j$:

$$\frac{1}{s + 0.5 - 0.867j} = A + \frac{A^*(s + 0.5 + 0.867j)}{s + 0.5 - 0.867j}$$

Setting $s = -0.5 - 0.867j$ and multiplying through by $s + 0.5 + 0.867j$ we have

$$\frac{1}{-0.5 - 0.867j + 0.5 - 0.867j} = A$$

$$+ \frac{A^*(-0.5 - 0.867j + 0.5 + 0.867j)}{-0.5 - 0.867j + 0.5 - 0.867j}$$

$$= A + 0$$

Therefore,

$$A = \frac{1}{-1.734j} = +0.577j$$

To solve next for A^*, multiply both equations by $s + 0.5 - 0.867j$:

$$\frac{1}{s + 0.5 + 0.867j} = \frac{A(s + 0.5 - 0.867j)}{s + 0.5 + 0.867j} + A^*$$

And setting $s = -0.5 + 0.867j$ we get

$$\frac{1}{-0.5 + 0.867j + 0.5 + 0.867j} = 0 + A^*$$

so

$$A^* = \frac{1}{1.734j} = -0.577j$$

Substituting the values of A and A^* into the $V_N(s)$ equation:

$$V_N(s) = \frac{0.577j - 0.577j}{s + 0.5 + 0.867j - s + 0.5 - 0.867j}$$

$$= \frac{0.577j(s + 0.5 - 0.867j) - 0.577j(s + 0.5 + 0.867j)}{(s + 0.5 + 0.867j)(s + 0.5 - 0.867j)}$$

$$= \frac{0.577(0.867) + 0.577(0.867)}{(s + 0.5)^2 + 0.867^2} = \frac{0.5 + 0.5}{(s + 0.5)^2 + 0.867^2}$$

Multiply both numerator and denominator by 0.867:

$$V_N(s) = 1/0.867 \left[\frac{0.867}{(s + 0.5)^2 + 0.867^2} \right]$$

This is in the desired form, which converts to

$$v_N(t)(1/0.867)\, \epsilon^{-0.5t}\, \sin(0.867t)$$
$$= 1.15\, \epsilon^{-0.5t}\, \sin(0.867t) \textbf{ ANS}$$

324

PROBLEM 16-4

What voltage appears across R_1 when $u(t)$ is a unit step?

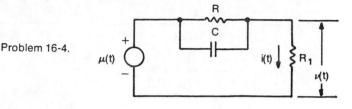

Problem 16-4.

Answer

The transformed loop equation can be written immediately as

$$U(s) - I(s)\left[\frac{R(1/sC)}{R + 1/sC}\right] - R_1 I(s) = 0$$

where $U(s)$, being the transform of $u(t)$, is equal to $1/s$.

Simplifying the equation,

$$1/s - I(s)\frac{R}{RsC + 1} - R_1 I(s) = 0$$

Multiply through by $s(RsC + 1)$:

$$RsC + 1 - I(s)Rs - I(s)R_1 s(RsC + 1) = 0$$
$$RsC + 1 = I(s)[Rs + R_1 s(RsC + 1)]$$

$$I(s) = \frac{RsC + 1}{Rs + R_1 s(RsC + 1)}$$

$$= \frac{RsC}{Rs + R_1 s(RsC + 1)} + \frac{1}{Rs + R_1 s(RsC + 1)}$$

$$= \frac{RC}{R + R_1 RsC + R_1} + \frac{1}{Rs + R_1 Rs^2 C + R_1 s}$$

$$= \frac{RC}{s(RR_1 C) + (R + R_1)} + \frac{1}{s^2(RR_1 C) + s(R + R_1)}$$

Divide through by $RR_1 C$:

$$I(s) = \frac{RC/RR_1 C}{s + R + R_1 /RR_1 C} + \frac{1/RR_1 C}{s^2 + s(R + R_1)/RR_1 C}$$

$$= 1/R_1 \left[\frac{1}{s + R + R_1 /RR_1} \right] C$$

$$+ 1/RR_1 C \left[\frac{1}{s(s + R + R_1 /RR_1 C)} \right]$$

$$= \frac{1}{R_1} \left[\frac{1}{s + \alpha} \right] + \frac{1}{RR_1 C} \left[\frac{1}{s(s + \alpha)} \right]$$

where $\alpha = (R + R_1)/RR_1 C$.

From the table of transforms,

$$i(t) = 1/R_1 \, \epsilon^{-\alpha t} + 1/RR_1 C$$
$$[1/(R + R_1 /RR_1 C)(1 - \epsilon^{-\alpha t})]$$

$$= 1/R_1 \, \epsilon^{-\alpha t} + 1/R + R_1 (1 - \epsilon^{-\alpha t})$$

For which

$$v(t) = i(t)R = \epsilon^{-\alpha t} + R_1 /R + R_1 - R_1 /R + R_1 \, \epsilon^{-\alpha t}$$

Factoring,

$$v(t) = \frac{R_1}{R + R_1} + \left(1 - \frac{R_1}{R + R_1} \right) \epsilon^{-\alpha t}$$

$$= \frac{R_1}{R + R_1} + \left(\frac{R + R_1 - R_1}{R + R_1} \right) \epsilon^{-\alpha t}$$

$$= \frac{R_1}{R + R_1} + \left(\frac{R}{R + R_1} \right) \epsilon^{-\alpha t}$$

$$= \frac{R_1}{R + R_1} + \left(\frac{R}{R + R_1} \right)$$

$$\exp\left(\frac{-(R + R_1)t}{RR_1 C} \right) \quad \textbf{ANS}$$

PROBLEM 16-5

For the network shown, prove that the input impedance at terminal pair 1 is $Z_{1N}(s) = 1$ ohm.

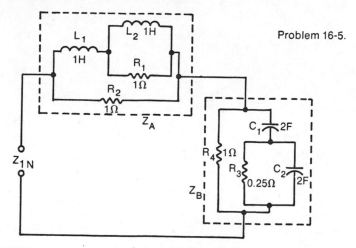

Problem 16-5.

Answer   *(ref. 1, p. 23-12; ref. 2, p. 198; ref. 5, p. 541)*

Now Z_{1N} is the impedance looking into the inputs, which is

$$Z_{1N} = Z_A + Z_B$$

$$Z_A = \frac{\left(sL_1 + \dfrac{sL_2 R_1}{sL_2 + R_1}\right)R_2}{sL_1 + \dfrac{sL_2 R_1}{sL_2 + R_1} \quad R_2}$$

$$Z_B = \frac{\left(1/sC_1 + \dfrac{R_3}{1 + sC_2 R_3}\right)R_4}{1/sC_1 + \dfrac{R_3}{1 + sC_2 R_3} + R_4}$$

Assigning the values for L, C, and R:

$$
\begin{aligned}
L_1 &= L_2 = 1 \\
R_1 &= R_2 = R_4 = 1 \\
C_1 &= C_2 = 2 \\
R_3 &= 1/4
\end{aligned}
$$

327

Therefore,

$$Z_A = \dfrac{\left(s(1) + \dfrac{s(1)\,(1)}{s(1) + 1}\right)(1)}{s(1) + \dfrac{s(1)\,(1)}{s(1) + 1} + 1}$$

$$= \frac{s(s + 1) + s}{s(s + 1) + s + (s + 1)}$$

$$= \frac{s^2 + 2s}{s^2 + 3s + 1}$$

$$Z_B = \frac{1/s(2) + \dfrac{1/4}{1 + s(2)\,(1/4)}\;(1)}{1/s(2) + \dfrac{1/4}{1 + s(2)\,(1/4)} + 1}$$

$$= \frac{\dfrac{1 + 1}{2s + 4 + 2s}\;\dfrac{1 + 1}{2s + 4 + 2s} + 1}$$

$$= \frac{4 + 2s + 2s}{4 + 2s + 2s + 2s(4 + 2s)}$$

$$= \frac{4 + 4s}{4s^2 + 12s + 4}$$

$$= \frac{s + 1}{s^2 + 3s + 1}$$

$$Z_{1N}(s) = \frac{s^2 + 2s}{s^2 + 3s + 1} + \frac{s + 1}{s^2 + 3s + 1}$$

$$= \frac{s^2 + 3s + 1}{s^2 + 3s + 1} = 1\Omega \text{ ANS}$$

Chapter 17

Power In AC Circuits

The Power Option P.E. candidate must be thoroughly familiar with the various theorems, conventions, transformations, and basic power circuits. The review material here has been produced by the Westinghouse Electric Corporation and serves to summarize the basic information required in working with power in AC circuits. Problems 17-15 through 17-22 cover electrical heating examples and have been provided by the General Electric Company and adapted for use as problems.

NETWORK THEOREMS

There are four basic network theorems, and these are explained as follows:

Superposition Theorem. This theorem states that each source of electromotive force (emf) produces currents in a linear network independently of those produced by any other emf. It follows that the emfs and currents of a given frequency can be treated independently of those of any other frequency, and of transients. The superposition theorem is a direct result of the fact that the fundamental simultaneous differential equations of the network are linear. (See any standard book on differential equations.)

Compensation Theorem. This theorem states that if the impedance of a branch of a network is changed by an amount ΔZ, the change in current in any branch is the same as would be produced by a compensating emf, $-\Delta ZI$, acting in series with the modified branch, where I is the original current in that branch. The term *compensating emf* means an emf which, if it were inserted, would neutralize the drop through ΔZ. This theorem follows directly from the superposition theorem.

Reciprocal Theorem. This theorem states that when a source of electromotive force is connected across one pair of terminals of a passive (no internal emfs) linear network and an ammeter is connected across a second pair of terminals, then the source of emf and the ammeter can be interchanged without altering the reading of the ammeter, provided neither the source nor receiver has internal impedance.

The proof for the reciprocal theorem may be deduced by considering an arbitrary network in which the solution for the current in any particular mesh (P) is

$$I_P = \frac{E_1 A_{P1}}{D} + \frac{E_2 A_{P2}}{D\ldots} + \frac{E_N A_{PN}}{D}$$

where D is the determinant of coefficients

$$D = \begin{matrix} Z_{11} & Z_{21} & Z_{31} & \cdot & \cdot & Z_{N1} \\ Z_{12} & Z_{22} & Z_{32} & \cdot & \cdot & Z_{N2} \\ Z_{13} & Z_{23} & Z_{33} & \cdot & \cdot & Z_{N3} \\ \cdot & \cdot & \cdot & & & \cdot \\ \cdot & \cdot & \cdot & & & \cdot \\ Z_{1N} & Z_{2N} & Z_{3N} & \cdot & \cdot & Z_{NN} \end{matrix}$$

and A_{PQ} is the cofactor of Z_{PQ} in the determinant.

If all emfs are zero except E_2 and E_3, for example, then if $E_2 = 0$, it follows that $I_2 = E_3 A_{23}/D$. While if $E_3 = 0$, then $I_3 = E_2 A_{32}/D$. So if $E_3 = E_2$, then I_3 must equal I_2 for conditions of the theorem.

Thevenin's Theorem. This theorem states that the current in any impedance Z_L, when connected to two terminals of a network, is the same as would be obtained if Z_L were supplied from a source voltage E' in series with an impedance Z', where E' is the open-circuit voltage at the terminals from which Z_L has been removed and Z' is the impedance that would be measured at these terminals after all generators have been removed and replaced by its internal impedance. (See Fig. 17-1.)

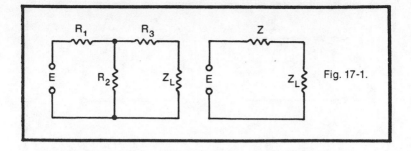

Fig. 17-1.

REFERENCE CURRENT AND VOLTAGE DIRECTIONS

To specify uniquely a vector current or voltage in a circuit, some system must be adopted to label the points between which the voltage is being described or the branch in which the current flows. This system must also indicate the *reference* or *positive direction*. Two common methods are the use of reference or positive direction arrows and the double-subscript notation.

Reference-Direction Arrows

When a network is to be solved to determine, for example, the current flow for a given set of impressed emfs, the network should first be marked with arrows (Fig. 17-2A and B) to indicate the reference of the positive direction for each current and voltage involved. These can be drawn arbitrarily, although if the predominant directions are known, their use as reference-direction arrows simplifies later interpretation.

The use of open-voltage arrowheads and closed (or solid) current arrowheads is sometimes used to avoid confusion in numerical work, where the E and I symbols are not used.

It must be decided at this point whether the voltage arrow is to represent a rise or a drop. In system calculations it is generally used as the rise in voltage. While this decision is arbitrary, once made, it must be adhered to consistently.

Finally, a vector value must be assigned to the voltage or current. It can be expressed as a complex number, or in polar form, or graphically and gives the magnitude and relative phase of the quantity with respect to some reference. A symbol, such as I_S, can be used to designate this vector quantity.

The known vectorial voltages or currents must be associated with the reference arrows in a manner consistent with the conditions of the problem. For example, suppose the problem in Fig. 17-2A is to determine the currents that would flow with the two voltage sources 180 degrees out of phase, and

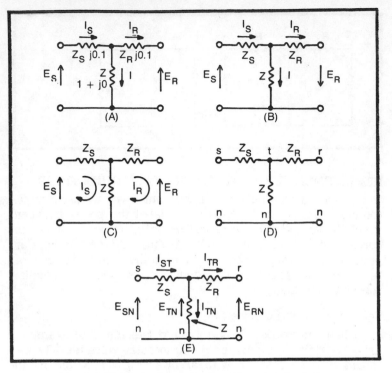

Fig. 17-2. Methods of notation used in network solution.

100 votts each, RMS, 60 Hz. Then if E_S is taken as $100 + j0$, E_R must be taken as $-100 + j0$. Had the arbitrary reference-direction arrows been taken as in Fig. 17-2B, then for the same problem a consistent set of voltages would be

$$E_S = 100 + j0, E_R = 100 + j0$$

Ordinarily, the reference-direction arrows for shunt voltages are directed from the neutral to the phase conductor as in Fig. 17-2A.

Summarizing then, the complete specification of a quantity in the reference-direction arrow system involves three elements:

1. The reference-direction arrow, drawn arbitrarily.
2. An agreement, consistently followed as to what the reference-direction arrow means; particularly whether the voltage arrow means the voltage of the point above the tail or the drop from tail to point.
3. A vector to represent the magnitude and relative phase of the quantity with respect to a reference.

332

Suggested convention: For voltages, the vector quantity shall indicate the voltage of the point of the arrow above the tail, that is, the rise in the direction of the arrow. It then is also the drop from point to tail.

Mesh Currents and Voltages

The mesh current system involves a somewhat different point of view. Here each current is continuous around a mesh and several currents may flow in the same branch. For example, in Fig. 17-2C, I_S and I_R flow in Z. The branch current is the vector sum of all the mesh currents in the branch, taken in the reference direction for the branch current. If such a network can be laid out "flat," it is most convenient to take the reference direction for mesh currents as simply "clockwise." Or circular arrows can be used as shown in Fig. 17-2C.

The same reference directions can be conveniently used for mesh emfs, which are the vector sum of all emfs acting around a particular mesh, taken in the reference direction.

Double-Subscript Notation

A double-subscript notation, as in Fig. 17-2D, is sometimes used and is, of course, equivalent to the drawing of reference arrows. Here again an arbitrary decision must be made as to what is intended. Suggested convention: I_{ST} means the current from S to T. E_{SN} means the voltage of S above N. It is apparent then that $E_{NS} = -E_{SN}$; $I_{ST} = -I_{TS}$, etc.

Setting Up Equations

If the work is analytical, by the method of equations, the equations must be set up consistent with the reference-direction arrows, regardless of the values of any known currents or voltages. Consistent equations for Figs. 17-2A, B, C, and D are as follows, using Kirchoff's laws. The voltage equations are written on the basis of adding all of the voltage rises in a clockwise direction around each mesh. The total must of course be zero. The current equations are written on the basis that the total of all the currents flowing up to a point must equal zero. Arrows and double subscripts have the meanings given in the preceding suggested conventions.

Referring to Fig. 17-2A,

$$E_S - I_S Z_S - IZ = 0$$
$$-I_R Z_R - E_R + IZ = 0$$
$$I_S - I - I_R = 0$$

Referring to Fig. 17-2B,

$$E_S - I_S Z_S - IZ = 0$$
$$-I_R Z_R + E_R + IZ = 0$$
$$I_S - I - I_R = 0$$

Referring to Fig. 17-2C,

$$E_S - I_S (Z_S + Z) + I_R Z = 0$$
$$I_S Z - I_R (Z + Z_R) - E_R = 0$$

Referring to Fig. 17-2D or,

$$E_{SN} - I_{ST} Z_S - I_{TN} Z = 0$$
$$I_{TN} Z - I_{TR} Z_R - E_{RN} = 0$$
$$I_{ST} - I_{TN} - I_{TR} = 0$$

SOLUTION BY EQUATIONS

A network of n meshes can be represented as having independent currents, I_1 to I_N, as shown in Fig. 17-3. The branch currents are combinations of these. (See *Branch Currents*.)

Mesh Impedances

Mesh impedances are defined generally as Z_{PQ}, which is a voltage drop in the reference direction in mesh Q per unit current in reference direction in mesh P. The curved arrows in Fig. 17-3 indicate reference directions in each mesh. In general,

$$Z_{PQ} = Z_{QP}$$

The impedance Z_{PP} is called the self-impedance, while the impedance Z_{PQ} is the mutual impedance.

Specifically in Fig. 17-3,

$$Z_{11} = R_A + R_B + j(X_{IA} + X_{IB} - X_{CA} - X_{CB})$$
$$Z_{12} = -R_B - j(X_{IB} - X_{CB})$$
$$Z_{13} = 0$$
$$Z_{14} = 0, \text{ etc.}$$
$$Z_{22} = R_B + R_C + R_D + R_M + j(X_{IB} + X_{IC} + X_{ID} - X_{CB} - X_{CC})$$

$$Z_{23} = -jX_{IDE}$$
$$Z_{24} = -R_M$$
$$Z_{25} = 0, \text{ etc.}$$
$$Z_{12} = Z_{21}$$
$$Z_{13} = Z_{31}, \text{ etc.}$$

where the first subscripts I and C indicate inductive and capacitive reactances respectively, and the polarity marks signify that the mutual flux links the two windings in a manner to produce maximum voltages at the same instant at the marked ends of the windings (Fig. 17-3B).

Mesh Emfs

Reference-positive directions for branch emfs, E_A , E_B , etc., are shown by associated arrows.

A mesh emf is the sum of the branch emfs acting around that particular mesh in the reference direction.

The same reference direction will be used for mesh emfs as for mesh currents.

Specifically in Fig. 17-3.

$$E_1 = E_A - E_T$$
$$E_2 = -E_C$$
$$E_3 = -E_B$$
$$E_4 = E_D + E_F + E_C \text{ , etc.}$$
$$E_N = E_G$$

Equations

Kirchoff's Law states that the voltage drop around each closed mesh must equal the emf impressed in that mesh.

$$I_1 Z_{11} + I_2 Z_{21} +$$

$$I_3 Z_{31} + \text{ . . . } + I_N Z_{N1} = E_1$$
$$I_1 Z_{12} + I_2 Z_{22} +$$

$$I_3 Z_{32} + \text{ . . . } + I_N Z_{N2} = E_2$$
$$I_1 Z_{13} + I_2 Z_{23} +$$

$$I_3 Z_{33} + \text{ . . . } + I_N Z_{N3} = E_S$$
$$\cdot \quad \cdot \quad \cdot \quad \cdot \quad \cdot$$
$$\cdot \quad \cdot \quad \cdot \quad \cdot \quad \cdot$$
$$\cdot \quad \cdot \quad \cdot \quad \cdot \quad \cdot$$

$$I_1 Z_{1N} + I_2 Z_{2N} +$$

$$I_3 Z_{3N} + \text{ . . . } + I_N Z_{NN} = E_N$$

Branch Currents

Branch currents can now be obtained by combination. The vector sum of all mesh currents flowing in a branch is the branch current.

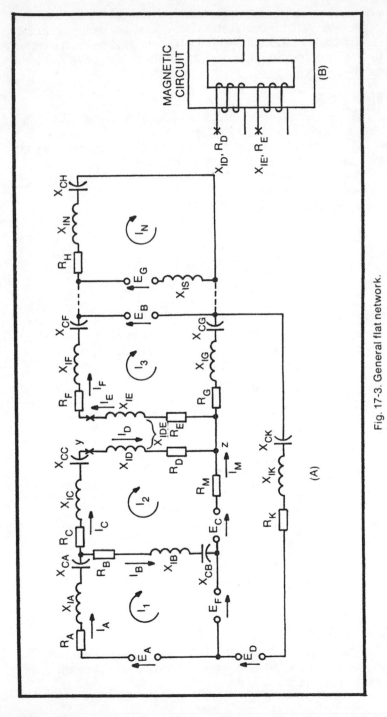

Fig. 17-3. General flat network.

Specifically with reference to Fig. 17-3,

$$I_A = I_1$$
$$I_B = I_1 - I_2$$
$$I_C = I_2$$
$$I_M = I_4 - I_2 \text{ , etc.}$$

Branch Voltages

The branch voltages, E_{YX}, E_{YZ}, etc., or the voltages between any two conductively connected points in the network, as E_{XZ}, can be obtained by vectorial addition of all voltages, both emfs and drops through any path connecting the two points.

The voltage drop from x to y, D_{XY}, and the voltage of point x above point y, E_{XY}, are the same. (Note that drop is measured from first subscript to second. The voltage of the first subscript is measured above the second.)

$$D_{XY} = E_{XY} = I_C R_C + jI_C (X_{IC} - X_{CC})$$
$$D_{YZ} = E_{YZ} = I_D R_D + jI_D X_{ID} - jI_E X_{IDE}$$
$$D_{WZ} = E_{WZ} = -E_C + I_M R_M$$
$$D_{XZ} = E_{XZ} = I_C R_C + jI_C (X_{IC} - X_{CC}) + I_D R_D + jI_D X_{ID} - jI_E X_{ID}$$

Note that

$$D_{XZ} = D_{XY} + D_{YZ}$$
$$E_{XZ} = E_{YZ} + E_{XY}$$

Example of Solution by Equations

(A) Given the impedances and emfs of the network in Fig. 17-4 required to find the currents. Note: The headings (B), (C), etc., refer to the corresponding paragraphs B, C, etc., in which the method and equations are given.

(B) Mesh impedances:

$$Z_{11} = 0 + j(7.55 + 136) = 0 + j143.55$$
$$Z_{12} = Z_{21} = -j113.0$$
$$Z_{22} = 5 + j(150.5 - 2.65) = 5 + j147.85$$

(C) Mesh emfs:

$$E_1 = 2.9 + j56.6$$
$$E_2 = -0.29 - j5.66$$

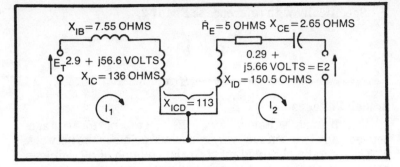

Fig. 17-4. Example of a solution by equations.

(D) Equations: It is unnecessary to write these out completely since only the solutions are desired. However, for completeness they are

$$I_1 (0 + j143.55) + I_2 (-j113.0) = 2.9 + j56.6$$
$$I_1 (-j113.0) + I_2 (5 + j147.85) = -0.29 - j5.66$$

(E) Mesh currents:

$$D = \begin{vmatrix} 0 + j143.6 & -j113.0 \\ -j113.0 & 5 + j147.9 \end{vmatrix} = -8400 + j718$$

$A_{11} = 5 + j147.9$
$A_{12} = A_{21} = +j113$
$A_{22} = 0 + j143.6$

$$I_1 = \frac{E_1 A_{11}}{D} + \frac{E_2 A_{12}}{D}$$

$$= \frac{(2.9 + j56.6)(5 + j147.9)}{-8400 + j718} + \frac{(-0.29 - j5.66)(j113)}{-8400 + j718}$$

$$= 0.921 + j0.007$$

$$I_2 = \frac{E_1 A_{21}}{D} + \frac{E_2 A_{22}}{D}$$

$$= \frac{(2.9 + j56.6)(j113)}{-8400 + j718} + \frac{(-0.29 - j5.66)(+j143.6)}{-8400 + j718}$$

$$= 0.662 + j0.034$$

Note that the term $A_{12}/D = A_{21}/D$ is the *transfer admittance* between meshes 1 and 2, or is the current in either

338

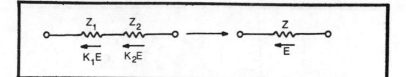

Fig. 17-5. Impedances in series.

one of these meshes per unit of emf impressed in the other. Thus the voltage $E_2 = -0.1E_1$ and likewise the current in mesh 1 resulting from E_2 is -0.1 times the current in mesh 2 due to E_1.

TRANSFORMATIONS IN IMPEDANCE FORM

Certain transformations are desirable and often necessary in solving network problems. These basic transformations occur between series and parallel networks, delta and star, Pi and T.

Impedances in Series

Refer to Fig. 17-5.

$$Z = Z_1 + Z_2 \qquad\qquad K_1 + K_2 = 1$$

$$K_1 = \frac{Z_1}{Z_1 + Z_2} = \frac{Z_1}{Z_2} \qquad K_2 = \frac{Z_2}{Z_1 + Z_2} = \frac{Z_2}{Z}$$

Impedances in Parallel

Refer to Fig. 17-6. The parallel of two impedances is the product divided by the sum.

$$Z = \frac{Z_1 Z_2}{Z_1 + Z_2} \qquad\qquad K_1 + K_2 = 1$$

$$K_1 = \frac{Z_2}{Z_1 + Z_2} \qquad\qquad K_2 = \frac{Z_1}{Z_1 + Z_2}$$

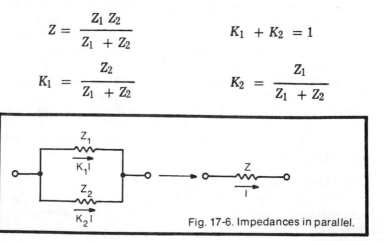

Fig. 17-6. Impedances in parallel.

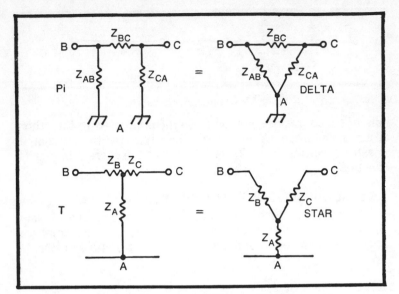

Fig. 17-7. Pi and delta; T and stars of Y are the same.

Suggested order of calculation

$D = Z_1 + Z_2$
$K_1 = Z_2 / D = $ current in Z_1 per unit current in Z.
$K_2 = 1 - K_1 = $ current in Z_2 per unit current in Z.
$Z = K_1 Z_1$

Delta and Star Networks

Delta and star forms used in general networks are identical with Pi and T forms used in specialized transmission forms of networks. See Fig. 17-7. The difference is simply in the manner of drawing the circuit. Thus the star-delta and delta-star transformations are at once, T-to-Pi and Pi-to-T transformations. The arrow between parts of the figures indicates that the figure on the left is being transformed to the figure on the right. It is assumed then that the currents are determined for the figure on the right and the equations under the figure on the left are for determining the resulting currents (or voltages) in it.

Delta to Star (or Pi-to-T) Transformation

The star impedances are the product of adjacent delta impedances divided by the sum of all delta impedances. (See Fig. 17-8).

340

$$D = Z_{AB} + Z_{BC} + Z_{CA}$$
$$Z_A = Z_{CA} Z_{AB} / D$$
$$Z_B = Z_{AB} Z_{BC} / D$$
$$Z_C = Z_{BC} Z_{CA} / D$$
$$I_{AB} = -(Z_{CA}/D)I_A + (Z_{BC}/D)I_B$$
$$= -K_{CA} I_A + K_{BC} I_B$$

$$I_{BC} = -(Z_{AB}/D)I_B + (Z_{CA}/D)I_C$$
$$= -K_{AB} I_B + K_{CA} I_C$$

$$I_{CA} = -(Z_{BC}/D)I_C + (Z_{AB}/D)I_A$$
$$= -K_{BC} I_C + K_{AB} I_A$$

where K_{AB}, K_{BC}, and K_{CA} are current distribution factors such that $K_{AB} + K_{BC} + K_{CA} = 1$.

Suggested order of calculation:

$$D = Z_{AB} + Z_{BC} + Z_{CA}$$
$$K_{AB} = Z_{AB}/D$$
$$Z_A = Z_{CA} K_{AB}$$
$$K_{BC} = Z_{BC}/D$$
$$Z_B = Z_{AB} K_{BC}$$
$$K_{CA} = Z_{CA}/D$$
$$Z_C = Z_{BC} K_{CA}$$

After I_A, I_B, and I_C have been found, then I_{CA}, I_{AB}, and I_{BC} can be determined.

Star to Delta (or T to Pi) Transformation

Refer to Fig. 17-9.

$$D = 1Z_A + 1/Z_D + 1/Z_C$$
$$Z_{AB} = DZ_A Z_B$$
$$Z_{BC} = DZ_B Z_C$$
$$Z_{CA} = DZ_C Z_A$$
$$I_A = I - I_{AB}$$
$$I_B = I_{AB} - I_{BC}$$
$$I_C = I_{BC} - I_{CA}$$

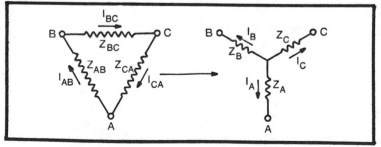

Fig. 17-8. Delta of star (impedance form).

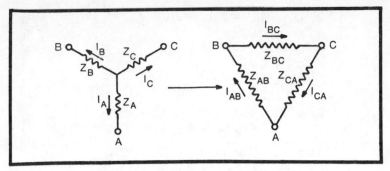

Fig. 17-9. Star of delta (impedance form).

Alternative forms of the transformation formulas follows:

$$N = Z_A Z_B + Z_A Z_C + Z_B Z_C$$
$$Z_{AB} = N/Z_C = Z_A + Z_B + Z_A Z_B /Z_C$$
$$Z_{BC} = N/Z_A = Z_B + Z_C + Z_B Z_C /Z_A$$
$$Z_{CA} = N/Z_B = Z_C + Z_A + Z_C Z_A /Z_B$$

Star With Mutuals to Star Without Mutuals

Refer to Fig. 17-10.

$$Z_1 = Z_A + Z_{BC} - Z_{AB} - Z_{CA}$$
$$Z_2 = Z_B + Z_{CA} - Z_{BC} - Z_{AB}$$
$$Z_3 = Z_C + Z_{AB} - Z_{CA} - Z_{BC}$$

Polarity marks require that with all reference directions from center outward as shown, all self and mutual drops are from center outward. That is, it is understood that with the polarity marks as shown, the voltage drop from the center to A will be written:

$$D_{NA} = I_A Z_A + I_B Z_{AB} + I_C Z_{CA}$$

and the numerical (vector) values and signs assigned to Z_{AB} and Z_{AC} must be such as to make this true. It follows that Z_{AB}

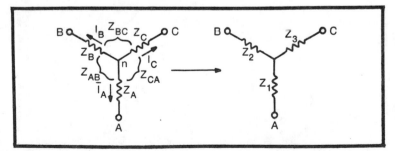

Fig. 17-10. Star with mutuals to star without mutuals (impedance form).

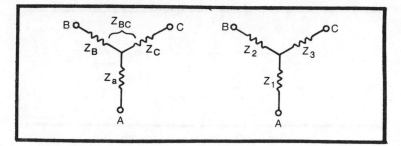

Fig. 17-11. Star with one mutual to star without mutual (impedance form).

is defined as the voltage drop from N to A divided by the current from N to B that causes the drop.

Special Case: Star with one mutual between two branches to star without mutual. (See Fig. 17-11.)

$$Z_1 = Z_A + Z_{BC}$$
$$Z_2 = Z_B - Z_{BC}$$
$$Z_3 = Z_C - Z_{BC}$$

Two Self-Impedances With Mutual to Star or T

This is also the equivalent circuit of a two-winding transformer (Fig. 17-12).

$$Z_1 = Z_{11} - Z_{12}$$
$$Z_2 = Z_{22} - Z_{12}$$

NOTE: This transformation involves bringing B and D to the same potential and is permissible only when these potentials are not otherwise fixed. Strictly, the form on the

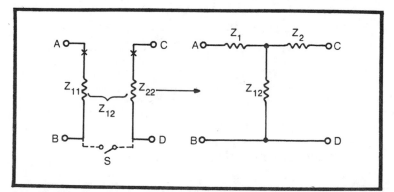

Fig. 17-12. Two self impedances and a mutual transformed to an equivalent star or T. Or the equivalent circuit of a two winding transformer.

343

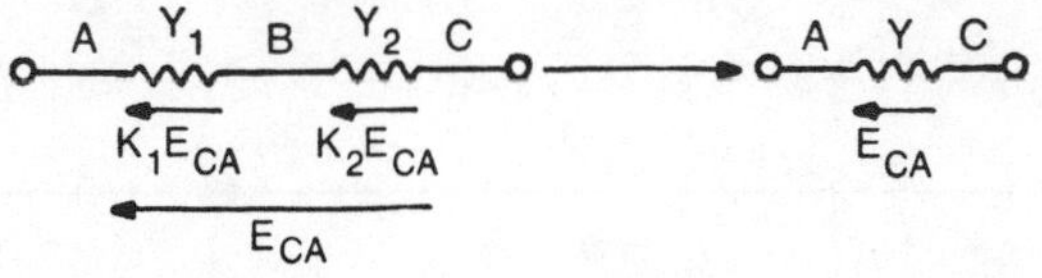

Fig. 17-13. Admittances in series.

right is equivalent to that on the left with switch S closed. However, if the closure of S would not alter the current division, it can be considered closed and the equivalent circuit used. The resulting potentials E_{AB} and E_{CD} will be correct but the potentials E_{CA} and E_{DB}, which are definite in the equivalent, are actually indeterminate in the original circuit and must not be construed as applying there. See note under E for meaning of polarity marks, considering B and D as point N.

This is the familiar equivalent circuit of a two-winding transformer, provided all impedances have first been placed on a common turns basis. In this case Z_{12} is the exciting impedance and $Z = Z_1 + Z_2$ the leakage impedance.

TRANSFORMATIONS IN ADMITTANCE FORM

Similar to the preceding transformations in impedance form, the basic transformations can also be written in admittance form.

Admittances in Series

Refer to Fig. 17-13.

$$Y = \frac{Y_1 \, Y_2}{Y_1 + Y_2}$$

$$K_1 + K_2 = 1$$

$$K_1 = \frac{Y_2}{Y_1 + Y_2}$$

$$K_2 = \frac{Y_1}{Y_1 + Y_2}$$

Suggested order of calcuation.

$$D = Y_1 + Y_2$$
$$K_1 = Y_2 / D$$
$$K_2 = 1 - K_1$$
$$Y = K_1 \, Y_1 = K_2 \, Y_2$$

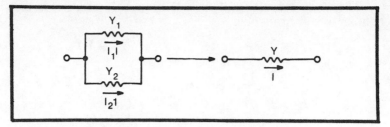

Fig. 17-14. Admittances in parallel.

Admittances in Parallel

Refer to Fig. 17-14.

$$Y = Y_1 + Y_2$$

$$K_1 K_2 = 1$$

$$K_1 = \frac{Y_1}{Y_1 + Y_2} = \frac{Y_1}{Y}$$

$$K_2 = \frac{Y_2}{Y_1 + Y_2} = \frac{Y_2}{Y}$$

General Star to Mesh Transformation

This is also known as the "elimination of a junction." See Fig. 17-15.

Note: A network can be solved by eliminating one junction point after another until a single-branch mesh remains.

Rule: A mesh branch is the product of adjacent star branches divided by the sum of all star branches.

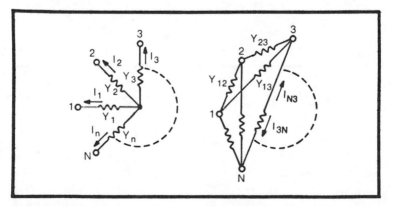

Fig. 17-15. General star to mesh transformation or elimination of a junction (admittance form).

345

The mesh contains $N/2(N-1)$ branches, where N is the number of star branches.

$$D = Y_1 + Y_2 + Y_3 + \ldots + Y_N$$
$$Y_{12} = Y_1 \, Y_2 \, /D$$
$$Y_{13} = Y_1 \, Y_3 \, /D, \text{ etc.}$$
$$Y_{PQ} = Y_P \, Y_Q \, /D$$
$$I_1 = I_{21} + I_{31} + \ldots + I_{NL}$$
$$I_2 = I_{12} + I_{32} + \ldots + I_{N2} \text{, etc.}$$
$$I_P = I_{1P} + I_{2P} + \ldots + I_{NP}$$

In which the positive reference direction for any mesh current I_{PQ} is toward terminal Q.

Suggested order of calculation:

$$D = Y_1 + Y_2 + Y_3 + \cdots + Y_N$$
$$Y_1 \, /D = K_1$$
$$Y_{12} = Y_1 \, Y_2 \, /D = K_1 \, Y_2$$
$$Y_{13} = Y_1 \, Y_3 \, /D = K_1 \, Y_3 \text{, etc.}$$
$$Y_2 \, /D = K_2$$
$$Y_{23} = Y_2 \, Y_3 \, /D = K_2 \, Y_3$$
$$Y_{24} = Y_2 \, Y_4 \, /D = K_2 \, Y_4 \text{, etc.}$$

Star to Delta (or T to Pi) Transformation

This is actually a special case of the preceding transformation. Referring to Fig. 17-16,

$$D = Y_A + Y_B + Y_C$$
$$Y_{AB} = Y_A \, Y_B \, /D$$
$$Y_{BC} = Y_B \, Y_C \, /D$$
$$Y_{CA} = Y_C \, Y_A \, /D$$
$$I_A = I_{CA} - I_{AB}$$
$$I_B = I_{AB} - I_{BC}$$
$$I_C = I_{BC} - I_{CA}$$

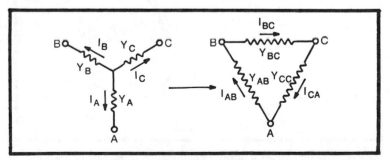

Fig. 17-16. Star to delta (admittance form).

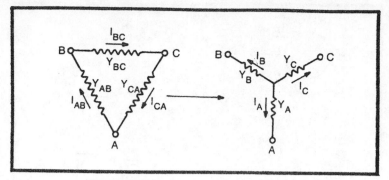

Fig. 17-17. Delta to star (admittance form).

Suggested order of calculation is the same as for the general transformation.

Delta to Star (or Pi to T) Transformation

Refer to Fig. 17-17.

$$D = 1/Y_{AB} + 1/Y_{BC} + 1/Y_{CA}$$
$$Y_A = DY_{CA} Y_{AB}$$
$$Y_B = DY_{AB} Y_{BC}$$
$$Y_C = DY_{BC} Y_{CA}$$

$$I_{AB} = -K_{CA} I_A + K_{BC} I_B$$
$$I_{BC} = -K_{AB} I_B + K_{CA} I_C$$
$$I_{CA} = -K_{BC} I_C + K_{AB} I_A$$

where $K_{AB} = 1/DY_{AB}$
$K_{BC} = 1/DY_{BC}$
$K_{CA} = 1/DY_{CA}$

UNBALANCE AND FAULT CONDITIONS

Equations for calculating the sequence quantities at the point of unbalance are given below for the unbalance conditions that occur frequently. In these equations E_{1F}, E_{2F}, and E_{0F} are components of the line-to-neutral voltages at the point of unbalance; I_{1F}, I_{2F}, and I_{0F} are components of the fault current I_F; Z_1, Z_2, and Z_0 are impedances of the system (as viewed from the unbalanced terminals) to the flow of the sequence currents; and E_A is the line-to-neutral positive-sequence generated voltage.

Three-Phase Fault

Refer to Fig. 17-18.

$$I_{1F} = I_F = \frac{E_{A1}}{Z_1}$$

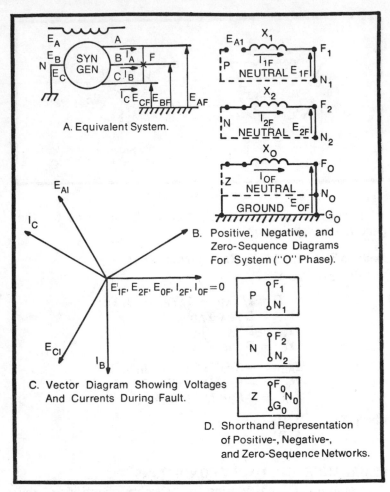

Fig. 17-18. Three-phase short circuit on generator.

Single-Line-to-Ground Fault

Refer to Fig. 17-19.

$$I_{1F} = I_{2F} = I_{0F} = \frac{E_{A1}}{Z_1 + Z_2 + Z_0}$$

$$I_F = I_{1F} + I_{2F} + I_{0F} = 3I_{0F}$$

$$E_{1F} = E_{A1} - I_{1F}Z_1 = E_{A1}\frac{(Z_2 + Z_0)}{Z_1 + Z_2 + Z_0}$$

$$E_{2F} = -I_{2F} Z_2 = -\frac{E_{A1} Z_2}{Z_1 + Z_2 + Z_0}$$

$$E_{0F} = -I_{0F} Z_0 = -\frac{E_{A1} Z_0}{Z_1 + Z_2 + Z_0}$$

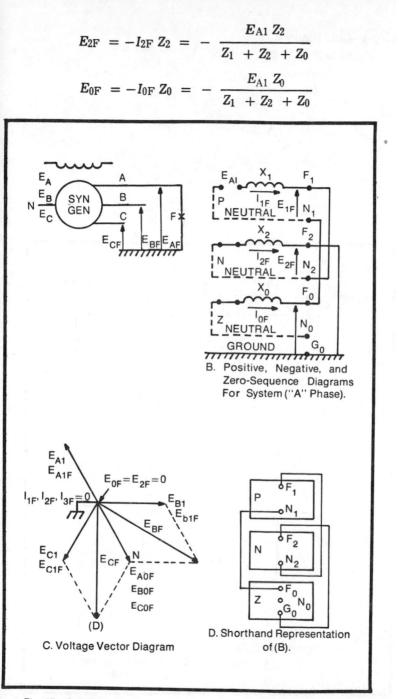

B. Positive, Negative, and Zero-Sequence Diagrams For System ("A" Phase).

C. Voltage Vector Diagram

D. Shorthand Representation of (B).

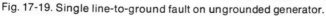

Fig. 17-19. Single line-to-ground fault on ungrounded generator.

Line-to-Line Fault

Refer to Fig. 17-20.

$$I_{1F} = -I_{2F} = \frac{E_{A1}}{Z_1 + Z_2}$$

$$I_F = \sqrt{3}I_{1F}$$

$$E_{1F} = E_{A1} - I_{1F}Z_1 = \frac{E_{A1}Z_2}{Z_1 + Z_2}$$

$$E_{2F} = -I_{2F}Z_2 = \frac{E_{A1}Z_2}{Z_1 + Z_2}$$

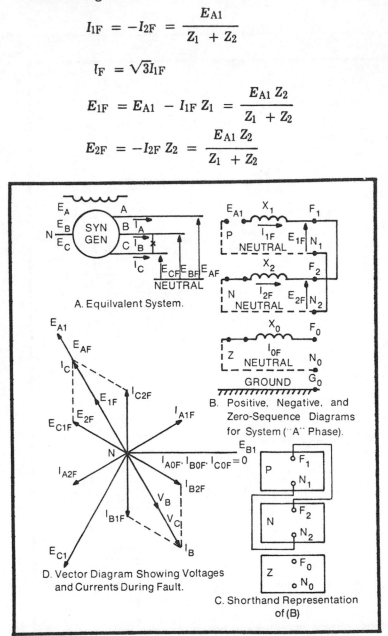

A. Equivalent System.

B. Positive, Negative, and Zero-Sequence Diagrams for System ("A" Phase).

D. Vector Diagram Showing Voltages and Currents During Fault.

C. Shorthand Representation of (B)

Fig. 17-20. Line-to-line fault on grounded or ungrounded generator.

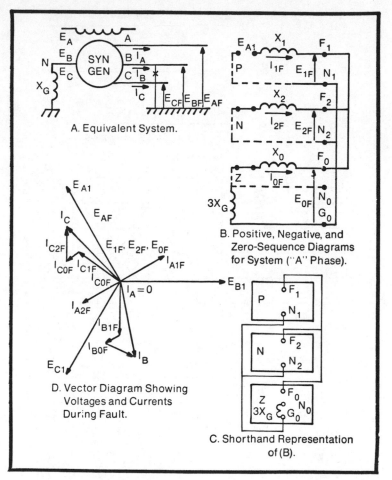

A. Equivalent System.

B. Positive, Negative, and Zero-Sequence Diagrams for System ("A" Phase).

D. Vector Diagram Showing Voltages and Currents During Fault.

C. Shorthand Representation of (B).

Fig. 17-21. Double line-to-ground fault on generator grounded through a neutral reactor.

Double Line-to-Ground Fault

Refer to Fig. 17-21.

$$I_{1F} = \frac{E_{A1}}{Z_1 + Z_2 Z_0 / (Z_2 + Z_0)} = \frac{E_{A1}(Z_2 + Z_0)}{Z_1 Z_2 + Z_1 Z_0 + Z_2 Z_0}$$

$$I_{2F} = -\frac{Z_0}{Z_2 + Z_0} I_{1F} = \frac{-Z_0 E_{A1}}{Z_1 Z_2 + Z_1 Z_0 + Z_2 Z_0}$$

$$I_{0F} = -\frac{Z_2}{Z_2 + Z_0} = \frac{-Z_2 E_{A1}}{Z_1 Z_2 + Z_1 Z_0 + Z_2 Z_0}$$

$$E_{1F} = E_{A1} - I_{1F} Z_1 = \frac{Z_2 Z_0 E_{A1}}{Z_1 Z_2 + Z_1 Z_0 + Z_2 Z_0}$$

$$E_{2F} = -I_{2F} Z_2 = \frac{Z_2 Z_0 E_{A1}}{Z_1 Z_2 + Z_1 Z_0 + Z_2 Z_0}$$

$$E_{0F} = -I_{0F} Z_0 = \frac{Z_2 Z_0 E_{A1}}{Z_1 Z_2 + Z_1 Z_0 + Z_2 Z_0}$$

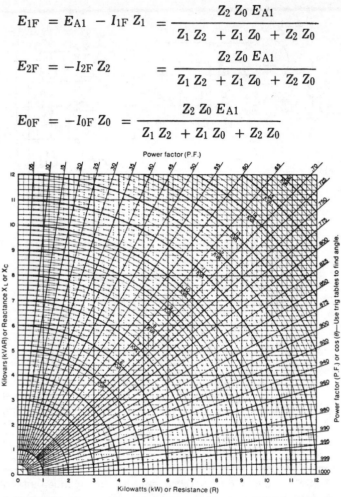

1. To find kVA and power factor, follow the horizontal line from the kVAR value and the vertical line from the kW value. At the intersection of these lines is the kVA arc and the power factor radius.

2. To find kW and kVAR. From the intersection of the kVA arc and the power factor radius; a horizontal to the left gives kVAR, a vertical line down gives kW.

3. This chart can be used for values of kW, kVA, and kVAR greater or less than 1 through 12 by using multiples of 10 on all quantities except power factor.

4. To convert from rectangular to polar, use vertical and horizontal lines, which are R and X_L or X_C, respectively. At the intersection of rectangular coordinates, follow the polar arc to either edge and read Z. Also, at the intersection, follow the angle line to its edge and read cos (θ). Use trig table to get angle.

5. To convert from polar to rectangular, at the intersection of the angle, cos (θ), and polar value Z, move down to read R and to the left to read X_L or X_C.

Fig. 17-22. Conversion chart for kW, kVA, kVAR, power factor cos (θ), rectangular and polar coordinates. (Courtesy EEM).

CONVERSION FACTORS AND NOMOGRAPHS

Table 17-1 lists a summary of conversion factors for converting admittance and impedance, and Pi and T networks.

Figure 17-22 is a useful nomograph for AC power, power factor, and reactance problems, and it will also permit rectangular to polar conversions.

Figure 17-23 is a general nomograph for power, voltage, current, and resistance, permitting all variables to be seen at

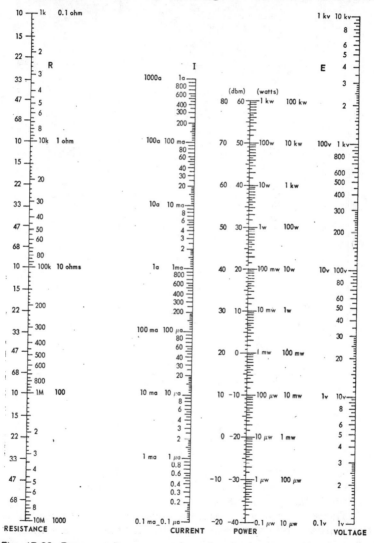

Fig. 17-23. Power, voltage, current, and resistance nomograph. To determine any two parameters, lay a straight-edge between any two knowns and read the other two unknowns. (Courtesy EDN)

Table 17-1. Conversion Formulas for Transmission Type Networks.

		To Convert From		
		A B C D	Admittance	Impedance
ABCD	$A=$	ABCD Constants	$\dfrac{Y_{11}}{-Y_{12}}$	$\dfrac{Z_{22}}{Z_{12}}$
	$B=$	$E_S = AE_R + BI_R$ $I_S = CE_R + DI_R$	$\dfrac{1}{Y_{12}}$	$\dfrac{Z_{11}Z_{22} - Z_{12}^2}{Z_{12}}$
	$C=$	$E_R = DE_S - BI_S$ $I_R = -CE_S + AI_S$	$\dfrac{Y_{12}^2 - Y_{11}Y_{22}}{Y_{12}}$	$\dfrac{1}{Z_{12}}$
	$D=$		$\dfrac{Y_{22}}{Y_{12}}$	$\dfrac{Z_{11}}{Z_{12}}$
Admittance	$Y_{11}=$	$\dfrac{A}{B}$	Admittance Constants	$\dfrac{Z_{22}}{Z_{11}Z_{22} - Z_{12}^2}$
	$Y_{12}=$	$-\dfrac{1}{B}$	$I_1 = Y_{11}E_1 + Y_{12}E_2$	$\dfrac{Z_{12}}{Z_{11}Z_{22} - Z_{12}^2}$
	$Y_{22}=$	$\dfrac{D}{B}$	$I_2 = Y_{12}E_1 + Y_{22}E_2$	$\dfrac{Z_{11}}{Z_{11}Z_{12} - Z_{12}^2}$
Impedance	$Z_{11}=$	$\dfrac{D}{C}$	$\dfrac{Y_{22}}{Y_{11}Y_{22} - Y_{12}^2}$	Impedance Constants
	$Z_{12}=$	$\dfrac{1}{C}$	$-\dfrac{Y_{12}}{Y_{11}Y_{22} - Y_{12}^2}$	$E_1 = Z_{11}I_1 + Z_{12}I_2$
	$Z_{22}=$	$\dfrac{A}{C}$	$\dfrac{Y_{11}}{Y_{11}Y_{22} - Y_{12}^2}$	$E_2 = Z_{12}I_1 + Z_{22}I_2$
Equiv. Pi	$Y_R=$	$\dfrac{A-1}{B}$	$Y_{11} + Y_{12}$	$\dfrac{Z_{22} - Z_{12}}{Z_{11}Z_{22} - Z_{12}^2}$
	$Z=$	B	$-\dfrac{1}{Y_{12}}$	$\dfrac{Z_{11}Z_{22} - Z_{12}^2}{Z_{12}}$
	$Y_S=$	$\dfrac{D-1}{B}$	$Y_{22} + Y_{12}$	$\dfrac{Z_{11} - Z_{12}}{Z_{11}Z_{22} - Z_{12}^2}$
Equiv. T	$Z_R=$	$\dfrac{D-1}{C}$	$\dfrac{Y_{22} + Y_{12}}{Y_{11}Y_{22} - Y_{12}^2}$	$Z_{11} - Z_{12}$
	$Y=$	C	$-\dfrac{Y_{11}Y_{22} - Y_{12}^2}{Y_{12}}$	$\dfrac{1}{Z_{12}}$
	$Z_S=$	$\dfrac{A-1}{C}$	$\dfrac{Y_{11} + Y_{12}}{Y_{11}Y_{12} - Y_{12}^2}$	$Z_{22} - Z_{12}$

the same time, which can be helpful in some transmission and load-matching problems.

The following conversion factors are also quite useful in the heating problems presented in the last half of the chapter.

$$1\,\text{Btu} = 252\,\text{calories} = 0.293\,\text{watt-hours}$$
$$1\,\text{Btu per pound} = 1.8\,\text{calories per gram}$$

Table 17-1 continued.

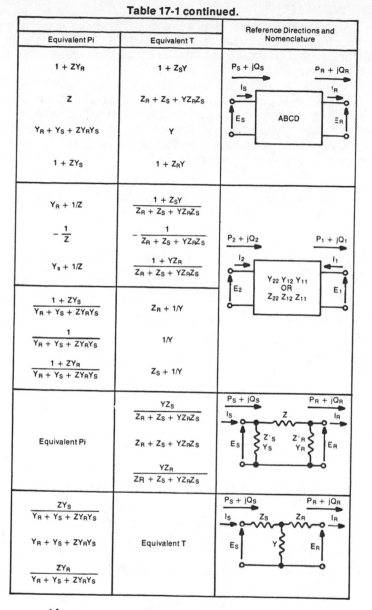

Equivalent Pi	Equivalent T	Reference Directions and Nomenclature
$1 + ZY_R$	$1 + Z_S Y$	$P_S + jQ_S$ → ; $P_R + jQ_R$ →
Z	$Z_R + Z_S + YZ_R Z_S$	I_S, I_R
$Y_R + Y_S + ZY_R Y_S$	Y	E_S, ABCD, E_R
$1 + ZY_S$	$1 + Z_R Y$	
$Y_R + 1/Z$	$\dfrac{1 + Z_S Y}{Z_R + Z_S + YZ_R Z_S}$	$P_2 + jQ_2$; $P_1 + jQ_1$
$-\dfrac{1}{Z}$	$-\dfrac{1}{Z_R + Z_S + YZ_R Z_S}$	I_2, I_1
$Y_s + 1/Z$	$\dfrac{1 + YZ_R}{Z_R + Z_S + YZ_R Z_S}$	E_2, $Y_{22}\, Y_{12}\, Y_{11}$ OR $Z_{22}\, Z_{12}\, Z_{11}$, E_1
$\dfrac{1 + ZY_S}{Y_R + Y_S + ZY_R Y_S}$	$Z_R + 1/Y$	
$\dfrac{1}{Y_R + Y_S + ZY_R Y_S}$	$1/Y$	
$\dfrac{1 + ZY_R}{Y_R + Y_S + ZY_R Y_S}$	$Z_S + 1/Y$	
Equivalent Pi	$\dfrac{YZ_S}{Z_R + Z_S + YZ_R Z_S}$	$P_S + jQ_S$ → ; $P_R + jQ_R$ →
	$Z_R + Z_S + YZ_R Z_S$	I_S, Z, I_R ; E_S, Z'_S Y_S, Z'_R Y_R, E_R
	$\dfrac{YZ_R}{Z_R + Z_S + YZ_R Z_S}$	
$\dfrac{ZY_S}{Y_R + Y_S + ZY_R Y_S}$		$P_S + jQ_S$ → ; $P_R + jQ_R$ →
$Y_R + Y_S + ZY_R Y_S$	Equivalent T	I_S, Z_S, Z_R, I_R ; E_S, Y, E_R
$\dfrac{ZY_R}{Y_R + Y_S + ZY_R Y_S}$		

1 horsepower = 745.2 watts
1 kilowatt-hour = 3412 Btu
1 gallon = 231 in. = 0.1337 cu ft = 3.785 liters

In addition, 1 kilowatt-hour represents the energy required to evaporate 3.5 pounds of water at 212°F.

PROBLEM 17-1

An aluminum cable, steel reinforced, contains 32 circular conductors of aluminum of which each is 0.1562 inches in diameter, and 20 strands of steel conductor of which each is 0.09 inches in diameter.

(A) What would be the resistance of this cable per mile?

(B) By what percentage does the presence of the steel core alter the resistance in comparison to what it would have been had the cable contained only the aluminum wires? Assume the resistivity of the aluminum to be 18Ω per circular-mil foot, and for the steel, 95Ω per circular-mil foot.

Answer

Aluminum cable: A diameter of 0.1562 inches corresponds to 156.2 mils. The area is then

$$(0.1562 \text{ mils})^2 = 24,500 \text{ CM}$$

Steel reinforced cable: A diameter of 0.09 inches is 90 mils, and

$$(90 \text{ mils})^2 = 8100 \text{ CM}$$

Resistance per mile, aluminum cable:

$$R_{AL} = \frac{(18)(5280)}{(32)(24.5 \times 10^3)} = 0.121\Omega \text{ per mile}$$

Resistance per mile, steel reinforcing:

$$R_S = \frac{(95)(5280)}{(20)(8100)} = 3.1\Omega \text{ per mile}$$

(A) Total resistance is the combined resistance of the aluminum and steel in parallel:

$$R_T = (0.121)(3.1)/(0.121 + 3.1)$$

$$= 0.1162\Omega \text{ per mile ANS}$$

(B) Percentage difference in resistance is

$$100(R_{AL} - R_T)/R_{AL} = 100(0.121 - 0.1162)/(0.121)$$
$$= 3.97\% \text{ less ANS}$$

PROBLEM 17-2

In an industrial plant, the primary source of power is polyphase alternating current. However, the industrial plant has need for direct-current energy for a demand of about 200 kW. To supply this direct-current energy, the engineer may use: (1) A synchronous motor, DC generator set. (2) An induction motor, DC generator set. (3) Mercury arc rectifiers. (4) Ignitron rectifiers. (5) Silicon controlled rectifiers.

Visualize this plant, then select one of the five possible methods for supplying the required direct current. Give the reasons governing your selection.

Answer

Method (1) consumes much power, is inefficient, and may need power-factor correction.

Method (2) is cheaper than method (1), but the energy losses are higher and so are the maintenance costs.

Method (3) is cheaper than the first two methods, but it has limited current capabilities, possibly making it inadequate.

Method (4) has lower power losses then the first three methods, and it has good current-handling capabilities.

Method (5) has the lowest losses, the highest efficiency, and occupies the smallest space of all. This method also requires less cooling energy than method (4).

Conclusion: Method (5) is the best choice.

PROBLEM 17-3

What are the commercially used electrical characteristics of the following items of equipment?

(A) Incandescent lamp for street lighting.
(B) Incandescent lamp for house lighting.
(C) Incandescent lamp for flood lighting.
(D) Lamp for movie projection.
(E) Sewing machine motor.
(F) Washing machine motor.
(G) Household refrigerator motor.
(H) Starting motor for 5-passenger automobile.
(I) Elevator motor for 20-story office building.
(J) Streetcar motor
(K) Neon sign
(L) Motor for a pump rated at 10,000 gpm and 200 ft head.
(M) Electrostatic dust collector.

Answer

(A) 120V to 240V; 7A; 2000 Cp. (Mercury vapor is preferred.)
(B) 50W to 100W; relatively long life.
(C) 500W to 750W; shorter life.

(D) 500W; high intensity; short life.

(E) AC/DC universal motor.

(F) Capacity, split-phase, induction motor.

(G) Induction motor (also capacity type).

(H) Series DC; high torque.

(I) Three-phase, induction, DC motor with an AC-to-DC converter.

(J) Series motor; 550V to 750V DC.

(K) 20 kV; high reactance.

(L) Squirrel cage motor.

(M) High voltage; low current DC; similar to neon type.

PROBLEM 17-4

An industrial load of 30,000 kW, 2300V, has a lagging power factor of 0.6. To receive the benefit of a preferential rate, the power factor must be improved to 0.8

(A) Specify the rating of the shunt capacitors to be used to accomplish this power-factor improvement.

(B) If the transformers were originally installed to just handle the load, how much additional capacity will be made available for plant expansion when the power factor is improved?

Answer   *(ref. 13, pp. 23-83)*

(A) Referring to the Westinghouse manual, the correction factor for converting from a power factor of 0.6 to a power factor of 0.8 is 0.583. The kVA rating of the capacitor is then

$$(0.583) (30,000) = 17.450 \text{ kVA ANS}$$

(B) The present kVA load is

$$30,000/0.6 = 50,000 \text{ kVA}$$

The new kVA load will be

$$30,000/0.8 = 37,500 \text{ kVA}$$

The additional kVA available to the plant is equal to the difference, which is

$$50,000 - 37,500 = 12,500 \text{ kVA ANS}$$

PROBLEM 17-5

Explain how the power factor of a given 60 Hz load can be measured with test apparatus consisting of a voltmeter, ammeter, and an unknown noninductive fixed resistance. Neglect meter losses.

Answer

$$\text{power factor} = \frac{\text{actual power}}{\text{apparent power}} = \frac{\text{watts}}{\text{volt-amperes}} = \frac{I^2 R}{VA}$$

Using these relationships and the circuit in Fig. 17-24, the power factor can be determined by the following list of steps.

1—Select a noninductive resistor R that is much less than load Z.

2—Insert R in series with the load and connect to the 60 Hz line.

3—Measure the voltage drop across Z, which is E_L .

4—Measure the current through the circuit, which is the load current I_L .

5—Calculate the value of Z using the formula $Z = E_L / I_L$.

6—Remove R from the circuit and connect Z to the 60 Hz line.

7—Remeasure the voltage and current, finding E_A and I_A .

8—Now that we know Z and I_A , calculate the power $W = I_A{}^2 / Z$.

9—Calculate the volt-amperes $VA = I_A E_A$.

10—The power factor is then equal to W/VA.

PROBLEM 17-6

(A) Given a 10 mH inductance in series with a 50Ω resistor and operating at 60 Hz, find the power factor angle ϕ and $\cos(\phi)$.

(B) Given a 400 Hz inverter supplying a 5 kΩ load connected in parallel with a $0.05\mu F$ capacitor, find the power factor angle ϕ and $\cos(\phi)$.

Answer

Use the nomograph in Fig. 17-25 to solve the problems.

(A) Draw a line from 60 Hz to 10 mH. Draw a second line from 50Ω (R_S) through the crossover line and read $\phi = 4°3'$ and $\cos(\phi) = 0.997$.

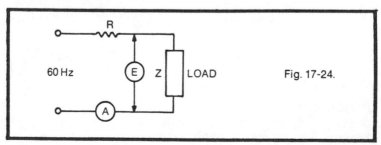

60 Hz   R   E   Z   LOAD   Fig. 17-24.

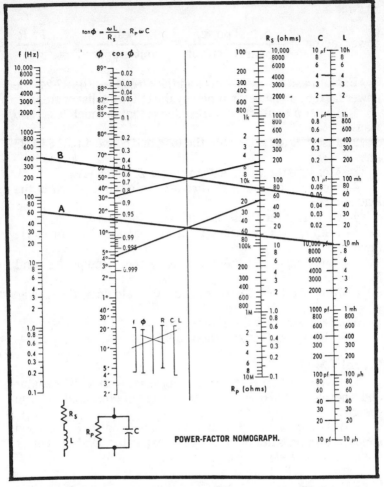

$$\tan\phi = \frac{\omega L}{R_s} = R_p\,\omega\,C$$

POWER-FACTOR NOMOGRAPH.

Fig. 17-25. (A) Given: A 10 mH inductance in series with a 50-ohm resistor and operating at 60 Hz, find ϕ and cos (ϕ). Solution: Draw a line from 60 Hz to 10 mH. Draw a second line from 50Ω (R_s) through the crossover line and read $\phi = 4°3'$, also cos (ϕ) = 0.997. (B) Given: A 400 Hz inverter supplying a 5000Ω load in parallel with a 0.05 μF capacitor. Find ϕ and cos (ϕ) (P_F). Solution: Draw a line from 400 Hz to 0.05 μF. Draw a second line from 5K (R_p), through the crossover line and read $\phi = 23°$, cos (ϕ) = 0.85. (Courtesy EDN)

(B) Draw a line from 400 Hz to 0.05μF. Draw a second line from 5K (R_P) through the crossover line and read $\phi = 32°$ and cos(ϕ) = 0.85.

Alternate Answer

For more accuracy, use your calculator to solve the problems. The governing formulas are

360

$$\tan(\phi) = 2\pi fL/R_S = 2\pi fR_p C$$
$$\phi = \tan^{-1}(2\pi fL/R_S) = \tan^{-1}(2\pi fR_p C)$$
$$P_F = \cos(\phi)$$

PROBLEM 17-7

At a hydro site with an average annual output of 400 million kWh, a firm capacity of 85,000 kW may be installed. The total cost is estimated at \$185/kW with fixed charges of 10%. The cost of the hydro operation is \$0.83/kW of installation. There could be 50 miles of transmission with transformer stations costing \$2,500,000 and with total annual charges of 13%. The transmission efficiency would be 95% for demand and energy.

An alternative steam plant at the load center would cost \$16/kW plus 2.5 mils/kWh.

(A) What is the annual saving in favor of the cheaper power supply?

(B) If the steam plant costs \$120/kW, what total rate of return could be earned on the extra investment required for the hydro (including transmission) project?

Answer

Develop a tabular chart showing the annual costs of the hydro plant vs the steam plant. Then total the cost of each approach and determine the difference in cost. See Table 17-2.

(A) The annual savings (difference in table amounts) is \$720,000 per year.

(B) If the steam plant is installed, the total installation cost would be equal to the power charged plus the transmission line, which is

$$(\$120/\text{kW})\ (85,000\ \text{kW}) + \$2,500,000 = \$12,700,000$$

Table 17-2. Cost Comparison.

ITEM	HYDRO (millions)	STEAM (millions)
1. Installation cost:		
(185)(85,000) =	\$15.7	
(120)(85,000) =		\$10.2
2. transmission line:	2.5	2.5
3. Fixed charges:		
10% of item (1) =	1.57	
(2.5 mils)(4 × 10⁸) =		1.0
4. Cost of operation:		
(0.83)(85,000) =	0.07	
(16)(85,000) =		1.36
5. Maintenance:		
13% of item (2) =	0.342	0.342
Annual cost of operation:	\$1.982	\$2.702

The initial cost of the hydro plant is the sum of items (1) and (2) in Table 17-2 which is $18,200,000. So the cost difference between hydro and steam is $5,500,000, with steam being the cheaper.

The percentage return on the extra investment is then

$$\frac{\$720,000}{\$5,500,000} = 13.1\% \text{ ANS}$$

PROBLEM 17-8

A power plant of 60,000 kW capacity can be built at a cost of $120 per kW and will produce energy at an operating cost of $0.0075 per kWh. A plant of same capacity but different type equipment can be built at a cost of $90 per kW and with operating cost of $0.01 per kWh. Assuming that interest, depreciation, insurance, and taxes amount to 16% of the first cost, annually, and that the average annual load factor will be 50%, which plant would you build? Give supporting data for your answer.

Answer

The operating costs are summarized in Table 17-3, and they show that the cost of operating the first plant would be less than the second, mainly because the cost to produce its energy is only $0.0075/kWh compared to $0.01/kWh.

PROBLEM 17-9

A 600-volt DC power station delivers energy to a transmission line of 0.6 ohms resistance, at the end of which is floated a storage battery whose emf is 520 volts and whose internal resistance is 1.1 ohms. What is the maximum power that can be taken from the line at a point halfway between the station and the battery?

Table 17-3. Cost Comparison.

ITEM	FIRST PLANT	SECOND PLANT
1. Output:	60,000 kW	60,000 kW
2. Operating cost:	$120/kWh	$90/kWh
3. Total price:	$7,200,000	$5,400,000
4. Cost of energy:	$0.0075/kWh	$0.01/kWh
5. Interest, etc:	16% of first year	16% of first year
6. Load factor	50%	50%
7. Average yearly output:	30,000 kWh	30,000 kWh
8. Cost of yearly output:	$225/hour	$300/hour
9. Burden at 16%	$1,153,000/yr	$866,000/yr
10. Cost of operation based on item (8):	$1,970,000/yr	$2,630,000/yr
11. Cost plus burden:	$3,123,000/yr	$3,496,000/yr

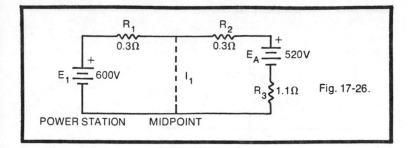

POWER STATION   MIDPOINT

Fig. 17-26.

Answer

Dividing the transmission line in half, as shown in Fig. 17-26, the line resistances and battery resistance comprise a series DC circuit whose total resistance is 1.7Ω. The maximum current delivered by the power station without the proposed midpoint tap is then

$$I_T = (600 - 520)/R_T$$
$$= 80/1.7$$
$$= 47.06A$$

Assuming that the line is broken at its midpoint and that the power station continues to deliver its load, the power available at the midpoint is

$$P = E_1 I_1 - I_1^2 R_1$$
$$= (600)(47.06) - (47.06)^2 (0.3)$$
$$= 27.57 \text{ kW ANS}$$

PROBLEM 17-10

Tabulate the approximate no-load losses of a 6900V, 60 Hz, single-phase line that is 10 miles long. The line uses No. 4 hard-drawn copper, with 24-inch spacing, five 2.5 kVA transformers per mile, and five 10-ampere watthour meters per mile.

Answer

No. 4 wire has a resistance of 1.52Ω per mile, which for the 20 miles of wire required for a 10-mile line is 30.4Ω. The reactance is 0.55Ω per foot spacing, of 1.1Ω per 2-foot spacing, and this amounts to 22Ω for the line. The total line impedance is then

$$Z = \sqrt{(30.4)^2 + (22)^2}$$
$$= 37.5\Omega$$

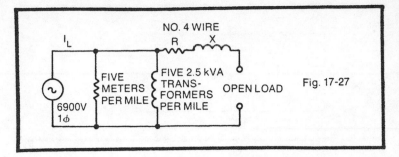

NO. 4 WIRE

I_L

6900V
1ϕ

FIVE
METERS
PER MILE

FIVE 2.5 kVA
TRANS-
FORMERS
PER MILE

R   X

OPEN LOAD

Fig. 17-27

Assuming a 2% loss in the transformers, each of the 50 transformers would have a loss of 50 VA, making a total loss of 2500 VA. At a 0.8 power factor, the power loss is 2000W. The losses involved here consist of copper, hysteresis, and eddy current losses. The copper losses would depend upon loading, but under no-load conditions this would be negligible compared to other losses. The hysteresis loss varies directly with frequency and is approximately $B^{1.6}$, where B is the maximum flux density. The eddy current loss varies with flux density squared, and can be considered constant for all loads, even at no-load.

The 50 watthour meters would consume about 0.4W per meter, or about 20W total, which is negligible compared to other losses.

Assuming that all transformers and watthour meters were lumped together at the end of the transmission line, the line current would be

$$I_L = (2500VA)/(6900V)$$
$$= 0.362A$$

And the line loss with transformers connected would then be approximately

$$P_L = I^2 Z$$
$$= (0.362)^2 (37.5)$$
$$= 4.9W \text{ ANS}$$

So, the line loss under no-load conditions is negligible compared to the 2000-watt loss or the 2500 VA loss of the transformers.

PROBLEM 17-11

Let us assume the typical transmission system shown in Fig. 17-28 to have a single line-to-ground fault on one end of the 66kV line as shown. The line construction is given in Fig.

17-28B and the generator constants in Fig. 17-28C. Calculate the following:

(A) Positive-sequence reactance to the point of fault.
(B) Negative-sequence reactance to the point of fault.
(C) Zero-sequence reactance to the point of fault.
(D) Fault current.
(E) Line currents, line-to-ground voltages, and line-to-line voltages at the breaker adjacent to the fault.
(F) Line currents, line-to-ground voltages, and line-to-line voltages at the terminals of G'.

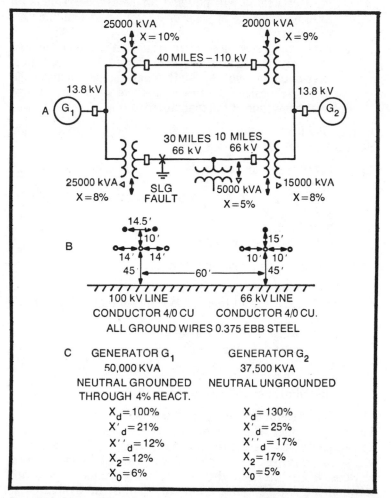

Fig. 17-28. Typical system assumed for fault calculation. (A) System single-line diagram. (B) Line construction. (C) Tabulation of generator constants.

(G) Line currents, line-to-ground voltages, and line-to-line voltages at the 110 kV breaker adjacent to the 25,000 kVA transformer.

Answer

Let's make the following assumptions:

1. That the fault currents are to be calculated using transient reactances.
2. A base of 50,000 kVA for the calculations.
3. That all resistances can be neglected.
4. That a voltage, positive-sequence, as viewed from the fault of j 100% will be used for reference. This is an assumed voltage of $j66,000/\sqrt{3}$ volts between line "A" and neutral.
5. That the reference phases on either side of the star-delta transformers are chosen such that positive-sequence voltage on the high side is advanced 30° in phase position from the positive-sequence voltage on the low side of the transformer.

(A) Positive-sequence reactance: For 4/0 copper conductors $X_A = 0.497$ ohms per mile.

$$X_D = (1/3) \ (X_D \text{ for 14 feet} + X_D \text{ for 14 feet} + X_D \text{ for 28 feet})$$
$$= (1/3) \ (0.320 + 0.320 + 0.404) = 0.348 \text{ ohms per mile}$$
$$X_1 = X_2 \ X_A + X_D$$
$$= 0.497 + 0.348$$
$$= 0.845 \text{ ohms per mile ANS}$$

(B) Negative-sequence reactance:

$$X_A = 0.497 \text{ ohms per mile.}$$
$$X_D = (1/3) \ (X_D \text{ for 10 feet} + X_D \text{ for 10 feet} + X_D \text{ for 20 feet})$$
$$= (1/3) \ (0.279 + 0.279 + 0.364)$$
$$= 0.307 \text{ ohms per mile}$$
$$X_1 = X_2 = X_A + X_D$$
$$= 0.497 + 0.307$$
$$= 0.804 \text{ ohms per mile ANS}$$

(C) Zero-sequence circuit: Since zero-sequence currents flowing in either the 110 or the 66 kV line will induce a zero-sequence voltage in the other line and in all three ground wires, the zero-sequence mutual reactances between lines, between each line and the two sets of ground wires, and between the two sets of ground wires, must be evaluated as well as the zero-sequence self-reactances. Indeed, the

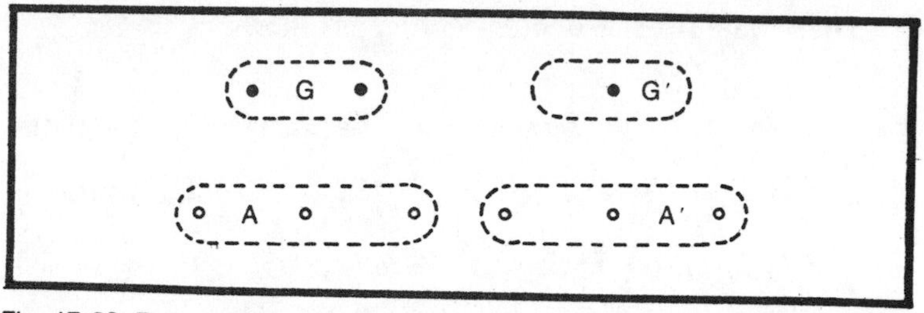

Fig. 17-29. Zero-sequence circuits formed by the 110 kV line (A), the 66 kV line ZA'), the two ground wires (G), and the single ground wire (G').

zero-sequence self-reactance of either the 110 or the 66 kV line will be affected by the mutual coupling existing with all of the ground wires. The three conductors of the 110 kV line, with ground return are assumed to form one zero-sequence circuit, denoted by A in Fig. 17-29; the two ground conductors for this line, with ground return, form the zero-sequence circuit denoted G; the three conductors for the 66 kV line, with ground return, form the zero-sequence circuit denoted A'; and the single ground wire for the 66 kV line, with ground return, forms the zero-sequence circuit denoted G'. Although not strictly correct, we assume the currents carried by the two ground wires of circuit G are equal. Then let:

E_O = zero-sequence voltage of circuit A

E_{G0} = zero-sequence voltage of circuit G = 0, since the ground wires are assumed to be continuously grounded.

E_0' = zero-sequence voltage of circuit A'

E'_{G0} = zero-sequence voltage of circuit G' = 0, since the ground wire is assumed to be continuously grounded.

I_0 = zero-sequence current of circuit A

I_G = zero-sequence current of circuit G

I_0' = zero-sequence current of circuit A'

I_G' = zero-sequence current of circuit G'

It should be remembered that unit I_0 is one ampere in each of the three line conductors with three amperes returning in ground; unit I_G is 3/2 amperes in each of the two ground wires with three amperes returning in the ground; unit I_0' is one ampere in each of the three line conductors with three amperes returning in the ground; and unit I_G' is three amperes in the ground wire with three amperes returning in the ground.

367

These quantities are inter-related as follows:

$$E_0 = I_0\,Z_0\,(A) + I_G\,Z_{0(AG)} + I_0\,'Z_0\,(AA') + I_G\,'Z_0\,(AG')$$

$$E_{G0} = I_0\,Z_{0(AG)} + I_G\,Z_0\,(G) + I_0\,'Z_0\,(A'G) + I_G\,'Z_0\,(GG') = 0$$

$$E_0\,' = I_0\,Z_{0(AA')} + I_G\,Z_{0(A'G)} + I_0\,'Z_{0(A')} + I_G\,'Z_0\,(A'G')$$

$$E'_{G0} = I_0\,Z_{0(AG')} + I_G\,Z_{0(GG')} + I'_0\,Z_{(A'G')} + I'_G\,Z_{0(G')} = 0$$

where

$Z_0\,(A)$ = zero-sequence self-reactance of the A circuit
$= X_A + X_E - (2/3)(X_D$ for 14 feet $+ X_D$ for 14 feet $+ X_D$ for 28 feet)
$= 0.497 + 2.89 - 2(0.348)$
$= 2.69$ ohms per mile

$Z_{0(A')}$ = zero-sequence self-reactance of the A' circuit
$= X_A + X_E - (2/3)(X_D$ for 10 feet $+ X_D$ for 10 feet $+ X_D$ for 20 feet)
$= 0.497 + 2.89 - 2(0.307)$
$= 2.77$ ohms per mile

$Z_0\,(G)$ = zero-sequence self-reactance of the G circuit
$= (3/2)X_A + X_E - (3/2)(X_D$ for 14.5 feet)
$= (3/2)\,(2.79) + 2.89 - (3/2)\,(0.324)$
$= 6.59$ ohms per mile.

$Z_{0(G')}$ = zero-sequence self-reactance of the G' circuit
$= 3X_A + X_E$
$= 3(2.79) + 2.89$
$= 11.26$ ohms per mile

$Z_0\,(AG)$ = zero-sequence mutual reactance between the A and G circuits
$= X_E - (3/6)\,(X_D$ for 12.06 feet $+ X_D$ for 12.06 feet $+ X_D$ for 12.35 feet $+ X_D$ for 12.35 feet $+ X_D$ for 23.5 feet $+ X_D$ for 23.5 feet)
$= 2.89 - 3(0.3303)$
$= 1.90$ ohms per mile

$Z_0\,(AA')$ = zero-sequence mutual reactance between the A and A' circuits
$= X_C - (3/9)(X_D$ for 60 feet $+ X_D$ for 50 feet $+ X_D$ for 70 feet $+ X_D$ for 46 feet $+ X_D$ for 36 feet $+ X_D$ for 56 feet $+ X_D$ for 74 feet $+ X_D$ for 84 feet)
$= 2.89 - 3(0.493)$
$= 1.411$ ohms per mile

$Z_{0\,(AG')}$ = zero-sequence mutual reactance between the A
and G' circuits

= $X_E - (3/3)(X_D$ for 75 feet + X_D for 62
feet + X_D for 48 feet

= $2.89 - 3(0.498)$

= 1.40 ohms per mile

$Z_{0(A'G')}$ = zero-sequence mutual reactance between the A'
and G' circuits

= $X_C - (3/3)(X_D$ for 15 feet + X_D for 18.03
feet + X_D for 18.03 feet)

= $2.89 - 3(0.344)$

= 1.86 ohms per mile

Similar definitions apply for $Z_{0(A'G)}$ and $Z_{0(GG')}$. In each
case the zero-sequence mutual reactance between two circuits
is equal to X_C minus three times the average of the X_D s for
all possible distances between conductors of the two circuits.

The zero-sequence self-reactance of the 110 kV line in the
presence of all zero-sequence circuits is obtained by letting I_0'
be zero in the above equations and solving for E_0/I_0. Carrying
out this rather tedious process, it will be found that

$$E_0/I_0 = 2.05 \text{ ohms per mile}$$

The zero-sequence self-reactance of the 66 kV line in the
presence of all zero-sequence circuits is obtained by letting I_0
be zero in the equations and solving for E_0'/I_0'. It will be
found that

$$E_0'/I_0' = 2.25 \text{ ohms per mile}$$

The zero-sequence mutual reactance between the 66 and
the 110 kV line in the presence of all zero-sequence circuits is
obtained by letting I_0' be zero and solving for E_0'/I_0'. When
this is done, it will be found that E_0'/I_0' (with $I_0' = 0$)

$(I'_0 = 0) = (E'_0/I_0')$ (with $I'_0 = 0$)

= (E_0/I'_0) (with $I_0 = 0$)

= 0.87 ohms per mile.

The sequence networks are shown in Figs. 17-30, 31, and 32
with all reactances expressed in percent on a 50,000 kVA base
and the networks set up as viewed from the fault. Illustrative
examples of expressing these reactances in percent on a 50,000
kVA base follow.

Positive-sequence reactance of G_2:

$$(25)\ \frac{(50\ 000)}{(37\ 500)} = 33.3\%$$

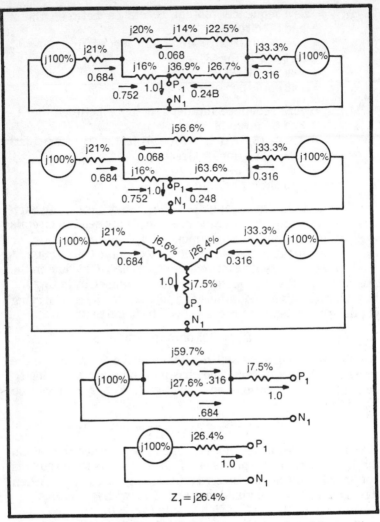

Fig. 17-30. Reduction of the positive-sequence network and the positive-sequence distribution factors.

Positive-sequence reactance of the 66 kV line:

$$\frac{(0.804)\ (40)\ (50\ 000)}{(66)\ (66)\ (10)} = 36.9\%$$

Positive-sequence reactance of the 110 kV line:

$$\frac{(0.845)\ (40)\ (50\ 000)}{(110)\ (110)\ (10)} = 14\%$$

370

Zero-sequence mutual reactance between the 66 kV and the 110 kV line for the 30 mile section:

$$\frac{(0.87)\,(30)\,(50\,000)}{(110)\,(66)\,(10)} = 18\%$$

The distribution factors are shown on each sequence network; obtained by finding the distribution of one ampere taken as flowing out at the fault.

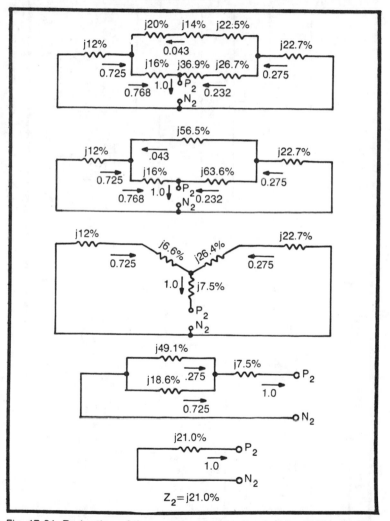

Fig. 17-31. Reduction of the negative sequence network and the negative sequence distribution factors.

371

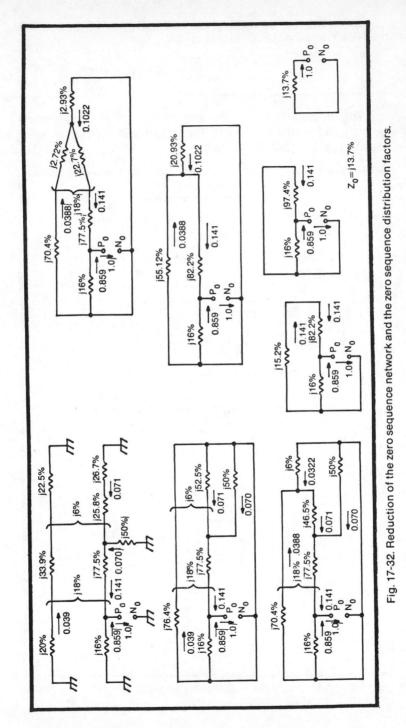

Fig. 17-32. Reduction of the zero sequence network and the zero sequence distribution factors.

Each network is finally reduced to one equivalent impedance as viewed from the fault.

(D) The sequence networks are connected in series to represent a single line-to-ground fault. The total reactance of the resulting single-phase network is

$$Z_1\% + Z_2\% + Z_0\% = 26.4\% + 21.0\% + 13.7\% = 61.1\%$$

Then

$$I_{0F} = I_{1F} = I_{2F} \frac{j100\%}{j61.1\%} \ 1.637 \text{ p.u.}$$

Since normal current for the 66kV circuit (for a base kVA of 50,000) is

$$\frac{50,000}{\sqrt{3} \times 66} = 437.5 \text{ amperes}$$

and

$$I_0 = I_1 = I_2 = (1.637) \ (437.5) = 715 \text{ amprees}$$

The total fault current is

$$I_0 + I_1 + I_2 = 4.911 \text{ p.u.} = 2145 \text{ amperes.}$$

The sequence voltages at the fault:

$$\begin{aligned}
E_1 &= E_{A1} - I_1 Z_1 \\
&= j100\% - j(1.637) \ (26.4)\% \\
&= j56.9\% \\
&= j21,700 \text{ volts}
\end{aligned}$$

$$\begin{aligned}
E_2 &= -I_2 Z_2 \\
&= -j(1.637) \ (21)\% \\
&= -j34.4\% \\
&= -j13,100 \text{ volts}
\end{aligned}$$

$$\begin{aligned}
E_0 &= -I_0 Z_0 \\
&= -j(1.637) \ (13.7)\% \\
&= -j22.5\% \\
&= -j8,600 \text{ volts}
\end{aligned}$$

$$\begin{aligned}
E_{AG} &= E_0 + E_1 E_2 = 0 \\
E_{BG} &= E_0 + A^2 E_1 + AE_2 \\
&= 30,200 - j12,900 \\
&= 32\ 800 \text{ volts.}
\end{aligned}$$

373

$$E_{CG} = E_0 + AE_1 + A^2 E_2$$
$$= -30,200 - j12,900$$
$$= 32\,800 \text{ volts}$$

$$E_{AB} = E_{AG} - E_{BG}$$
$$= -30,200 + j12,900$$
$$= 32,800 \text{ volts}$$

$$E_{BC} = E_{BG} - E_{CG}$$
$$= 60,400 \text{ volts}$$

$$E_{CA} = E_{CG} - E_{AG}$$
$$= -30,200 - j12,900$$
$$= 32,800 \text{ volts}$$

(E) Using the distribution factors in the sequence networks at this point:

$$I_1 = (0.752)\,(1.637) = 1.231 \text{ p.u.} = 540 \text{ amperes}$$
$$I_2 = (0.768)\,(1.637) = 1.258 \text{ p.u.} = 550 \text{ amperes}$$
$$I_0 = (0.859)\,(1.637) = 1.407 \text{ p.u.} = 615 \text{ amperes}$$
$$I_A = I_0 + I_1 + I_2 = 1705 \text{ amperes}$$
$$I_B = I_0 + A^2 I_1 + AI_2 = 70 + j8.6 = 70.5 \text{ amperes}$$
$$I_C = I_0 + AI_1 + A^2 I_2 = 70 - j8.6 = 70.5 \text{ amperes}$$

The line-to-ground and line-to-line voltages at this point are equal to those calculated for the fault.

(F) The base, or normal, voltage at this point is 13800 volts line-to-line, or 7960 volts line-to-neutral.

The base, or normal, voltage at this point is 13800 volts line-to-line, or 7960 volts line-to-neutral.

The base, or normal, current at this point is

$$\frac{50\,000}{\sqrt{3} \times 13.8} = 2090 \text{ amperes}$$

Since a star-delta transformation is involved, there will be a phase shift in positive- and negative-sequence quantities.

$$I_1 = (0.684)\,(1.637)\,(2090)\epsilon^{-j30}$$
$$= 2340$$
$$= 2030 - j1170$$

$$I_2 = (0.725)\,(1.637)\,(2090)\epsilon^{+j30}$$
$$= 2480$$
$$= 2150 + j1240$$

$$I_0 = 0$$
$$I_A = I_0 + I_1 + I_2$$
$$= 4180 + j70$$
$$= 4180 \text{ amperes}$$

374

$$I_B = I_0 + A^2 I_1 + AI_2$$
$$= -4180 + j70$$
$$= 4180 \text{ amperes}$$

$$I_C = I_0 + AI_1 + A^2 I_2$$
$$= -j140$$
$$= 140 \text{ amperes}$$

The sequence voltages at this point are:

$$E_1 = (j100\% - j0.684 \times 21 \times 1.637\%)\epsilon^{-j30}$$
$$= -A^2 \, 76.5\%$$
$$= 3045 + j5270 \text{ volts}$$

$$E_2 = (-j0.725 \times 12 \times 1.637\%)\epsilon^{j30}$$
$$= -A14.2\%$$
$$= 565 - j980 \text{ volts}$$

$$E_0 = 0$$
$$E_{AG} = E_1 + E_2$$
$$= 3610 + j4290$$
$$= 5600 \text{ volts}$$

$$E_{BG} = A^2 E_1 + AE_2$$
$$= 3610 - j4290$$
$$= 5600 \text{ volts}$$

$$E_{CG} = AE_1 + A^2 E_2$$
$$= -7220$$
$$= 7220 \text{ volts}$$

$$E_{AB} = +j8580$$
$$= 8580 \text{ volts}$$

$$E_{BC} = 10\,830 - j4290$$
$$= 11\,650 \text{ volts}$$

$$E_{CA} = -10\,830 - j4290$$
$$= 11\,650 \text{ volts}$$

(G) The base, or normal, voltage at this point is 110,000 volts line-to-line; or 63,500 volts line-to-neutral.

The base, or normal, current at this point is

$$\frac{50\,000}{\sqrt{3} \times 110} = 262 \text{ amperes}$$

The sequence currents at this point are:

$$I_1 = (-0.068)\,(1.637)\,(262)$$
$$= -29.2 \text{ amperes}$$

375

$$I_2 = (-0.043)(1.637)(262)$$
$$= -18.4 \text{ amperes}$$

$$I_0 = (0.039)(1.637)(262)$$
$$= 16.7 \text{ amperes}$$

$$I_A = I_0 + I_1 + I_2$$
$$= -30.9$$
$$= 30.9 \text{ amperes}$$

$$I_B = I_0 + A^2 I_1 + A I_2$$
$$= 40.5 + j9.35$$
$$= 41.6 \text{ amperes.}$$

$$I_C = I_0 + A I_1 + A^2 I_2$$
$$= 40.5 - j9.35$$
$$= 41.6 \text{ amperes.}$$

The sequence voltages at this point are:

$$E_1 = j100\% - j(0.684)(1.637)$$
$$(21)\% - j(-0.068)(1.637)(20)\%$$
$$= j78.7\%$$
$$= j50\,000 \text{ volts}$$

$$E_2 = -j(0.725)(1.637)(12)\% - j(-0.043)(1.637)(20)\%$$
$$= j12.8\%$$
$$= -j8130 \text{ volts}$$

$$E_0 = -j(0.039)(1.637)(20)\%$$
$$= -j1.3\%$$
$$= -j825 \text{ volts}$$

$$E_{AG} = E_0 + E_1 + E_2$$
$$= j41.000$$
$$= 41\,000 \text{ volts}$$

$$E_{BG} = E_0 + A^2 E_1 + A E_2$$
$$= 50\,300 - j21\,750$$
$$= 54\,800 \text{ volts}$$

$$E_{CG} = E_0 + A E_1 + A^2 E_2$$
$$= -50\,300 - j21\,750$$
$$= 54\,800 \text{ volts}$$

$$E_{AB} = -50\,300 + j62\,750$$
$$= 80\,400 \text{ volts}$$

$$E_{BC} = 100\,600$$
$$= 100\,600 \text{ volts}$$

$$E_{CA} = -50\,300 - j62\,750$$
$$= 80\,400 \text{ volts}$$

PROBLEM 17-12

A single line-to-ground fault occurs on a 3-phase transmission line with grounded neutral. The system impedances looking in at the point of the fault are:

$$\text{Positive sequence: } Z_1 = j25\%$$
$$\text{Negative sequence: } Z_2 = j20\%$$
$$\text{Zero sequence: } Z_0 = j15\%$$

Find the percent fault current. Assume no fault impedance and neglect all system resistances. The system voltage to neutral is to be taken as 100%.

Answer

A diagram showing the line-to-ground fault is given in Fig. 17-33, with the positive, negative, and zero sequence impedances in series. The total fault impedance is then

$$Z_T = j25\% + j20\% + j15\%$$
$$= j60\%$$

So the fault current per line is

$$I_1 = 100\%/60\%$$
$$= 167\%$$

The total fault current is then three times the line current, so

$$I_F = 3I_1$$
$$= 3(167\%)$$
$$= 500\% \text{ ANS}$$

PROBLEM 17-13

A wye-connected generator rated at 2400V has 0.2Ω resistance and 2.5Ω reactance per phase. The generator is connected by lines, each having an impedance of $2\Omega \angle 30°$ to a wye-wye stepup transformer bank having a transformation

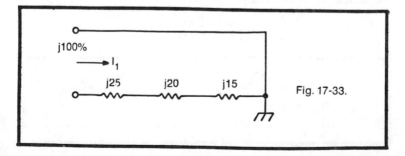

Fig. 17-33.

377

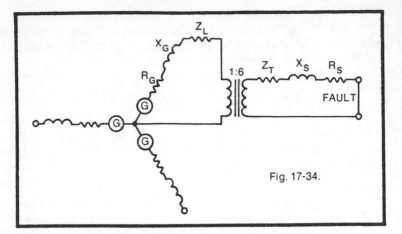

Fig. 17-34.

ratio of 6. Each transformer has a total equivalent impedance, referred to the high side, of $100\Omega \angle 60°$, and the transformer bank is connected to a load by lines, each having a resistance of 25Ω and an inductive reactance of 50Ω. Calculate the fault current for a 3-phase symmetrical short-circuit at the load.

Answer

For a 3-phase fault, start by examining the simpler 1-phase fault shown in Fig. 17-34. Taking the transformation ratio into account, the equivalent circuit with all parameters referred to the load side of the transformer is shown in Fig. 17-35. Inserting values:

$$36R_G = (36)(0.2) = 7.2$$
$$36X_G = (36)(2.5) = 90$$
$$36Z_L = (36)(2 \angle 30°) = 72 \angle 30° = 62.35 + j36$$
$$Z_T = 100 \angle 60° = 50 + j86.60$$
$$X_S = 50$$
$$R_S = 25$$

The sum of which is $144.55 + j262.6 = 299.8 \angle 61.2°$. Therefore, the fault current in the 3-phase circuit is

$$I_F = 14,400/299.8\sqrt{3}$$
$$= 27.73A \angle -61.2° \text{ ANS}$$

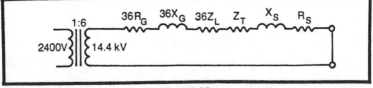

Fig. 17-35.

PROBLEM 17-14

Two power stations, A and B, are located 20 miles from each other and are each connected to a common substation located 40 miles from each power station. Power station A has a capacity of 50,000 kW, and power station B has a capacity of 100,000 kW. Assuming representative reactances for the generating equipment, transmission lines, and transformers, what would be the duty on the oil circuit breakers if a fault occurred midway between power station A and the substation?

Answer

Assumptions: generators have 15% Z; transformers, 5% Z; line is 138 kV; power factor, 100%. Under these conditions, station B with its 100,000 kW capacity would provide 418A line current.

From a wire handbook, at 418A and 15-foot spacing, the line impedance is

$$\begin{aligned}
Z_L &= R_A + j(X_A + X_D) \\
&= (0.33\Omega/\text{mi}) + j(0.477 + 0.329\Omega/\text{mi}) \\
&= 0.33 + j0.806\Omega/\text{mi} \\
&= 0.871 \angle 68°\Omega/\text{mi} \\
&= 34.84 \angle 68° \text{ ohms per 40 miles}
\end{aligned}$$

The resistance to a fault occurring at the midpoint of the line would be one-half, or 17.42 ohms.

Assuming 100% power factor,

$$Z = (10)(\%Z)(kV)^2/(kVA)$$

so the Z power phase at station A is

$$Z_A = (10)(15)(138)^2/(50,000) = 57.1\Omega$$

and at power station B,

$$Z_B = (10)(15)(138)^2/(100,000) = 28.6\Omega$$

and at the transformer of A,

$$Z_{TA} = (10)(5)(138)^2 (50,000) = 19.0\Omega$$

and at the transformer of B,

$$Z_{TB} = (10)(5)(138)^2/(100,000) = 9.5\Omega$$

Redrawing the system as in Fig. 17-36 shows that the impedance from power station A to the fault is

$$\begin{aligned}
Z_{AF} &= Z_A + Z_{TA} + Z_L (20 \text{ mi}) \\
&= 57.1 + 19.0 + 17.4 \\
&= 93.5\Omega
\end{aligned}$$

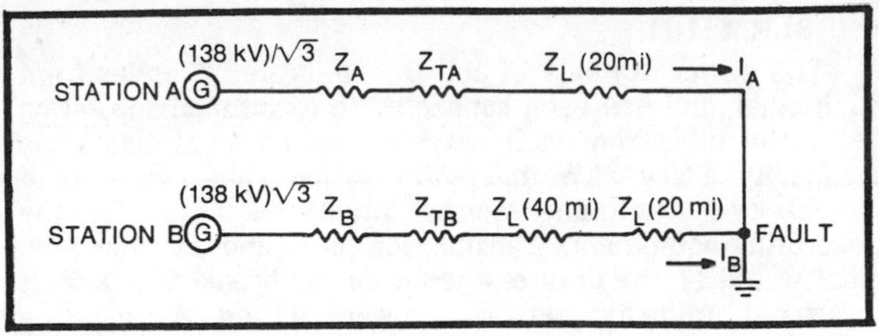

Fig. 17-36.

And the impedance from power station B to the fault is

$$Z_{BF} = Z_B + Z_{TB} + Z_L \text{ (40 mi)} + Z_L \text{ (20 mi)}$$
$$= 28.6 + 9.5 + 34.8 + 17.4$$
$$= 90.3\Omega$$

The equivalent impedance seen by the fault is the parallel combination of Z_{AF} and Z_{BF}, which is 45.9Ω. Therefore, the fault current is

$$I_F = (138,000)/(45.9) = 1736A$$

which is the sum of the currents delivered by each power station; that is $I_A = 852A$ and $I_B = 884A$.

The recommended circuit breaker rating would be 138 kV at 600A, assuming peak handling capabilities. The MVA rating for this case would be $(884)(138,000) = 122$ MVA for this fault condition. Ordinarily, a worst-case fault condition would be taken into account when making such a breaker rating.

PROBLEM 17-15

(A) How much would it cost to electrically heat the water in a pool measuring 20 feet by 40 feet by 6 feet deep? The cost of electrical energy is 3.2 cents per kilowatt-hour. The water is supplied at 40°F. Assume heat losses to be 25%. The water is to be heated to 68°F.

(B) How long would it take to heat the pool from 40°F to 68°F if power is supplied to the heater at the rate of 240 kW?

Answer

(A) The volume of the pool is

$$V = (20)(40)(6) = 4800 \text{ ft}^3$$

Since water weighs about 62.4 lb/ft^3, the pool contains almost 300,000 lb of water.

To raise the temperature of one pound of water by one degree Fahrenheit, one Btu is required. And for the required

28-degree increase over the supply temperature, a total of 8.4 million Btu is needed. Allowing for a 25% heat loss, the energy requirement is pushed up to 10.5 million Btu.

Now, there are 3412 Btu per kWh, so the electrical energy required is

$$kWh = (10.5 \times 10^6)/3412$$
$$= 3077$$

And the cost of this energy at 3.2 cents per kWh is

$$(3077)(0.032) = \$98.46 \text{ ANS}$$

(B) The time required to heat the pool would be

$$(3077 \text{ kWh})/(240 \text{ kW}) = 12.82 \text{ hours ANS}$$

PROBLEM 17-16

An open steel tank, 2.3 ft in diameter, 2 ft deep, weighing 186 lb, and covered with 2 inches of insulation, contains 60 gallons of olive oil, which must be heated from 70°F to 220°F within one hour and maintained at that temperature.

(A) What power rating would you recommend for the electric heating element?

(B) How much power is required to maintain the desired temperature after the oil has reached 220°F?

Answer   (ref. 28)

(A) From Table 17-4, the specific heat of steel is 0.12 Btu/lb-F. From Table 17-5, the specific heat of olive oil is 0.471 Btu/lb-F. From Table 17-6, the weight of olive oil is 7.75 lb/gal. From Fig. 17-37, the oil surface loss at 220°F is 145 W/ft^2. And from Fig. 17-38, the insulated wall loss for a temperature rise of 150 degrees is 10 W/ft^2.

To heat the oil initially, the power required is

$$\frac{(60 \text{ lb})(7.75 \text{ lb/gal})(0.471 \text{ Btu/lb-F})(150°)}{(3412 \text{ Btu/kWh})} = 9.63 \text{ kWh}$$

To heat the tank initially,

$$\frac{(186 \text{ lb})(0.12 \text{ Btu/lb-F})(150°)}{(3412 \text{ Btu/kWh})} = 0.98 \text{ kWh}$$

The average oil surface loss is

$$\frac{(4.15 \text{ ft}^2)(145 \text{ W/ft}^2)(1 \text{ hr})}{(2)(1000 \text{ W/kW})} = 0.30 \text{ kWh}$$

381

Table 17-4. Properties of Solids.

SUBSTANCE	AVERAGE SPECIFIC HEAT (BTU/LB-F)	HEAT OF FUSION (BTU PER LB)	MELTING POINT (F)	WEIGHT (LB PER CU FT)
Aluminum	0.23	138	1216	160
Antimony	0.52	25	1166	423
Asphalt	.40	40	250 ±	65
Beeswax		75	144	60
Bismuth	.031	23	520	610
Brass	.10		1700 ±	525
Brickwork and Masonry	.220	—	—	140
Carbon	.204	—	—	—
Copper	.10	75	1981	550
Glass	.20	—	2200 ±	165
Graphite	.20	—	—	130
Ice	.504	—	32	56
Iron, cast	.13	—	2300 ±	450
Iron, wrought	.12	—	2800 ±	480
Lead, solid	.031	10	621	710
Lead, melted	.04	—	—	—
Magnesia, 85%	.222	—	—	19
Nickel	.11	—	2642	550
Paper	.45	—	—	58
Paraffin	.70	63	133	56
Pitch, hard	—	—	300 ±	83
Rubber	.40	—	—	95 ±
Silver	.057	38	1761	655
Solder (50% Pb, 50% Sn)	.04	17	415	580
Steel	.12	—	2550 ±	490
Sugar	.30	—	320	105
Sulphur	.203	17	230	125
Tallow	—	—	90 ±	60
Tin, solid	.056	25	450	455
Tin, melted	.064	—	—	—
Type metal (85% Pb, 15% Sb)	.040	—	500	670
Wood, pine	.45 ±	—	—	34
Wood, oak	.45 ±	—	—	50
Zinc	.095	51	787	445

(Courtesy General Electric)

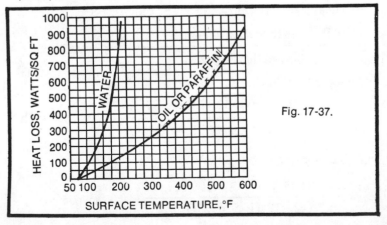

Fig. 17-37.

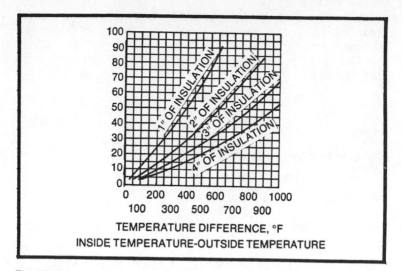

Fig. 17-38. Heat losses from insulated walls. Curves are for standard high-grade material, such as 85% magnesia, Rockwool, etc. (Courtesy General Electric)

The average tank surface loss is

$$\frac{[(2)(2.3)(\pi) + (1.15)^2 (\pi) \text{ ft}^2](10 \text{ W/ft}^2)(1 \text{ hr})}{(2)(1000 \text{ W/kW})} = 0.09 \text{ kWh}$$

The total of these initial heat values is 11.0 kWh, but to assure rapid heating, allow a 20% safety factor. This brings the initial heating requirement up to 13.2 kWh, so the heater rating should be 13.2 kWh **ANS**.

Table 17-5. Properties of Liquids.

SUBSTANCE	AVERAGE SPECIFIC HEAT (BTU/LB-F)	HEAT OF VAPORIZATION (BTU PER LB)	BOILING POINT (F)	WEIGHT	
				(LB PER CU FT)	(LB PER GAL)
Acetic acid	0.472	153	245	66	8.81
Alcohol	.65	365	172	55	7.35
Benzine	.45	166	175	56	7.49
Ether	.503	160	95	46	6.15
Glue (2/3 water)	.895	—	—	69	9.21
Glycerine	.58	—	554	79	10.58
Mercury	.0333	117	675	845	112.97
Oil, cottonseed	.47	—	—	60	7.76
Oil, machine	.40	—	—	58	7.75
Oil, olive	.471	—	570±	58	7.75
Paraffin, melted	.71	—	750±	56	7.49
Petroleum	.51	—	—	56	7.49
Sulphur, melted	.234	652	601	—	—
Turpentine	.41	133	319	54	7.22
Water	1.0	965	212	62.5	8.34

(Courtesy General Electric)

(B) Once up to temperature, there is no additional material to heat, nor is there any heat of fusion or vaporization.

The heat loss through the oil surface is double the average loss, or 0.60 kW.

$$\frac{(4.15 \text{ ft})^2 \ (145 \text{ W/ft}^2)}{(1000 \text{ W/kW})} = 0.60 \text{ kW}$$

The heat loss through the tank surface is

$$\frac{(18.6 \text{ ft}^2)(10 \text{ W/ft}^2)}{(1000 \text{ W/lW})} = 0.19 \text{ kW}$$

Again, allowing for a 20% safety factor, the power required to maintain the temperature is

$$(1.2)(0.60 + 0.19) = 0.95 \text{ kW ANS}$$

PROBLEM 17-17

An electric press has two steel platens, each measuring 52 inches by 104 inches by 1.5 inches. After initial heat-up from 70°F to 400°F in 1 hour, 50 lb sheets of fiberboard (specific heat of 0.64 Btu/lb-F) are processed by drying and compressing to 3/16-inch thickness, at a rate of 4 per hour. The platens are

Table 17-6 Properties of Gases and Vapors.

SUBSTANCE	AVERAGE SPECIFIC HEAT, CONSTANT PRESSURE (BTU/LB-F)	WEIGHT AT APPROX 70 F AND ATMOSPHERIC PRESSURE (LB PER CU FT)
Acetylene	0.35	0.073
Air	.237	0.80
Alcohol	.453	
Ammonia	.520	.048
Argon	1.24	.1037
Carbon dioxide	.203	.123
Carbon monoxide	.243	.078
Chlorine	.125	.20
Ethylene	.40	.0728
Helium	1.25	.0104
Hydrochloric acid	.195	.102
Hydrogen	3.41	.0056
Methane	.60	.0447
Methyl chloride	.24	.1309
Nitric oxide	.231	0.779
Nitrogen	.245	.078
Oxygen	.218	.09
Sulphur dioxide	.155	.179

(Courtesy General Electric)

closed during the initial heat-up and open 1.5 minutes of each 15-minute working cycle. The nonworking horizontal surface of each platen is insulated from the press, but the edges are exposed.

(A) Determine the initial heat-up requirement.
(B) Determine the power required for operation.

Answer *(ref. 28)*

From Table 17-4, the specific heat of steel is 0.12 Btu/lb-F, and the weight of steel is 490 lb/ft^3. Thus the weight of each platen is

$$\frac{(52 \times 104 \times 1.5 \text{ in}^3)(490 \text{ lb/ft}^3)}{(1728 \text{ in}^3/\text{ft}^3)} = 2300 \text{ lb}$$

From Fig. 17-39, the steel surface loss at 400°F is 300 W/ft^2. The heat required to heat the platens is then

$$\frac{(4600 \text{ lb})(0.12 \text{ Btu/lb-F})(330°)}{(3412 \text{ Btu/kWh})} = 53.5 \text{ kWh}$$

The average platen edge loss during heating is

$$\frac{(6.6 \text{ ft}^2)(300 \text{ W/ft}^2)(1 \text{ hr})}{(2)(1000 \text{ W/kWh})} = 1.0 \text{ kWh}$$

Allowing a 20% safety factor, the total heat requirement is

$$(1.2)(53.5 + 1.0 \text{ kWh}) = 65.4 \text{ kWh ANS}$$

The power required for initial heat-up is then

$$(65.4 \text{ kWh})/(1 \text{ hr}) = 65.4 \text{ kW ANS}$$

(B) To heat the fiberboard requires

$$\frac{(50 \text{ lb})(0.64 \text{ Btu/lb-F})(330°)}{(3412 \text{ Btu/kWh})} = 3.1 \text{ kWh}$$

The continuous loss at the platen edges is

$$\frac{(6.6 \text{ ft}^2)(300 \text{ W/ft}^2)(0.25 \text{ hr})}{(1000 \text{ W/kW})} = 0.5 \text{ kWh}$$

The loss during the time the platens are open is

$$\frac{(75\,\text{ft}^2\,)(300\,\text{W/ft}^2\,)(0.025\,\text{hr})}{(1000\,\text{W/kW})} = 0.6\,\text{kWh}$$

Allowing a 20% safety factor, the total heat requirement during each 15-minute cycle is

$$(1.2)(3.1 + 0.5 + 0.6) = 5.0\,\text{kWh}$$

The power required for operation is then

$$(5.0\,\text{kWh})/(0.25\,\text{hr}) = 20\,\text{kW ANS}$$

PROBLEM 17-18

A steel crucible is 1 ft high, has an outside diameter of 1.5 ft and inside diameter of 1.34 ft. weighs 230 lb, and contains 800 lb of lead. The lead is to be heated from 70°F to 720°F in one hour and held at that temperature for 30 minutes before pouring.

(A) What is the initial electrical power required for heat-up?

(B) What power is required for normal operation after heat-up?

(C) What type of electrical heaters would you recommend?

Answer   *(ref. 28)*

(A) From Table 17-4, the specific heat of steel is 0.12 Btu/lb-F, the specific heat of solid lead is 0.031 Btu/lb-F, the

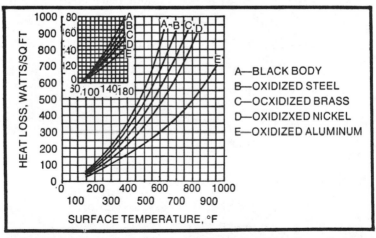

A—BLACK BODY
B—OXIDIZED STEEL
C—OCXIDIZED BRASS
D—OXIDIZXED NICKEL
E—OXIDIZED ALUMINUM

Fig. 17-39. Heat losses from uninsulated smooth solid surfaces. Assumed external ambient temperature of 70° F. (Courtesy General Electric)

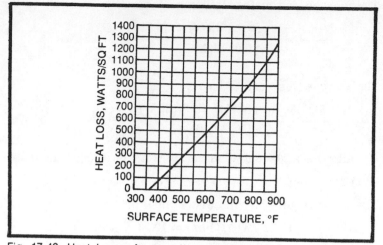

Fig. 17-40. Heat losses from molten metal surfaces. Curve is for soft metals, such as lead, tin, solder, babbitt, type metal, etc. Assumed external ambient temperature of 70° F. (Courtesy General Electric)

melting point of lead is 621°F, the heat of fusion of lead is 10 Btu/lb, and the specific heat of melted lead is 0.04 Btu/lb-F. From Fig. 17-39, the steel surface loss at 720°F is 920 W/ft². And from Fig. 17-40, the lead surface loss at 720°F is 715 W/ft².

To heat the crucible initially, the power required is

$$\frac{(230 \text{ lb})(0.12 \text{ Btu/lb-F})(720 - 70°F)}{3412 \text{ Btu/kWh}} = 5.3 \text{ kWh}$$

To heat the solid lead to the melting point,

$$\frac{(800 \text{ lb})(0.031 \text{ Btu/lb-F})(621 - 70°F)}{3412 \text{ Btu/kWh}} = 4.0 \text{ kWh}$$

The power required to melt the lead is

$$\frac{(800 \text{ lb})(10 \text{ Btu/lb})}{3412 \text{ Btu/kWh}} = 2.3 \text{ kWh}$$

To heat the melted lead,

$$\frac{(800 \text{ lb})(0.04 \text{ Btu/lb-F})(720 - 621°F)}{3412 \text{ Btu/kWh}} = 0.9 \text{ kWh}$$

387

The average lead surface loss is

$$\frac{[(\pi)(0.67)^2 \text{ ft}^2](715 \text{ W/ft}^2)(1 \text{ hr})}{(2)(1000 \text{ W/kW})} = 0.5 \text{ kWh}$$

The average steel surface loss is

$$\frac{[(\pi)(1.5)(1) + (\pi)(0.75)^2 \text{ ft}^2](920 \text{ W/ft}^2)(1 \text{ hr})}{2(1000 \text{ W/kW})} = 3.0 \text{ kWh}$$

Allowing a 20% safety factor, the total heat requirement is

$$(1.2)(5.3 + 4.0 + 2.3 + 0.9 + 0.5 + 3.0 \text{ kWh}) = 19.2 \text{ kWh}$$

The power required for initial heat-up is then

$$(19.2 \text{ kWh})/(1 \text{ hr}) = 19.2 \text{ kW ANS}$$

(B) Once up to temperature, there is no additional material to heat, nor is there any heat of fusion or vaporization.

The lead surface loss is

$$\frac{[(\pi)(0.67)^2 \text{ ft}^2](715 \text{ W/ft}^2)(0.5 \text{ hr})}{1000 \text{ W/kW}} = 0.5 \text{ kWh}$$

The steel surface loss is

$$\frac{[(\pi)(1.5)(1) + (\pi)(0.75)^2 \text{ ft}^2](920 \quad \text{W/ft}^2)(0.5 \quad \text{hr})}{1000 \text{ W/kW}} = 3.0 \text{ kWh}$$

Again applying a 20% safety factor, the total heat requirement is

$$(1.2)(0.5 + 3.0 \text{ kWh}) = 4.2 \text{ kWh}$$

The power required for operation is then

$$(4.2 \text{ kWh})/(0.5 \text{ hr}) = 8.4 \text{ kW ANS}$$

(C) Since the initial heat-up requirements are the greatest, cast-in immersion heaters totaling 20 kW would be best for the application.

PROBLEM 17-19

Steel drills, each weighing 0.157 lb, are to be placed in a 60 lb rack and dip coated in melted paraffin, at the rate of 1500 drills per hour with 20 lb of paraffin. For this process an open-top uninsulated steel tank is to be used, measuring 1.5 ft wide, 2 ft long, 1.5 ft deep, weighing 140 lb, and containing 168 lb of paraffin to be heated from 70°F to 150°F in 2 hours.

(A) What is the initial heat-up requirement?

(B) What is the normal operating power required?

(C) What recommendations would you make regarding the electrical heaters?

Answer   *(ref. 28)*

(A) From Table 17-4, the specific heat of steel is 0.12 Btu/lb-F, the specific heat of solid paraffin is 0.70 Btu/lb-F, the melting point of paraffin is 133°F, and the heat of fusion of paraffin is 63 Btu/lb. From Table 17-5, the specific heat of melted paraffin is 0.71 Btu/lb-F. From Fig. 17-37, the paraffin surface loss at 150°F is 70 W/ft^2. And from Fig. 17-39, the steel surface loss at 150°F is 55 W/ft^2

To heat the steel tank initially, the power required is

$$\frac{(140\ lb)(0.12\ Btu/lb\text{-}F)(150 - 70°F)}{3412\ Btu/kWh} = 0.39\ kWh$$

To heat the solid paraffin to the melting point,

$$\frac{(168\ lb)(0.70\ Btu/lb\text{-}F)(133 - 70°F)}{3412\ Btu/kWh} = 2.17\ kWh$$

The power required to melt the paraffin is

$$\frac{(168\ lb)\ (63\ Btu/lb)}{3412\ Btu/kWh} = 3.10\ kWh$$

To heat the melted paraffin,

$$\frac{(168\ lb)(0.71\ Btu/lb\text{-}F)(150 - 133°F)}{3412\ Btu/kWh} = 0.59\ kWh$$

The average paraffin surface loss is

$$\frac{(3\ ft^2)(70\ W/ft^2)\ (2\ hrs)}{(2)(1000\ W/kW)} = 0.21\ kWh$$

The average tank surface loss is

$$\frac{(13.5\ ft^2)(55\ W/ft^2)(2\ hrs)}{(2)(1000\ W/kW)} = 0.74\ kWh$$

Allowing a 20% safety factor, the total initial heat requirement is

$$(1.2)(0.39 + 2.17 + 3.10 + 0.59 + 0.21 + 0.74\,\text{kWh})$$
$$= 8.64\,\text{kWh}$$

The power required for initial heat-up is then

$$(8.64\,\text{kWh})/(2\,\text{hrs}) = 4.32\,\text{kW}\ \textbf{ANS}$$

(B) Power required to heat the drills and rack is

$$\frac{[60 + (1500)(0.157)\,\text{lb/hr}](0.12\,\text{Btu/lb-F})80°\text{F})}{3412\,\text{Btu/kWh}} = 0.83\,\text{kW}$$

To heat additional solid paraffin to the melting point,

$$\frac{(20\,\text{lb/hr})(0.70\,\text{Btu/lb-F})(133 - 70°\text{F})}{3412\,\text{Btu/kWh}} = 0.26\,\text{kW}$$

Heat of fusion, to melt the additional paraffin,

$$\frac{(20\,\text{lb/hr})(63\,\text{Btu/lb})}{3412\,\text{Btu/kWh}} = 0.37\,\text{kW}$$

To heat the melted paraffin,

$$\frac{(20\,\text{lb/hr})(0.71\,\text{Btu/lb-F})(150 - 133°\text{F})}{3412\,\text{Btu/kWh}} = 0.07\,\text{kW}$$

The paraffin surface loss is

$$\frac{(3\,\text{ft}^2)(70\,\text{W/ft}^2)}{1000\,\text{W/kW}} = 0.21\,\text{kW}$$

The tank surface loss is

$$\frac{(13.5\,\text{ft}^2)(55\,\text{W/ft}^2)}{1000\,\text{W/kW}} = 0.74\,\text{kW}$$

Allowing a 20% safety factor, the power required for operation is

$$(1.2)(0.83 + 0.26 + 0.37 + 0.07 + 0.21 + 0.74\ \text{kW}) = 2.98\ \text{kW}$$
$$\textbf{ANS}$$

(C) Since the initial heat-up requirements are greater, heaters rated 5 kW may be selected for the application. If immersion heaters rated higher than 16 W/in^2 are selected, they should initially be operated at half voltage until the paraffin is melted. Refer to Table 17-7 for maximum watt densities to be used for various materials.

Table 17-7. Maximum Watts Density Ratings.

These are suggested ratings only and will differ when flow velocity. heat transfer rate. or operating temperature vary.

MATERIAL BEING HEATED		OPERATING TEMPERATURE (F)	MAXIMUM WATTS PER SQUARE INCH
Acid Solution		180	40
Alkaline Solution		212	40
Ammonia Plating Solution		50	25
Asphalt. Tar. Heavy Compounds		200 300 400 500	10 8 7 6
Caustic Soda	2% 10% 75%	210 210 180	45 25 25
Citrus Juices		185	20
Degreasing Solution. Vapor		275	20
Electro-plating Solution		180	40
Ethylene Glycol		300	10
Fatty Acids		150	20
Freon		300	3
Gasoline		300	3 – 5
Glue (heat indirectly with water bath)			
Glycerine		50	40
Kerosene		300	3 – 5
Lead-stereotype Pot		600	35
Linseed Oil		150	50
Metal Melting Pot		500 – 900	20 – 27
Molasses		100	4 – 5
Oakite Solution		212	40
Oils	Bunker C Fuel Dowtherm A Dowtherm E Fuel Pre-heating Machine (SAE 30) Mineral Vegatable	160 600 400 180 250 200 400 400	10 20 12 9 18 20 16 30
Paraffin or Wax		150	16
Perchlorethylene		200	20
Potassium Hydroxide		160	25
Prestone		300	30
Propylene Glycol		150	20
Salt Bath. Molten		800-950	25 – 30
Sodium Cyanide		140	40
Sodium Hydride		720	28
Steel Tubing. cast into aluminum cast into iron		500-750 750-1000	50 55
Sulphur. Molten		600	10
Tin. Molten		600	20
Trichlorethylene		150	20
Water		212	55

(Courtesy General Electric)

PROBLEM 17-20

An oven with inside measures of 2 ft wide, 3 ft deep, and 3 ft high weighs 800 lb and contains a 50 lb steel tray. The oven is insulated with 2 inches of magnesia insulation weighing 154 lb. After being heated from 70°F to 250°F in one hour, the oven is to be used for drying 150 lb motors, one every 30 minutes. After heat-up, air is vented into the oven at the rate of 4 complete changes per working cycle. Assume that there is 2 lb of water in each motor.

(A) Determine the initial heat-up requirement of the oven.

(B) Find the normal operating power requirement.

(C) Recommend suitable heating elements.

Answer *(ref. 28)*

(A) From Table 17-4, the specific heat of steel is 0.12 Btu/lb-F and the specific heat of magnesia is 0.222 Btu/lb-F. From Table 17-5, the specific heat of water is 1.0 Btu/lb-F, the boiling point of water is 212°F, and the heat of vaporization of water is 965 Btu/lb. From Table 17-6, the specific heat of air is 0.237 Btu/lb-F, and the weight of air is 0.080 lb/ft^3.

The weight of the air in the oven is

$$[(2)(3)(3) \text{ft}^3](0.080 \text{ lb/ft}^3) = 1.44 \text{ lb}$$

From Fig. 17-38, insulated wall loss at a 180°F rise is 12 W/ft^2.

Power required to heat the oven initially is

$$\frac{(860 \text{ lb})(0.12 \text{ Btu/lb-F})(250 - 70°\text{F})}{3412 \text{ Btu/kWh}} = 5.45 \text{ kWh}$$

To heat the insulation,

$$\frac{(154 \text{ lb})(0.222 \text{ Btu/lb-F})(250 - 70°\text{F})}{3412 \text{ Btu/kWh}} = 1.80 \text{ kWh}$$

To heat the tray,

$$\frac{(50 \text{ lb})(0.12 \text{ Btu/lb-F})(250 - 70°\text{F})}{3412 \text{ Btu/kWh}} = 0.32 \text{ kWh}$$

To heat the air,

$$\frac{(1.44 \text{ lb})(0.237 \text{ Btu/lb-F})(250 - 70°\text{F})}{3412 \text{ Btu/kWh}} = 0.02 \text{ kWh}$$

The average oven surface loss is

$$\frac{(42\,\text{ft}^2)\,(12\,\text{W/ft}^2)\,(1\,\text{hr})}{(2)\,(1000\,\text{W/kW})} = 0.25\,\text{kWh}$$

Allowing a typical 30% safety factor for oven applications to cover door losses and other contingencies, the total initial heat requirement is

$$(1.3)(5.45 + 1.80 + 0.32 + 0.02 + 0.25\,\text{kWh}) = 10.20\,\text{kWh}$$

Power required for the initial heat-up is

$$(10.20\,\text{kWh})/(1\,\text{hr}) = 10.2\,\text{kW ANS}$$

(B) To heat each motor,

$$\frac{(150\,\text{lb})\,(0.12\,\text{Btu/lb-F})\,(250 - 70°\text{F})}{3412\,\text{Btu/kWh}} = 0.95\,\text{kWh}$$

To heat the additional air,

$$\frac{(4)\,(1.44\,\text{lb})\,(0.237\,\text{Btu/lb-F})\,(250 - 70°\text{F})}{3412\,\text{Btu/kWh}} = 0.07\,\text{kWh}$$

To heat the water in each motor,

$$\frac{(2\,\text{lb})\,(1.0\,\text{Btu/lb-F})\,(212 - 70°\text{F})}{3412\,\text{Btu/kWh}} = 0.08\,\text{kWh}$$

The heat of vaporization needed to evaporate the water is

$$\frac{(2\,\text{lb})\,(965\,\text{Btu/lb})}{3412\,\text{Btu/kWh}} = 0.57\,\text{kWh}$$

The oven surface loss is

$$\frac{(42\,\text{ft}^2)\,(12\,\text{W/ft}^2)\,(0.5\,\text{hr})}{1000\,\text{W/kW}} = 0.25\,\text{kWh}$$

Allowing a 30% safety factor, the heat required per working cycle is

$$(1.3)(0.95 + 0.07 + 0.08 + 0.57 + 0.25\,\text{kWh}) = 2.50\,\text{kWh}$$

Power required for operation is

$$(2.5\,\text{kWh/cycle})/(0.5\,\text{hr/cycle}) = 5.0\,\text{kW}\ \textbf{ANS}$$

(C) Since the initial heat-up requirements are greater, tubular, strip, oven, or duct heaters totaling 10 kW may be selected for the application. These may be connected to give half power for the 5 kW operating requirement. Alternatively, by changing the time allowed for initial heat-up to 2 hours, thereby reducing the hourly requirement to 5.1 kW, a 5 kW heater may be satisfactory for both heat-up and operation.

PROBLEM 17-21

A room is 11 feet wide, 13 feet long, and 8 feet high. It is to be maintained at 70°F. The room is of frame construction with window area of 65 square feet. Air is changed twice per hour. The lowest outside temperature is expected to be 10°F, which is 15 degrees higher than the record low of −5°F. A natural-convection air heater is to be installed, with associated controls. Determine the required kilowatt capacity of the heater.

Answer *(ref. 28)*

For comfort heating, the basic calculations for determining power requirements have been combined into a simple formula. For average construction, and assuming no loss through the floor, the power required for comfort heating is

$$P = \frac{(0.02NC + 1.13G + KA)(T_1 - T_0)}{(3412\ \text{Btu/kW})}$$

where
P = power requirement (kW)
0.02 = heat absorption of air (Btu/ft^3 −F)
1.13 = heat transmission of glass (Btu/h-ft^2 -F)
N = number of complete air changes per hour (usually 2 or 3)
C = cubical contents of room (ft^3)
G = exposed glass surface area (ft^2)
K = heat transmission of exposed walls and ceiling (Btu/h-ft^2 -F)
A = exposed wall and ceiling surface area (ft^2)
$(T_1 - T_0)$ = temperature difference between inside and outside (F), which should be taken as 15°F higher than the lowest recorded temperature for the given locality.

Table 17-8. Heat Transmission of Construction Materials.

CONSTRUCTION OF WALLS AND CEILINGS	K (BTU/HR-SQ FT-F)
Average frame construciton	0.25
12-inch brick construction	0.36
8-inch brick construction	0.50
10-inch concrete construction	0.62
6-inch concrete construction	0.79

Using the K value of 0.25 found in Table 17-8 and inserting values, the power requirements would be

$$P = \frac{[(0.02)(2)(1144) + (1.13)(65) + (0.25)(527)](70 - 10)}{(3412 \text{ Btu/kWh})}$$

$$= 4.42 \text{ kW ANS}$$

In this instance, a 4.5 kW natural-convection air heater would meet the heating requirements.

PROBLEM 17-22

Water at a temperature of 130°F flows through a 2-inch pipeline 35 ft long, at a rate of 10 gallons per minute. The steel pipe is covered with 2 inches of insulation, and the temperature of the air surrounding the pipe is 70°F. It is desired to heat the water to 180°F before it leaves the pipe.
(A) What is the initial heat-up requirement?
(B) What is the normal operating power required?

Answer *(ref. 28)*

From Table 17-4, the specific heat of steel is 0.12 Btu/lb-F. From Table 17-5, the specific heat of water is 1.0 Btu/lb-F, and the weight of water is 8.3 lb/gal. From Fig. 17-41, insulated pipe loss at a temperature rise of 110°F is 40 W/linear ft. And from Fig. 17-42, the insulation multiplier for 2-inch pipe is 0.59.

(A) In most applications of this type, water is allowed to flow until brought up to temperature by the installed operation power. Therefore, there is no initial heat-up requirement.

(B) The weight of water flowing in the pipe is

$$(10 \text{ gal/min})(60 \text{ min/hr})(8.3 \text{ lb/gal}) = 5000 \text{ lb/hr}$$

Power required to heat the flowing water is

$$\frac{(5000 \text{ lb/hr})(1.0 \text{ Btu/lb-F})(180 - 130°F)}{3412 \text{ Btu/kWh}} = 73.4 \text{ kW}$$

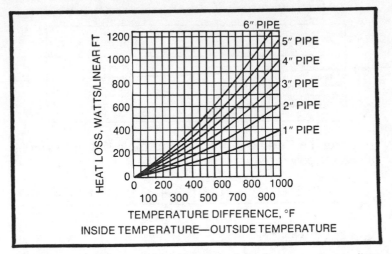

Fig. 17-41. Heat losses from insulated pipe lines. Curves are for pipelines covered with 1 inch of standard high-grade insulation. (Courtesy General Electric)

The pipe surface loss is

$$\frac{(35\,\text{ft})\,(40\,\text{W/ft})\,(0.59)}{1000\,\text{W/kW}} = 0.8\,\text{kW}$$

Allowing a 20% safety factor, the total power requirement is

$$(1.2)\,(73.4 + 0.8\,\text{kW}) = 89.0\,\text{kW}$$

The power per foot required for operation is

$$(89.0\,\text{kW})/(35\,\text{ft}) = 2.5\,\text{kW/ft}\ \textbf{ANS}$$

PROBLEM 17-23

Oil at a temperature of 200°F enters a 1-inch pipeline (1.315 inch OD) that is 29 feet long. The temperature of the air surrounding the steel pipe is 70°F. The pipe is to be heated to offset heat losses and to prevent a temperature drop in the oil.

(A) Assume an uninsulated pipe is used and determine the heating power requirements.

(B) What would be the percentage reduction in power required if the pipe were covered with 2 inches of insulation?

Answer   (ref. 28)

(A) From Fig. 17-39, the steel surface loss at 200°F is 90 W/ft^2.

Assume the pipeline is heated by the oil at 200°F, there is no initial heat-up requirement, and no power is needed to heat any additional material. Since oil enters and leaves the pipe at 200°F, no temperature change occurs so long as heat losses are replaced.

The pipe surface loss is then

$$\frac{(\pi)(1.315\text{ in.})(29\text{ ft})(90\text{ W/ft}^2)}{12\text{ in./ft}} = 900\text{ W}$$

Allowing a 20% safety factor, the total power requirement is

$$(1.2)(900\text{ W}) = 1080\text{ W}$$

Using an electric heat tape running the length of the pipeline, the power per foot required for operation is

$$(1080\text{ W})/(29\text{ ft}) = 37\text{ W/ft ANS}$$

(B) From Fig. 17-41, the insulated pipe loss at a 130°F rise is 25 W/linear ft. From Fig. 17-42, the insulation multiplier for 1-inch pipe is 0.63.

Pipe surface loss now becomes

$$(25\text{ W/ft})(0.63) = 15.8\text{ W/ft}$$

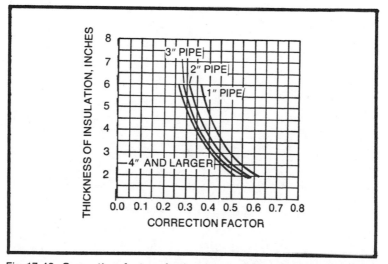

Fig 17-42. Correction factors for insulated pipeline heat losses. Where more than 1 inch of insulation is applied to pipe lines, obtain the heat loss from Fig. 2 and multiply by the appropriate factor from the above curves to determine the proper heat loss. (Courtesy General Electric)

And the safety factor raises the power per foot required for operation to

$$(1.2)(15.8 \text{ W/ft}) = 19 \text{ W/ft}$$

Comparing this figure with the power requirement established in (B) for an uninsulated pipeline indicates the relative operating economy of insulating pipes. In this case the percentage reduction in power consumption is

$$100(37 - 19)/37 = 48.6\% \text{ ANS}$$

Chapter 18
Power Transformers

Figure 18-1 illustrates a simple transformer. If the transformer is ideal, the voltage and current relationships are dependant solely upon the ratio of primary turns N_1 to secondary turns N_2. The following basic equations apply:

$$V_1/V_2 = N_1/N_2$$
$$I_2/I_1 = N_1/N_2$$
$$V_1 I_1 = V_2 I_2$$

The last equation represents a truly ideal (lossless) transformer in which the input power equals the output power.

Using these basic equations, it can easily be shown that the primary and secondary impedances are

$$Z_1 = V_1/I_1 = Z_2 (N_1/N_2)^2$$
$$Z_2 = V_2/I_2 = Z_1 (N_2/N_1)^2$$

That is,

$$Z_1/Z_2 = (N_1/N_2)^2$$

which is to say that the impedances are proportional to the square of the turns ratio. An equivalent circuit is shown in Fig. 18-2.

LUMPED PARAMETERS

The lumped parameters of a transformer are shown in Figs. 18-3 and 18-4. Resistances R_1 and R_2 represent the primary and secondary resistances of the transformer. Similarly, X_{L1} and X_{L2} represent the transformer

399

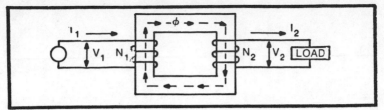

Fig. 18-1. The simple transformer.

reactances. Impedance Z_M in Fig. 18-4 is the complex impedance of the magnetizing branch of the transformer, but in power transformers the power loss and effects of this parameter are relatively small in comparison to the others and in most cases can be neglected.

For convenience, the secondary parameters can be moved to the primary side of the transformer, provided that the turns ratio is taken into account. In a 1:1 transformer the primary and secondary parameters can be lumped directly, adding the resistances and reactances in series. This leads to the representation of the equivalent circuit shown in Fig. 18-3. In most transformers, equivalent reactance X_{EQ} is much greater than equivalent resistance R_{EQ} and so is ignored.

When the turns ratio is not unity, the secondary parameters are seen in the primary circuit as modified by the turns ratio. This is also shown in Fig. 18-3.

OPEN-CIRCUIT SHORT-CIRCUIT PARAMETERS

Transformer values are often determined by open-circuit and short-circuit testing. The short or open is usually made in the secondary, as in Fig. 18-4. The shunting magnetic circuit (Z_M) is neglected.

The procedure is to measure V_1, I_1, and short-circuit power P_{SC}, then solve for the following short-circuit values:

$$Z_{SC} = V_1 / I_1$$
$$R_{SC} = P_{SC} / I_{SC}^2$$
$$X_{SC} = \sqrt{Z_{SC}^2 - R_{SC}^2}$$

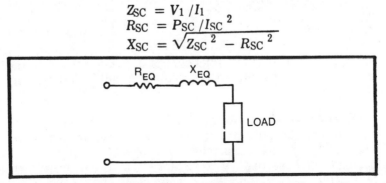

Fig. 18-2. Lumped primary with unity turns ratio.

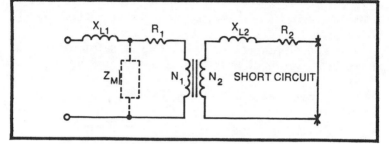

Fig. 18-3. Lumped primary of transformer having unequal turns ratio.

Applying the transformer turns ratio and lumping the parameters in the primary, and equivalent circuit can be drawn as in Fig. 18-5 for the she $\ast$ -circuit test condition. In a

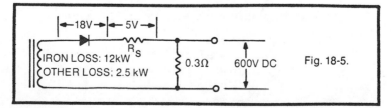

Fig. 18-4. Short-circuited secondary.

1:1 transformer, R_{EQ} and R_{SC} will be approximately equal to $R_1 + R_2$, while X_{EQ} and X_{SC} will be about equal $X_{L1} + X_{L2}$.

The short-circuit test is usually made with stepdown transformers, and it is the secondary which is shorted.

The open-circuit test is performed in a similar manner, leaving the secondary open circuited. Again measure V_1, I_1, and P_{OC}. The open-circuit parameters are then obtained by solving the following equations:

$$Z_{OC} = V_{OC}/I_{OC}$$
$$R_{OC} = P_{OC}/I_{OC}{}^2$$
$$X_{OC} = \sqrt{Z_{OC}{}^2 - R_{OC}{}^2}$$

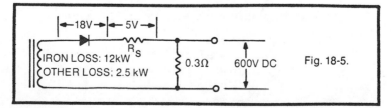

Open-circuit tests are usually performed on transformers by opening the high-voltage side and measuring the parameters on the low-voltage side.

POWER TRANSFORMERS

The efficiency of a power transformer is

$$\eta = P_{OUT}/P_{IN}$$
$$= (P_{IN} - P_{LOSS})/P_{IN}$$
$$= 1 - (P_{LOSS}/P_{IN})$$

where P_{LOSS} includes all sources of power loss within the transformer.

The percentage regulation is

$$\text{Reg. \%} = 100(V_{NL} - V_{FL})/V_{FL}$$

where V_{NL} and V_{FL} are the no-load and full-load output voltages.

The kilovolt-ampere (kVA) rating of a transformer is referenced to the input, or base, side of the transformer:

$$kVA = B_{BASE} I_{BASE}/1000$$

The percentage impedance of a transformer is

$$Z\% = 100(Z_{OUT}/Z_{BASE})$$

where

$$Z_{BASE} = 1000(kV_{BASE})^2/kVA$$

Also,

$$Z\% = \frac{100(Z_{OUT} kVA)}{1000(kV_{BASE})^2}$$

or

$$Z_{OUT} = \frac{10(kV_{BASE})^2 (Z\%)}{kVA}$$

Useful star and delta equivalents include the following:

$$\Delta(\text{volts}) = \sqrt{3}\, Y(\text{volts})$$
$$Y(\text{volts}) = \Delta(\text{volts})/\sqrt{3}$$
$$\Delta(\text{amps}) = Y(\text{amps})/\sqrt{3}$$
$$Y(\text{amps}) = \sqrt{3}\, \Delta(\text{amps})$$

Standard transformer types include:

SINGLE PHASE	3-PHASE DISTRIBUTION	POWER TRANSFORMER
1.5 kVA	112.5 kVA	1500 kVA
3 kVA	150 kVA	2500 kVA
5 kVA	300 kVA	3750 kVA
7.5 kVA	500 kVA	5000 kVA
10 kVA	750 kVA	7500 kVA
15 kVA	1000 kVA	10,000 kVA
25 kVA	1500 kVA	
37.5 kVA	2000 kVA	
50 kVA		
75 kVA		
100 kVA		
167 kVA		

Note: Single-phase types can be combined and used in 3-phase connections, such as three 50 kVA units to make a 150 kVA 3-phase unit.

TRANSFORMER EQUIVALENT CIRCUITS

The information in this section has been provided by courtesy of the Westinghouse Electric Corporation. The basic procedure to be followed in calculating the impedance values for a transformer equivalent circuit depends on the form of the original data and whether the final values are to be expressed in ohms or percent.

Procedure I is convenient for the simpler cases, when the original impedances are expressed in percent on a circuit base and when the final values are to be expressed in percent.

Procedure II is generally recommended for the more complicated cases, particularly for the ones involving neutral impedances or series transformers.

PROCEDURE I

The impedances of two- and three-winding transformers are normally given in percent on a circuit kVA base. With the basic data in this form, it is convenient to calculate the equivalent-circuit impedance values directly in percent. The equivalent circuits and equations for calculating the sequence quantities are given in Table 18-1 for the more common transformer connections. The following notation is employed in the table:

1. Terminal designations.

Circuit 4—abc terminals.
Circuit 5—a'b'c' terminals.
Circuit 6—a''b''c'' terminals.

2. Impedances.

Z_{45} %—impedance circuit 4 to circuit 5 in percent on 3-phase rated kVA of circuit 4.

Z_{46} %—impedance circuit 4 to circuit 6 in percent on 3-phase rated kVA of circuit 4.

Z_{56} %—impedance circuit 5 to circuit 6 in percent on 3-phase rated kVA of circuit 5.

Z_1 %, Z_0 %, Z_{H1} %, Z_{M1} %, Z_{L1} %, Z_{H0} %, Z_{M0} %, and Z_{L0} % are all in percent on the 3-phase rated kVA of circuit 4.

U_4, U_5, and U_6 designate the 3-phase kVA ratings of circuits 4, 5 and 6, respectively.

The impedances can be converted from one base to another by the relations:

$$Z_{46} \% = \frac{U_4}{U_6} \ Z_{64} \%.$$

$$Z_{56} \% = \frac{U_5}{U_6} \ Z_{65} \%.$$

$$Z_{45} \% = \frac{U_4}{U_5} \ Z_{54} \%.$$

PROCEDURE II

In many cases, particularly the ones involving neutral impedances or series transformers, less confusion results if the equivalent-circuit impedance values are calculated in ohms, rather than in percent. However, as the basic data are normally in percent, it is first necessary to convert to ohms using the following relations:

$$Z_{45} = \frac{10 Z_{45} \% E_4{}^2}{U_4}$$

$$Z_{46} = \frac{10 Z_{46} \% E_4{}^2}{U_4}$$

$$Z_{56} = \frac{10 Z_{56} \% E_5{}^2}{U_5}$$

where Z_{45} %, Z_{46} %, Z_{56} %, are as defined in Procedure I.

E_4, E_5 and E_6 are the line-to-line voltages, in kilovolts, in circuits 4, 5, and 6, respectively.

U_4, U_5 and U_6 are the 3-phase kVA ratings of circuits 4, 5, and 6, respectively.

Z_{45} is the impedance between circuits 4 and 5, in ohms, on circuit 4 voltage base.

Z_{46} is the impedance between circuits 4 and 6, in ohms, on circuit 4 voltage base.

Z_{56} is the impedance between circuits 5 and 6, in ohms, on circuit 5 voltage base.

PROBLEM 18-1

Determine the rating of a 3-phase transformer bank that is required to supply power to an electric pump lifting 300 gallons of water per minute to an elevation of 1000 feet. The pump motor has an efficiency of 65 percent.

Answer

Since water weighs 8.34 lb/gal, the pump's capacity is

$$(8.34 \text{ lb/gal})(300 \text{ gal/min}) = 2503.9 \text{ lb/min}$$

If the water is raised to a height of 1000 ft, the work required is

$$(2503.9 \text{ lb/min})(1000 \text{ ft}) = 2.5 \times 10^6 \text{ ft-lb/min}$$

The pump's horse power is then

$$\frac{(2.5 \times 10^6 \text{ ft-lb/min})}{(33,000 \text{ ft-lb/min})} = 75 \text{ hp}$$

Since 1 hp = 746 watts, the kilowatt rating of the pump is

$$(75 \text{ hp})(746 \text{ W/hp}) = 56 \text{ kW}$$

The pump's efficiency is 0.65, so the input power to the pump is

$$(56 \text{ kW})/(0.65) = 86.1 \text{ kW}$$

Assuming a typical power factor of 80%, the kVA rating of the pump would be

$$(86.1 \text{ kW})/(0.80) = 108 \text{ kVA}$$

With a required power capacity of 108 kVA, select a standard 112.5 kVA 3-phase power transformer, or use three 37.5 kVA single-phase transformers.

Table 18-1. Transformer Equivalent Circuits Used in Procedure I.

	TWO-CIRCUIT TRANSFORMERS		
DESCRIPTION	DIAGRAM OF CONNECTIONS	POSITIVE-SEQUENCE EQUIVALENT CIRCUIT	ZERO-SEQUENCE EQUIVALENT CIRCUIT
A-1 STAR/STAR SOLIDLY GROUNDED NEUTRALS| (FOR 3 PHASE CORE TYPE SEE TABLE 7)	5 ... 4	$Z_1\%$ $Z_1\% = Z_{45}\%$	$Z_0\%$ $Z_0\% = Z_{45}\%$
A-4 STAR/STAR NEUTRALS CONNECTED BUT UNGROUNDED (FOR 3 PHASE CORE TYPE SEE TABLE 7)	5 ... 4	SAME AS A-1	$Z_0\% = \infty$
A-5 STAR/DELTA SOLIDLY GROUNDED NEUTRAL	5 ... 4	$Z_1\%$ $j \cdot \epsilon$ $j30^4$ $Z_1\% = Z_{45}\%$	$Z_0\%$ $j Z_C\% = Z_{45}\%$

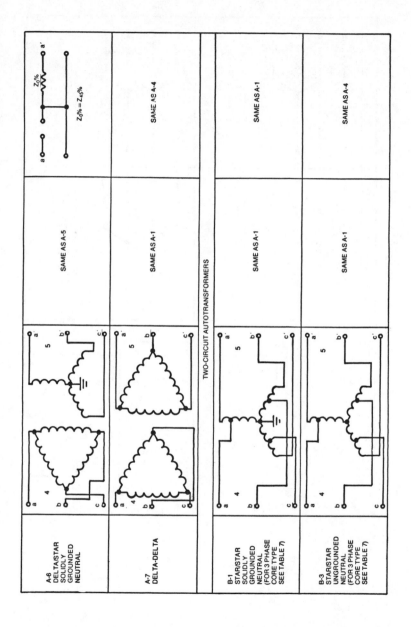

| A-6 DELTA/STAR SOLIDLY GROUNDED NEUTRAL | | SAME AS A-5 | $Z_0\% = Z_{45}\%$ |
| A-7 DELTA-DELTA | | SAME AS A-1 | SAME AS A-4 |

TWO-CIRCUIT AUTOTRANSFORMERS

| B-1 STAR/STAR SOLIDLY GROUNDED NEUTRAL (FOR 3 PHASE CORE TYPE SEE TABLE 7) | | SAME AS A-1 | SAME AS A-1 |
| B-3 STAR/STAR UNGROUNDED NEUTRAL (FOR 3 PHASE CORE TYPE SEE TABLE 7) | | SAME AS A-1 | SAME AS A-4 |

Table 18-1 continued.

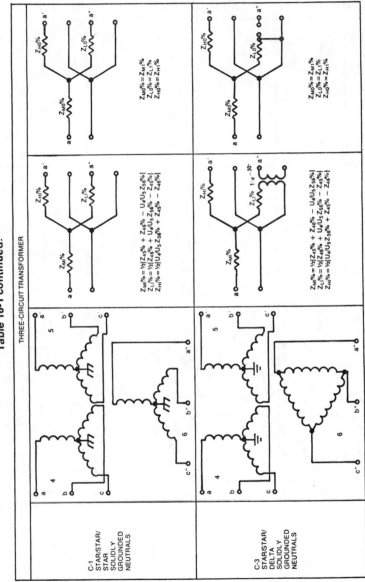

THREE-CIRCUIT TRANSFORMER

C-1 STAR/STAR/ STAR SOLIDLY GROUNDED NEUTRALS

$Z_{M1}\% = \frac{1}{2}[Z_{45}\% + Z_{46}\% - U_4/U_5 Z_{56}\%]$
$Z_{L1}\% = \frac{1}{2}[Z_{46}\% + U_4/U_5 Z_{56}\% - Z_{45}\%]$
$Z_{H1}\% = \frac{1}{2}[U_4/U_5 Z_{56}\% + Z_{45}\% - Z_{46}\%]$

$Z_{M0}\% = Z_{M1}\%$
$Z_{L0}\% = Z_{L1}\%$
$Z_{H0}\% = Z_{H1}\%$

C-3 STAR/STAR/ DELTA SOLIDLY GROUNDED NEUTRALS

$Z_{M1}\% = \frac{1}{2}[Z_{45}\% + Z_{46}\% - U_4/U_5 Z_{56}\%]$
$Z_{L1}\% = \frac{1}{2}[Z_{46}\% + U_4/U_5 Z_{56}\% - Z_{45}\%]$
$Z_{H1}\% = \frac{1}{2}[U_4/U_5 Z_{56}\% + Z_{45}\% - Z_{46}\%]$

$Z_{M0}\% = Z_{M1}\%$
$Z_{L0}\% = Z_{L1}\%$
$Z_{H0}\% = Z_{H1}\%$

408

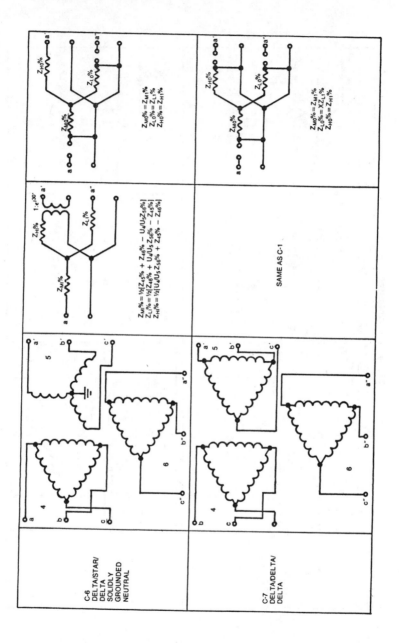

| C-6 DELTA/STAR/ DELTA SOLIDLY GROUNDED NEUTRAL | | $Z_{M1}\% = \frac{1}{2}[Z_{45}\% + Z_{46}\% - U_4/U_5 Z_{56}\%]$ $Z_{L1}\% = \frac{1}{2}[Z_{46}\% + U_4/U_5 Z_{56}\% - Z_{45}\%]$ $Z_{H1}\% = \frac{1}{2}[U_5/U_4 Z_{56}\% + Z_{45}\% - Z_{46}\%]$ | $Z_{M0}\% = Z_{M1}\%$ $Z_{L0}\% = Z_{L1}\%$ $Z_{H0}\% = Z_{H1}\%$ |
| C-7 DELTA/DELTA/ DELTA | | SAME AS C-1 | $Z_{M0}\% = Z_{M1}\%$ $Z_{L0}\% = Z_{L1}\%$ $Z_{H0}\% = Z_{H1}\%$ |

409

Table 18-1 continued.

THREE-CIRCUIT AUTOTRANSFORMERS

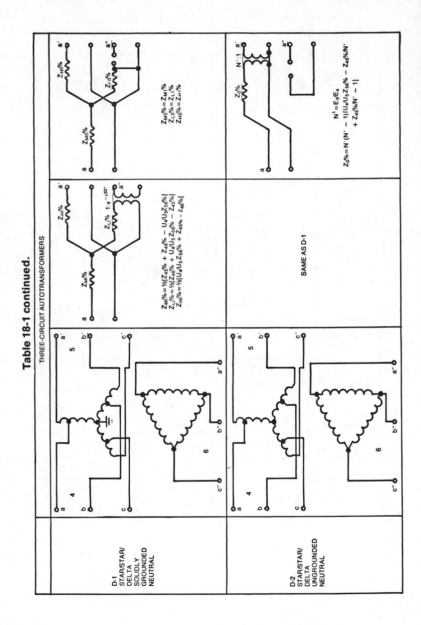

$$Z_{MO}\% = Z_{MI}\%$$
$$Z_{LO}\% = Z_{LI}\%$$
$$Z_{HO}\% = Z_{HI}\%$$

$$Z_{MI}\% = \tfrac{1}{2}[Z_{45}\% + Z_{46}\% - U_4/U_5 Z_{56}\%]$$
$$Z_{LI}\% = \tfrac{1}{2}[Z_{46}\% + U_4/U_5 Z_{56}\% - Z_{45}\%]$$
$$Z_{HI}\% = \tfrac{1}{2}[U_4/U_5 Z_{56}\% + Z_{45}\% - Z_{46}\%]$$

$$N' = E_5/E_4$$
$$Z_0\% = N'(N'-1)[U_4/U_5 Z_{56}\% - Z_{46}\%/N' + Z_{45}\%/N' - 1]$$

SAME AS D-1

D-1
STAR/STAR/
DELTA
SOLIDLY
GROUNDED
NEUTRAL

D-2
STAR/STAR/
DELTA
UNGROUNDED
NEUTRAL

410

PROBLEM 18-2

A transformer feeds a half-wave rectified power supply that delivers 600 volts DC to a 0.3-ohm resistive load operating at full load. Transformer iron losses are 12 kW and the auxiliary losses total 2.5 kW. The voltage drop across the power rectifier is 18 volts, and the transformer winding loss is 5 volts.

(A) Determine the overall efficiency of this circuit when operating at full load and draw a circuit showing all components.

(B) What is the overall efficiency at 20% of full load?

(C) Explain the difference in your answers for (A) and (B), if any.

Answer

(A) The circuit is as shown in Fig. 18-5. The output current delivered to the load is

$$I_0 = 600V/0.3\Omega$$
$$= 2000A$$

which means that the total output power is

$$P_0 = (600V)(2000A)$$
$$= 1200 \text{ kW}$$

The efficiency of this circuit is P_0/P_{IN}, where

$$P_{IN} = P_0 + P_L$$

with P_L representing the various losses. The iron loss (12 kW) and auxiliary loss (2.5 kW) are given. The copper loss is determined from the voltage drop at full load; that is,

$$P_{CU} = (5V)(2000A)$$
$$= 10 \text{ kW}$$

In addition, the rectifier loss is

$$P_R = (18V)(2000A)$$
$$= 36 \text{ kW}$$

Thus, the total loss is

$$P_L = 12 + 2.5 + 10 + 36$$
$$= 60.5 \text{ kW}$$

and

$$P_{IN} = 1200 + 60.5$$
$$= 1260.5 \text{ kW}$$

The circuit efficiency at full load is then

$$1200/1260.5 = 95.3\% \text{ ANS}$$

(B) At 20% of full load, only the iron loss will remain the same since this is due to excitation current flowing in the primary. All other losses, though, will be reduced to 20% of their full-load value. Therefore

$$P_L = 12 + 0.2(2.5 + 10 + 36)$$
$$= 21.7 \text{ kW}$$

The output load is also reduced, so

$$P_0 = (0.2)(1200 \text{ kW})$$
$$= 240 \text{ kW}$$

making

$$P_{IN} = 240 + 21.7$$
$$= 261.7 \text{ kW}$$

The efficiency is then

$$P_0/P_{IN} = 240/261.7$$
$$= 91.7\% \text{ ANS}$$

(C) Operating at 20% of full load reduces losses through a reduction of load current. The efficiency, however, does not remain constant because of the iron loss, which remains essentially unchanged at all loads. Consequently, the efficiency tends to decrease as the load is reduced.

PROBLEM 18-3

A two-winding transformer, nominally 240:24 volts, is connected as an autotransformer and installed as a booster on a 208-volt single-phase line. The load consists of ten 5000-watt, 230-volt, resistance-heating strips.

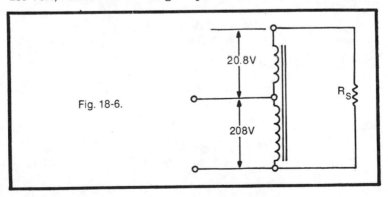

Fig. 18-6.

(A) What will be the open-circuit boosted voltage?

(B) What is the rating in kilowatts of the booster transformer required to serve the load? (Neglect transformer impedance.)

(C) If the booster transformer were removed from the system and the 208-volt source connected directly to the load, what would be the reduction in the heating effect?

Answer

(A) The transformer ratio is 240:24 or 10:1. When connected as an autotransformer (Fig. 18-6), the output voltage becomes

$$E = 208(10 + 1)/10$$
$$= 228.8\text{V ANS}$$

(B) Since $P = E^2/R$, the resistance of the strip heaters is

$$R = E^2/P$$
$$= (230)^2/5000$$
$$= 10.58\ \Omega$$

Operating at 228.8V, the heater and transformer capacity is now

$$P = E^2/R$$
$$= (228.8)^2/10.58$$
$$= 4948\text{W ANS}$$

(C) With the load operating at 208V, the heaters would consume

$$P = (208)^2/10.58$$
$$= 4089\text{W}$$

The heaters would thus be operating at only

$$4089/4948 = 82.6\%$$

of full load; a reduction of

$$100 - 82.6 = 17.4\%\text{ ANS}$$

PROBLEM 18-4

Show how to connect a 5 kVA, 2300:230 volt transformer to give a 10% boost on a 2300V line. Compute the load that can safely be put through this booster transformer.

Answer

The booster circuit is shown in Fig. 18-7. The current capacities of the windings determine the maximum safe load:

$$I_{LINE} = I_{PRI} + I_{SEC}$$

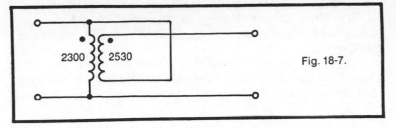

2300 ⎰⎱ 2530

Fig. 18-7.

The transformer is normally able to supply a secondary current of

$$I_{SEC} = (5000 \text{ VA})/230\text{V}$$
$$= 21.74\text{A}$$

and a primary current of

$$I_{PRI} = (5000 \text{ VA})/2300$$
$$= 2.174\text{A}$$

The total line current is then

$$I_{LINE} = 21.74 + 2.174$$
$$= 23.9\text{A}$$

An ideal transformer would therefore be able to deliver

$$P = (2300\text{V})(23.9\text{A})$$
$$= 54.97 \text{ kVA ANS}$$

PROBLEM 18-5

The following results were obtained from a test on a 20 kVA transformer: With the low-voltage winding short circuited, the rated current of 8.7A flowed in the high-voltage winding when 105V was applied. The power input under these conditions was 263 W. The voltage ratio is 2300:230. Find the voltage regulation (A) at 100% power factor and (B) at 75% pf leading.

Answer

The power loss is 263W at 105V, and this is attributable to copper losses. This makes the equivalent resistance referred to the primary

$$R_{01} = 263\text{W}/(8.7\text{A})^2$$
$$= 3.475\Omega$$

The equivalent impedance referred to the primary is then

$$Z_{01} = E_P /I_{SC}$$
$$= 105\text{V}/8.7\text{A}$$
$$= 12.07\Omega$$

The equivalent reactance referred to the primary is

$$X_{01} = \sqrt{(Z_{01})^2 - (R_{01})^2}$$
$$= \sqrt{(12.07)^2 - (3.475)^2}$$
$$= 11.56\Omega$$

The equivalent circuit of the power transformer is shown in Fig. 18-8. For this circuit, the rated primary current would be

$$I_P = (20\,\text{kVA})/105\text{V}$$
$$= 190.5\text{A}$$

The regulation is given by the formula

$$\text{Reg. \%} = 100(V_{NL} - V_{FL})/V_{FL}$$
$$= 100(E_1 - E_2)/E_2$$

where

$$E_1 = \sqrt{[E_2\cos(\theta) + IR]^2 + [E_2\sin(\theta) + IX_L]^2}$$

(A) At 100% pf, $\cos(\theta) = 1$ and $\sin(\theta) = 0$, so

$$E_1 = \sqrt{[E_2 + IR]^2 + [IX_L]^2}$$
$$= \sqrt{[2300 + (8.7)(3.475)]^2 + [(8.7)(12.07)]^2}$$
$$= 2332.6\text{V}$$

Therefore,

$$\text{Reg. \%} = 100(2332.6 - 2300)/2300$$
$$= 1.417\% \text{ ANS}$$

(B) At 75% pf leading, $\cos(\theta) = 0.75$ and $\sin(\theta) = 0.6614$. Thus,

$$E_1 = \sqrt{[(2300)(0.75) + (8.7)(3.475)]^2 +}$$

$$[(2300)(0.6614) + (8.7)(12.07)]^2$$

$$= 2393\text{V}$$

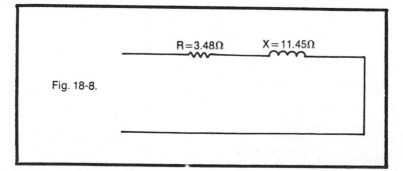

Fig. 18-8.

$R = 3.48\Omega$ $X = 11.45\Omega$

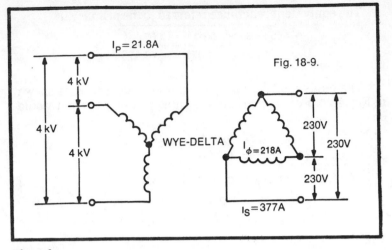

Fig. 18-9.

Therefore,

$$\text{Reg. } \% = 100(2393 - 2300)/2300$$
$$= 4.04\% \text{ ANS}$$

PROBLEM 18-6

Specify the primary and secondary current and voltage values and the kVA rating for the three transformers needed to supply a 3-phase balanced load of 120 kW and 80% power factor at 230V from a 4000V 3-phase line. Draw the wiring diagram and indicate all voltages and currents.

Answer

A wye-delta configuration is shown in Fig. 18-9. With an input of 4000V from the supply line, each of the wye-connected windings of the primary will be $4000/\sqrt{3}$, or about 2300V. To obtain 230V across each delta-connected secondary winding, the transformer turns ratio would be $2300 : 230 = 10 : 1$.

The transformer power requirements are determined from the formula

$$P = E_S \, I_S \, \cos(\theta) \, \sqrt{3}$$

where E_S and I_S are the secondary current and voltage. Solving for the secondary current rating,

$$I_S = \frac{P}{E_S \, \cos(\theta) \, \sqrt{3}} = \frac{120{,}000}{(230)(0.8)(1.732)} = 377\text{A}$$

The phase current flowing in the transformer secondaries would then be

$$I_\phi = 377/\sqrt{3}$$
$$= 218A$$

Assuming ideal transformers, the current flowing in the transformer primaries would then be related to the turns ratio, or

$$I_P = I_\phi /10$$
$$= 218/10$$
$$= 21.8A$$

PROBLEM 18-7

The transformers of a 3-phase city distribution system are rated at 2300:230 volts. The transformers are connected delta–delta and, in some cases, open delta. Discuss the changes in transformer connections that should be made if the primary is to be changed to 4-wire 4000 volts.

Answer

The present distribution system is shown in Fig. 18-10. In the case of an open delta, the voltages are the same as in the delta–delta except that the transformer must be rated at 86.6% or normal.

To achieve 4-wire 4000-volt operation in the primary hook-up, the primary should be connected in wye as shown in Fig. 18-11 and the fourth wire grounded. The secondary will remain in delta.

The primary phase voltage then becomes

$$E_\phi = E_{LINE} /\sqrt{3}$$
$$= 4000/\sqrt{3}$$
$$= 2310V$$

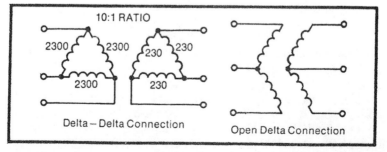

Delta – Delta Connection   Open Delta Connection

Fig. 18-10.

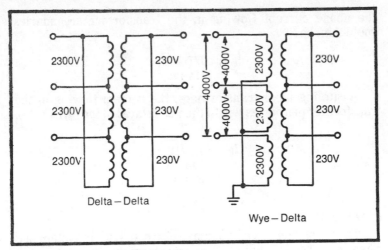

Fig. 18-11.

With a 10:1 turns ratio, the secondary voltage then becomes 231V.

In making the connections, it is essential to observe the polarity. The open delta should read zero volts before closing the delta loop.

PROBLEM 18-8

If three transformers of a 3-phase system have their connections changed from delta to wye, with no change in the primary circuit, what is the ratio of the new line voltage to the original line voltage? Explain the reason for this ratio.

Answer

The delta-to-wye configuration (Fig. 18-12) is a stepup transformer; that is,

$$\Delta(\text{volts}) = \sqrt{3} \times Y(\text{volts})$$

Consequently,

$$\Delta(\text{volts})/Y(\text{volts}) = \sqrt{3}\,\text{ANS}$$

The ratio comes about because the delta connection energizes the secondaries from line to line, whereas the wye connection energizes each secondary from line to ground. The resultant phase differences not only changes the output voltage, but it means that the different configurations cannot be operated in parallel. Delta—delta and wye—wye configurations do not produce a phase shift in the output, so these could be operated in parallel.

PROBLEM 18-9

Assume that you are in an isolated location where an 11,000-volt, 60 Hz service is available and that you require a lower voltage for a certain process requirement. The only transformer available is one that has the following data available:

> 500 kVA
> 13,200V : 400V
> 50 Hz
> copper loss: 4600W at full load
> iron loss: 2800W at full load
> eddy currents account for 22% of iron loss

If you connect the transformer to your available power supply, to what value should you change the full-load rating on the basis of the same heating loss?

Answer

The iron losses at 50 Hz are said to be 2800W, so the eddy-current losses are then $0.22 \times 2800 = 616$W, leaving a hysteresis loss of $2800 - 616 = 2184$W.

Now changing from 50 Hz to 60 Hz will cause the hysteresis loss to change, but the eddy-current loss will remain the same. The new hysteresis loss will then be

$$P_{H\,(60)} = P_{H(50)} \, (50/60)^{0.6}$$
$$= 2184(0.8333)^{0.6}$$
$$= 1958W$$

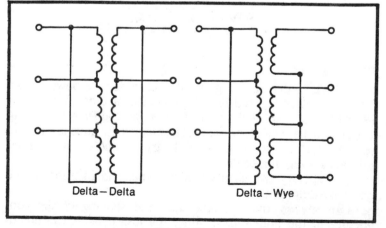

Delta — Delta   Delta — Wye

Fig. 18-12.

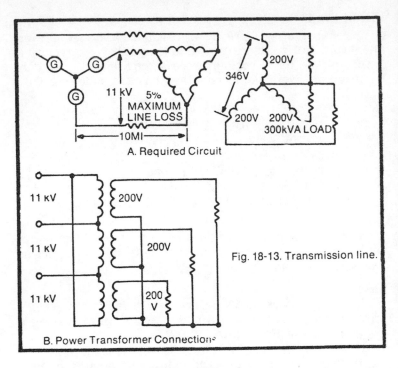

A. Required Circuit

Fig. 18-13. Transmission line.

B. Power Transformer Connections

The copper loss at 60 Hz will now be

$$P_{CU(60)} = P_{CU(50)} + P_E - P_{H(60)}$$
$$= 4600 + 616 - 2184$$
$$= 3032W$$

The rating at 60 Hz is

$$500\sqrt{P_{CU(60)}/P_{CU(50)}}$$
$$= 406\,kVA\ ANS$$

PROBLEM 18-10

A turbine-driven AC generator is to deliver power to an electric furnace located 10 miles from the turbine. The turbine generator is a 60 Hz, 3-phase, star-connected type producing 11,000 volts between terminals. The electric furnace is 3-phase, operating at 200 volts between electrodes and consuming 300 kVA at 80% lagging power factor.

(A) Specify the number, size, type, and location of the transformers required.

(B) Specify the voltage, size, and spacing of the conductors in the high tension line, the loss of which is not to exceed 5% of the delivered power.

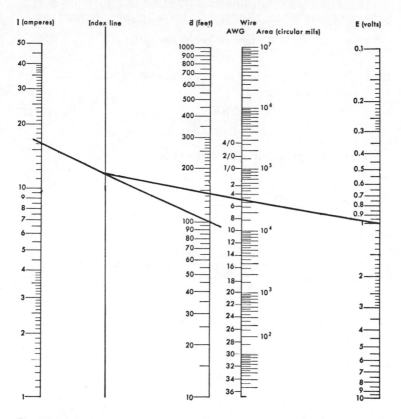

Fig. 18-14. Voltage drop vs wire size. To use nomograph, connect wire current to d(feet), which is the length of wire. From index line draw a line through wire size, if known, and read the voltage drop, or if desired voltage drop is known, read wire size. The problem can also be worked in reverse. To find the length of wire or the maximum current per given wire size and desired voltage drop, draw a line through E (volts) and wire size to establish crossing point at index line. If length of line is known draw a line through it and the index line and read (I+amps) or if I+amps is knwon, draw a line through it and the index line and read length of line to give referenced voltage drop. (Courtesy EDN)

Answer *(ref. 64)*

(A) The circuit arrangement should be as in Fig. 18-13A, with the power transformer connected as in Fig. 18-13B. Either a single, 300 kVA, 3-phase transformer or three, 100 kVA, single-phase transformers may be used. The transformer turns ratio is 11,000:200, or 55:1. To minimize line losses, the transformer(s) should be located as near the furnace as possible, but care must be taken to prevent the furnace heat from affecting the transformer(s) or power lines.

(B) The maximum line loss must be less than 5% of the furnace consumption, allowing for the 0.8 power factor. Thus,

$$P_{L(MAX)} = (300 \text{ kVA}) (0.8) (0.05)$$
$$= 12 \text{ kW}$$

This amounts to a loss of 4 kW per phase.

The line current is

$$I_L = \frac{300 \text{ kVA}}{(11 \text{ kV}) \sqrt{3}} = 15.75 \text{A}$$

The wire size in circular mills (CM) is

$$CM = \frac{\text{amps} \times 10.8 \times \text{feet} \times 2}{\text{voltage drop}}$$

$$CM = \frac{15.8 \times 10.8 \times 5280 \ (10) \ (2)}{11 \times 10^3 \ \times 0.05} = 32,763 \text{ CM}$$

Refer to wire tables (*ref. 64*) to get 33,100 CM = #5 AWG (or B&S) hard drawn, 97.3% conductivity. (Also, 0.226 in. diameter, 534 lb/mi.)

The resistance will be 1.91 Ω/mi, X_a = 0.613 Ω/mi at 1 ft spacing.

Alternate Answer

Determine the voltage drop per 100 ft. (See Fig. 18-14.)

$$\frac{550\text{V}}{5280 \ (10) \text{ ft}} \quad \therefore$$

$$\frac{X\text{V}}{100 \text{ ft}} = 1.04 \text{ volts per 100 ft}$$

Draw a line from 15.8A to 100 ft. Draw a second line from the Index line to 1.04V at the E (volts) line. Read 33,000 CM at AWG #5 wire size.

Chapter 19
Motors and Generators

The following is a summary of useful motor and generator formulas and design information. For further information, the reader should consult references 18 and 21−27 as listed in the appendix to this book.

SYMBOLS

D = damping constant in dyne-cm-sec/rad
E = voltage in volts
E_A = applied voltage in volts
E_C = rated control phase voltage in volts
E_B = brush voltage drop
E_F = back emf due to field winding
f = frequency in hertz
I_L = line current in amperes
I_S = stall current in amperes
I_T = tuned current in amperes
J = rotor moment of inertia in gm-cm^2
T_T = torque constant in dyne-cm/volt
K_V = motor gain in rad/sec-volt
N_L = loaded speed in rpm
N_O = no-load speed in rpm
N_S = synchronous speed in rpm
P = number of poles
P_F = power factor
P_O = power output in watts
P_{HP} = power output in horsepower
R = resistance in ohms

R_{OFF} = effective resistance in ohms
S = slip speed ratio
t = mechanical time constant in seconds
T = Torque
T_{FL} = full-load torque in oz-in.
T_L = loaded torque in oz-in.
t_R = reversing time in seconds
T_S = stall torque in oz-in.
W_{IN} = input wattage
W_O = output wattage
W_R = rotor wattage
W_S = shaft wattage
Z = impedance in ohms
η = efficiency
θ = angular position in radians
ϕ = magnetic flux

INDUCTION MOTORS

Synchronous speed:

$$N_S = 120 \, f/P$$

Shaft wattage:

$$W_S = T \, N_S \, (3/4000)$$

Torque:

$$T_L = (W_R/N_S)(4000/3)$$

Slip:

$$S = (N_S - N_O)/N_S$$

Output horsepower:

$$P_{HP} = (T_{FL} \, N_S)/(84{,}000)$$
$$P_{HP} = P_W/746$$

Torque:

$$T_L = (P_{HP}/N_S)/(84{,}000)$$

Efficiency:

$$\eta = 746 \, P_{HP}/(E_A \, I_L \, P_F)$$

Output wattage:

$$P_W = 746 \, P_{HP}$$

Input wattage:

$$P_{IN} = E_A \, I_L \, P_F$$

Power factor:

$$P_F = W_{IN}/(E_A I_L)$$

Theoretical acceleration at stall:

$$d^2 \theta/dt^2 = (7.062 \times 10^4)T_S/J$$

Reversing time:

$$t_R = (2.52 \times 10^{-6})J N_O/T_S$$
$$= 1.69T$$

Maximum power output:

$$P_O = (1.85 \times 10^{-4})T_S N_O$$

Effective resistance:

$$R_{EFF} = Z^2/R = Z/P_F = E/I_T$$

SERVO MOTOR TRANSFER FUNCTION

The Laplace transfer of a servo motor is given by the formula

$$0/E = \frac{K_V}{s(sT + 1)}$$

where:

motor gain $K_V = T_F/D$
torque constant $T_F = (7.062 \times 10^4)T_S/E_C$
damping constant $D = (6.75 \times 10^6)T_S/N_O$
mechanical time constant $t = J/D = (1.48 \times 10^{-6})J N_O/T_S$

UNITY POWER FACTOR

The tuning capacitance C required for parallel tuning to obtain unity power factor is given by the formula

$$C = \frac{\sqrt{1 - P_F^2}}{\omega Z}$$

where C = capacitance in farads
ω = powerline frequency in radians/second
P_F = winding power factor
Z = winding impedance

PHASE SHIFT

Simultaneous adjustment for stall conditions of the voltage and phase of the main motor winding is possible by using two

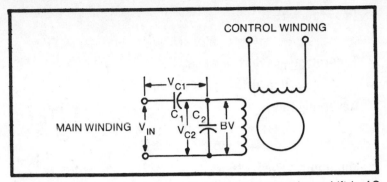

Fig. 19-1. In the two-capacitor method of correcting phase shift in AC motors, the voltage across the motor winding is also compensated to obtain zero phase shift. In the single-capacitor method, capacitor C_2 is removed.

capacitors as shown in Fig. 19-1. The following equations may be used to provide any magnitude ratio and for a phase shift of 90 degrees.

$$C_1 = B I P_F / \omega E$$

$$C_2 = \frac{I(\sqrt{1 - P_F^2} - B P_F)}{\omega E}$$

where P_F = power factor of the winding
E = rated voltage of main winding
B = magnitude ratio of line voltage to winding voltage
ω = frequency in rad/sec

For a 90-degree phase shift and unity voltage ratio ($B = 1$), the following formulas can be used for the special cases in which the powerline frequency is 60 Hz and 400 Hz.

For 60 Hz operation:

$C_1 (\mu F) = 2652 P_F I/E$
$C_2 (\mu F) = 2652 I(\sqrt{1 - P_F^2} - P_F)/E$

For 400 Hz operation:

$C_1 (\mu F) = 397.8 P_F I/E$
$C_2 (\mu F) = 397.8 I(\sqrt{1 - P_F^2} - P_F)/E$

The voltages across the capacitors are

$$E_{C1} = E\sqrt{1 + B^2 - 2B \cos(\theta)}$$
$$E_{C2} = BE$$

426

When the power factor exceeds 0.707, a single series capacitor should be used in series with the motor winding. Consequently, the voltage applied to the winding will be reduced, and magnitude ratio B will no longer equal unity. By removing capacitor C_2 from the circuit in Fig. 19-1, the numerator in the equation for calculating C_2 must become zero, for which

$$B = \sqrt{1 - P_F{}^2}/P_F$$

For a single-capacitor application at 60 Hz

$$C_1 \ (\mu F) = 2652 \ B \ I \ P_F \ /E$$

while the applied voltage across the winding becomes $E = B \ E_{IN}$.

DC SHUNT-WOUND MOTORS

The following equations apply to the DC shunt-wound motor as illustrated in Fig. 19-2. K_1 , K_2 , and K_3 are motor constants.

Torque:

$$T_L = K_1 \ I_A \ \phi$$

Speed:

$$N = K_2 \ (E_A - I_A \ R_A \ )/\phi$$

Power:

$$P_O = K_3 \ T_L \ N_L$$
$$= K_1 \ K_2 \ K_3 \ I_A \ (E_A - I_A \ R_A \ )$$

PERMANENT MAGNET, FIELD-TYPE MOTORS

PM field-type motors have linear speed/torque curves as shown in Fig. 19-3.

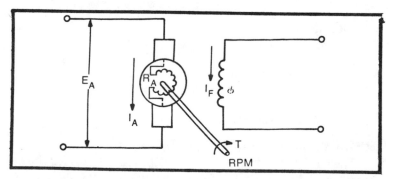

Fig. 19-2.

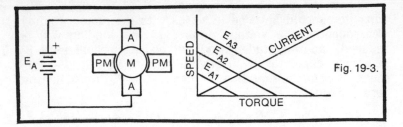

Fig. 19-3.

Output wattage:

$$W_O = N_L\, T_L\, (746 \times 10^{-6})$$

Input wattage:

$$W_{IN} = E_A\, I_A$$

Speed:

$$N_L = K_1\, E_F\, /n\phi$$

where n = number of armature conductors. Considering armature resistance R_A and brush voltage drop E_B, the reverse emf, denoted by E_F, is

$$E_F = E_A - E_B - I_A\, R_A$$

Efficiency:

$$\eta = W_O\, /W_{IN}$$

Output horsepower:

$$P_{HP} = W_O\, /746 \approx N_L\, T_L \times 10^6$$

Stall torque: The stall torque is directly proportional to the applied voltage and current, field strength, or number of armature conductors.

$$T_S = K_2\, N_L\, I_S\, \phi$$

Average torque:

$$T_A = I\alpha = T_S - T_L$$

where I = inertia in oz-in.-sec^2
   α = angular acceleration in rad/sec^2
   T_S = stall torque
   T_L = loaded torque

NOMOGRAMS AND CONVERSION FACTORS

The following material should prove useful for quick reference when working problems and estimating performance.

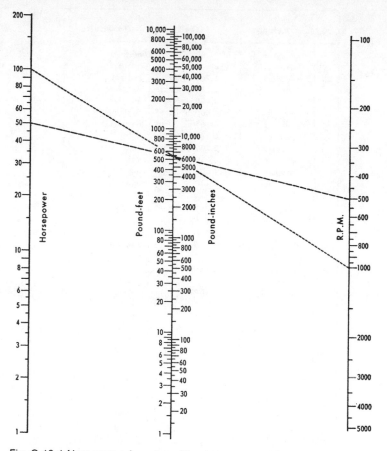

Fig. C-19-1 Nomogram for computing torque at various horsepowers and speeds. Example: A 50-hp motor at 500 rpm produces about 550 lb-ft (or 6500 lb-in.) of torque. (Courtesy EDN)

PROBLEM 19-1

A single-phase induction motor has the following name-plate ratings: 1 hp, 125 volts, 60 Hz, 40°C rise. Tests show the full-load efficiency to be 86 percent. Assume this motor is used at full load for 100 hours per month, and energy costs 3 cents per kWh.

(A) What will be the cost per month to operate this motor?

(B) If the full-load current is 9 amperes, find the power factor in percent.

Answer

(A) For the data given, the energy consumption is

$$(1 \text{ hp})(746 \text{ W/hp})(100 \text{ h})/(0.86 \text{ eff.}) = 86.7 \text{ kWh}$$

	Multiply:	By:	To Obtain:
ANGULAR MEASURE	degrees	17.45	mils
	degrees	60	minutes
	degrees	1.745×10^{-2}	radians
	mils	5.730×10^{-2}	degrees
	mils	3.438	minutes
	mils	1.000×10^{-3}	radians
	minutes	1.667×10^{-2}	degrees
	minutes	0.2909	mils
	minutes	2.909×10^{-4}	radians
	radians	57.30	degrees
	radians	1.000×10^3	mils
	radians	3.438×10^3	minutes
ANGULAR VELOCITY	deg/sec	1.745×10^{-2}	rad/sec
	deg/sec	0.1667	rpm
	deg/sec	2.778×10^{-3}	rps
	rad/sec	57.30	deg/sec
	rad/sec	9.549	rpm
	rad/sec	0.1592	rps
	rpm	6.0	deg/sec
	rpm	0.1047	rad/sec
	rpm	1.667×10^{-2}	rps
	rps	360	deg/sec
	rps	6.283	rad/sec
	rps	60	rpm
	ARMY MIL	$1/6400$	revolutions
DAMPING	$\dfrac{\text{ft-lb}}{\text{rad/sec}}$	20.11	$\dfrac{\text{oz-in}}{\text{rpm}}$
	$\dfrac{\text{oz-in}}{\text{rpm}}$	4.974×10^{-2}	$\dfrac{\text{ft-lb}}{\text{rad/sec}}$
	$\dfrac{\text{oz-in}}{\text{rpm}}$	6.75×10^{-2}	newton-m/rad/sec
DENSITY	g/cm^3	10^3	kg/m^3
	lb/ft^3	16.018	kg/m^3
DISTANCE	cm	10^{-2}	meters
	in	2.5400×10^{-2}	meters
	ft	0.30480	meters
	yd	0.91440	meters
	km	10^3	meters
	miles	1609.4	meters
ENERGY	ergs	10^{-7}	joules
	kwhr	3.6×10^6	joules
	calories	4.182	joules
	ft-lb	1.356	joules
	Btu	1055	joules

	Multiply:	By:	To Obtain:
FORCE AND WEIGHT	dynes	10^{-5}	newtons
	poundals	0.13826	newtons
	lb (force)	4.4482	newtons
INERTIA	g-cm^2	10^{-7}	kg-m^2
	g-cm^2	5.468×10^{-3}	oz-in^2
	g-cm^2	7.372×10^{-8}	slug-ft^2
	oz-in^2	1.829×10^2	g-cm^2
	oz-in^2	1.348×10^{-5}	slug-ft^2
	slug-ft^2	1.357×10^7	g-cm^2
	(lb-ft-sec^2)	7.419×10^4	oz-in^2
	slug-ft^2	1.357	kg-m^2
	lb-in^2	2.925×10^{-4}	kg-m^2
	oz-in^2	1.829×10^{-5}	kg-m^2
MASS	g	10^{-3}	kilograms
	slug	14.594	kilograms
POWER	ergs/sec	10^{-7}	watts
	cal/sec	4.182	watts
	Btu/hr	0.2930	watts
	joules/sec	1.00	watts
	horsepower	746	watts
	ft-lb/sec	1.356	watts
PRESSURE	dynes/cm^2	10^{-1}	newton/m^2
	psi	6.895×10^3	newton/m^2
	atmospheres	1.013×10^5	newton/m^2
	cm Hg	1333	newton/m^2
TORQUE	ft-lb	1.383×10^4	g-cm
	ft-lb	192	oz-in
	g-cm	7.235×10^{-5}	ft-lb
	g-cm	1.389×10^{-2}	oz-in
	oz-in	5.208×10^{-3}	ft-lb
	oz-in	72.01	g-cm
	oz-in	7.0612×10^{-3}	newton-m (joules)
TORQUE ERROR	$\dfrac{\text{oz-in}}{\text{minute}}$	0.0558	$\dfrac{\text{lb-ft}}{\text{rad}}$
	$\dfrac{\text{lb-ft}}{\text{rad}}$	17.9	$\dfrac{\text{oz-in}}{\text{minute}}$
VELOCITY	ft/sec	0.30480	m/sec
	miles/hr	0.44704	m/sec
	knots	1.152	miles/hr

Fig. C-19-2. Servomechanism Conversion factors. (Courtesy General Precision, Kearfott. Div.)

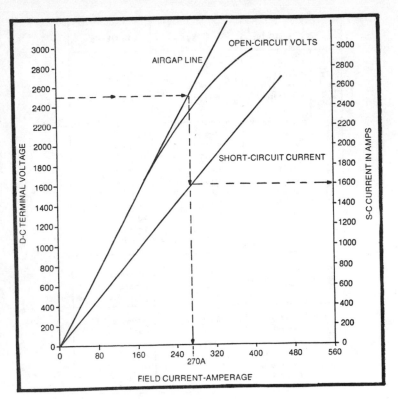

Fig. 19-4.

The cost at $0.03/kWh is then

$$(86.7)(0.03) = \$2.60 \text{ per month } \textbf{ANS}$$

(B) If the full-load input current is 9 amperes, the power factor is then

$$P_F = (746\,P_{HP})/E_{IN}\,I_{IN}\,\eta$$
$$= (746)(1)/(125)(9)(0.86)$$
$$= 77.1\% \textbf{ ANS}$$

PROBLEM 19-2

A water-wheel type of alternator is rated at 1500 kVA, 2500 volts, 3 phase, 60 Hz, wye connected. Tests run at rated speed to determine the open- and short-circuited characteristics gave the following results:

field current	0	80	160	240	320	400
open-circuit terminal voltage	0	750	1530	2150	2650	2970
short-circuit line current	0	480	920	1400	—	—

The DC resistance between terminals is 0.14 ohms. The effective resistance can be considered to be 1.4 times the ohmic resistance.

(A) Determine the synchronous reactance of the alternator per phase and calculate its full-load regulation at 80% power factor, lagging. Use the synchronous-impedance method.

(B) Would the actual voltage regulation be greater or less than the value obtained by the method in (A)?

(C) What modification of this method does the American Institute of Electrical Engineers recommend in order that the calculated results will come closer to the regulation existing under actual load conditions?

Answer

A plot of the short-circuit current and open-circuit voltage versus the filed current is given in Fig. 19-4. A straight line is drawn over the linear portion of the open-circuit voltage line and is called the airgap line. At 2500 rated volts, a line is drawn to the airgap line, and by dropping vertically, a field current is found that is 270A. At the intersection of the short-circuit current line, move horizontally to read a short-circuit current of 1600A. The rated current under these conditions is then

$$I = (1500\,\text{kVA})/(2.5\,\text{kV})\sqrt{3}$$
$$= 346\text{A}$$

The total resistance per phase (see Fig. 19-5) is half of the effective resistance, or

$$R_T = (1/2)\,(1.4)\,(0.14)$$
$$= 0.098\Omega$$

The short-circuit impedance is then

$$Z_S = (2500\text{V})/(1600\text{A})/\sqrt{3}$$
$$= 0.902\Omega$$

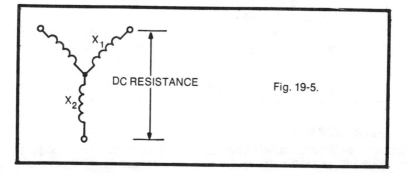

Fig. 19-5.

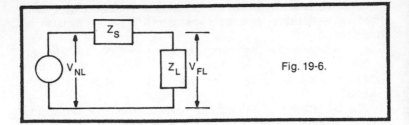

Fig. 19-6.

(A) The voltage per phase is $2500/\sqrt{3} = 1443$V; the load impedance is $Z_L = 1443$V$/346$A $= 4.17\Omega$. With a power factor of 0.8 lagging, $\cos(\theta) = R/Z = 0.8$, so $R = 3.336\Omega$. The synchronous reactance is then

$$X = \sqrt{Z^2 - R^2}$$
$$= 2.5\Omega \text{ ANS}$$

(B) The regulation is equal to $(V_{NL} - V_{FL})/V_{FL}$. Referring to Fig. 19-6, this can be stated as

$$\text{Reg.} = \frac{Z_S + Z_L}{|Z_L|} - 1$$

From part (A) of the problem, we have values for X and R, so

$$Z = \sqrt{X^2 + R^2}$$
$$= 4.17\Omega$$

We know that $Z_S = 0.902\Omega$ at $30°$ and that $Z_L = 4.17\Omega$ at $36.87°$, so the regulation equation breaks down as

$$\text{Reg.} = \left[\frac{(0.781 + 0.451j) + (3.336 + 2.5j)}{4.17} \right] - 1$$
$$= [(4.117 + 2.961j)/4.17] - 1$$
$$= [5.065/4.17] - 1$$
$$= 0.215$$
$$\% \text{Reg.} = 21.5\% \text{ ANS}$$

The actual value would be greater than the above answer because the synchronism is never 100 percent under full load.

(C) The AIEE procedure would be to determine the amount of slip at full load, using this realistic lag angle for making the calculations in part (B) of the problem.

PROBLEM 19-3

Two 3-phase alternators are operating in parallel to supply 9000 kW at 6600 volts and 0.90 power factor, lagging current.